MR. CHURCHILL'S TANK

MR. CHURCHILL'S TANK

The British Infantry Tank Mark IV

David Fletcher

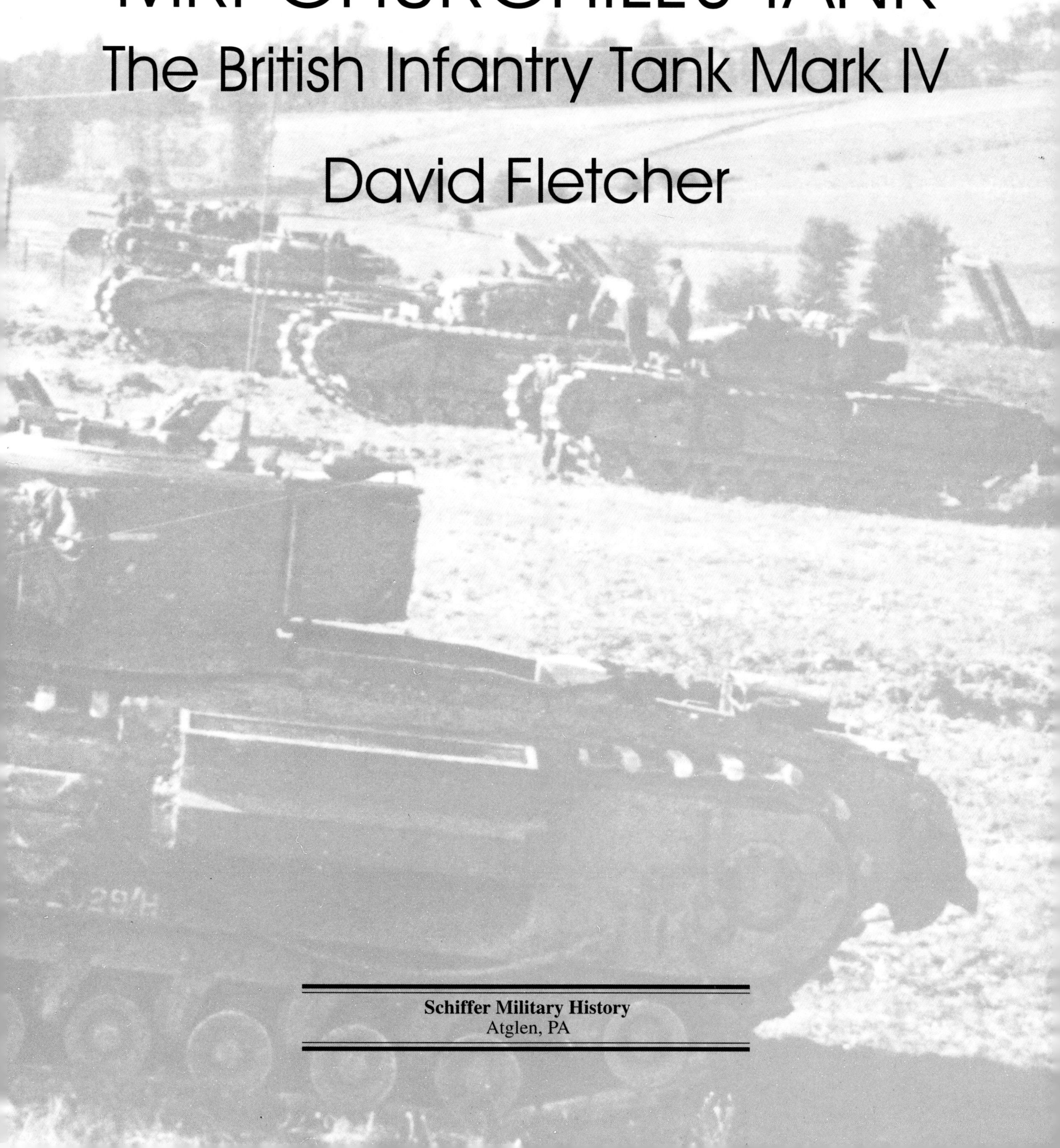

Schiffer Military History
Atglen, PA

Dustjacket artwork by David Pentland/2D Art
The scene depicted shows a Mark IV Churchill of the 6th Guards Tank Brigade during "Operation Bluecoat", Normandy, 1944.

Book Design by Robert Biondi.

Library of Congress Catalog Number: 98-86735.

Printed in China.
ISBN: 0-7643-0679-0

We are interested in hearing from authors with book ideas on military topics.

Published by Schiffer Publishing Ltd.
4880 Lower Valley Road
Atglen, PA 19310
Phone: (610) 593-1777
FAX: (610) 593-2002
E-mail: Schifferbk@aol.com.
Please write for a free catalog.
This book may be purchased from the publisher.
Please include $3.95 postage.
Try your bookstore first.

Contents

Introduction

One of the enduring legends of British history concerns the self-styled Lord Protector, Oliver Cromwell, from that short period when Britain sampled the delights of life as a republic. Old Noll, as he was known, sat for his portrait although he was, as they say, no oil painting. The artist did what he could to improve upon nature, only to be told by Cromwell that he wished to be portrayed "Warts an' all".

It may be true, but probably not, however this story is typical of the man and the fact that it was told at all shows that it was regarded as commendable and extremely unusual. And therein lies the problem. We are all enjoined to promote a positive image in all aspects of life, sport and work, but it ill becomes the would-be historian. Indeed anyone who attempts to chronicle the development of British tanks in an entirely positive way should really move over to writing novels.

Any researcher who sets out to study the development of British tanks in depth is liable to develop a jaundiced view. Yet it is probably true to say that similar studies of other nations, by fair minded historians, would reveal a similar tale. Even so the story of the Churchill tank is exceptional. Uninspired as it undoubtedly was, the programme of tank development devised by the British General Staff before World War II at least had some sort of logic to it. So how was it that a new design, quite outside that programme, was injected into the system against a good deal of professional advice? The answer seems to lie with the illustrious person with whom it shared its name. Winston Churchill, like President Lincoln in 1861, simply could not restrain himself from interfering in military matters. The difference is that whereas Lincoln managed to confine himself to matters of policy, Churchill also intervened in matters technical and he did this largely by instinct. Whether he was right or wrong it would be difficult to say, even after reading this book. Yet the fact remains that a tank designed and built in a hurry, in 1940, was still considered fit to go to war ten years later and it acquitted itself surprisingly well.

This is the fourth or fifth book to be published on the history of the Churchill tank, and undeniably the biggest. It has been written by someone who has studied the subject for years, has great admiration for the Churchill and hopefully a keen eye for its faults. Even so it is only half the story at best. The temptation to cover every version was immense but common sense, not to mention publishing limitations, indicated otherwise and in the end we elected to confine our study to the gun tanks. By eliminating all the special purpose versions we have been able to study the development of the gun tank in full and give a reasonable amount of space to its service history, for after all the two are inseparable. Much of what appears here is new, at least in the sense that it has not been published for public consumption before, but the sources from which it is drawn are extensive.

In the main these are found in the archives at the Tank Museum, which contain a wealth of official documents. These range from user manuals through trials reports and Government papers to regimental war diaries, combat reports and individual reminiscences. These have been supplemented by material obtained elsewhere, primarily with the assistance of friends whose contributions are acknowledged at the end of this piece. Where opinions are expressed it goes without saying that these are the responsibility of the author alone.

Even with this wealth of documentary evidence certain matters have proved impossible to resolve. Some might have been cleared up if the time for research was unlimited; although the odds are that if a writer waits until he knows everything then nothing will ever get done, and in any case, in fairness to one's publisher, a time limit has to be set. At the time of writing the book has already over run by three months and this can only be taken as a tribute to the patience of everyone at Atglen, Pennsylvania.

It could never have been done alone and thanks are due to Peter Gudgin in Leamington, Richard Harley in Edinburgh, Denis McCarthy in Dublin, Paul Handel in Sydney and particularly Tom Jentz in Pennsylvania for getting the whole project started. And indeed to Peter Schiffer and Robert Biondi in Atglen for publishing a book by a British author, on a British tank in the United States. May I say to American readers that I have not gone out of my way to explain our peculiar British way of doing things for their benefit. I believe in this age of instant communications these things will be understood well enough and any attempt to adopt American terminology would be patronising.

It is demonstrably futile to try and justify everything but by way of explanation it should be said that in recounting the Churchill's adventures some regiments have fared better than others. No favouritism is implied. It is simply that one cannot cover everything and the aim has been to bring out factors peculiar to the Churchill that are particularly relevant. If supporters of 9th and 48th Royal Tank Regiments feel hard done by it is only because both of them have excellent histories, currently in print, which do not need duplicating*. In the same way much of the wider background to this period has been covered in two of the author's earlier titles which, together, chronicle the development of British armour in the Second World War; *The Great Tank Scandal* (HMSO London 1989) and *The Universal Tank* (HMSO London 1993).

Like the remarkable individual for which it was named the Churchill tank has a fascinating history. It lasted a lot longer than its most optimistic designers could have imagined and, as is often the case, performed a lot better under the pressures of active service than anyone had a right to expect. Especially in view of its poor beginnings. That this was due more to the devotion and courage of those who served in it than any particular virtue of the tank itself is undeniable. Yet it managed to carve a reputation for itself that has endured and even earned a great deal of affection from those who fought in it and with it.

Author Tom Jentz wanted me to title this book 'The British Tiger' but I couldn't do it. As to why, I can only fall back on a story I have told elsewhere. In April 1944, during a secret debate in the House of Commons, two members of parliament confronted Prime Minister Winston Churchill with the undeniable inferiority of British tanks compared with their German counterparts. At one stage Richard Stokes MP requested that a Churchill and Tiger tank should be brought into New Palace Yard, just by Parliament Square, so that members of parliament could compare them. The arguments grew so heated, and the comments so personal that another MP stepped in to suggest an alternative method of deciding the matter. Perhaps Mr Churchill would care to take command of a Churchill tank while Mr Stokes took over the Tiger in order to settle the argument as honourable men should? All things being equal such a duel would have deprived Britain of an inspired leader at a critical time.

In truth there is little to be gained by making such comparisons and they will not be attempted in the following pages. What you have here is the Churchill tank; Warts an' all.

---

*Beale P., *Tank Tracks,* (Sutton Publishing).

Gudgin P., *With Churchills to War,* (Sutton Publishing).

1

The Winds of Change

In 1926 the 4th Battalion, Royal Tank Corps, moved from Bovington Camp in Dorset to Catterick in Yorkshire. It must have been quite a traumatic experience. Until that time all five tank battalions had been concentrated in the south of England, which enabled them to train together, but now the Fourth was banished to the north, to a bleak camp which the majority of soldiers regarded as one of the worst in Britain.

Things did not change much beyond that. Ever since the end of the First World War the Royal Tank Corps had been educating itself in the art of mobile, armoured warfare and in doing so had rather become divorced from the rest of the Army. This was not intentional. It resulted from a kind of persecution complex which led the tank soldiers to believe that they were outcasts, progressive outcasts perhaps, but misfits all the same. The more they preached the creed of armoured warfare the more they irritated the traditional arms, so they went their own way and encouraged one another with talk of mobile divisions in which tanks and armoured cars predominated.

The inevitable reaction to this isolationism was that other arms began to discuss tanks in relation to themselves and sometimes their thoughts were adopted by politicians particularly if, as a result, defence expenditure could be reduced. Thus when, in 1933, a senior infantry officer remarked that tanks would be more useful supporting the foot soldiers than chasing off across the countryside to outflank the enemy there were those who jumped at his words. It should not be surprising when one realises that a year later the Secretary of State for War was extolling the virtues of horsed cavalry! So, if tanks were not to replace cavalry they must support the infantry, primarily in direct assault operations against prepared defences. The formations required to operate tanks in this way would be classed as army tank battalions and their role would be little different from that originally devised for tanks in the First World War. Enthusiasts within the Royal Tank Corps could either go along with this thinking or ignore it and expect even greater isolation; they chose the former course. Major-General Percy Hobart, Inspector of the Royal Tank Corps, outlined the options in a paper prepared for the Chief of the Imperial General Staff's Research Committee in April 1934.

Hobart, who based his theories of armoured warfare on Sherman's campaign in Georgia, was happier promoting the concept of an independent tank brigade but he acknowledged the purpose, if not the value of army tank battalions which were "normally employed only after an opportunity for decisive action has been found and fixed by other arms". In other words infantry support tanks could only go into action once artillery had softened up the target and infantry had secured the flanks. The question then was - what kind of tanks should an army tank battalion employ? Hobart acknowledged the existence of two schools of thought.

One advocated small, heavily armoured tanks to be used in large numbers. Small meant inconspicuous; hard to locate and difficult to hit. Quantity meant overwhelming numbers and thick armour implied battlefield survival at a time when light anti-tank guns were regarded by many as the infantry's convincing answer to the tank threat. Such tanks would be armed only with machine-guns. Indeed they were in effect little more than mobile, protected machine-gun posts. As an alternative there were those who believed that a tank intended for infantry support should carry a cannon, and that by definition meant a large machine, but what precisely was its role? Some, who still viewed the infantry support tank in First World War terms, considered that the gun should be used for neutralising enemy machine-guns and destroying fortified structures such as pill boxes; for them a gun capable of dealing out high explosive was favoured. Others saw the role of such tanks as protection of the infantry from enemy tanks so they preferred a high velocity, flat trajectory weapon. It is instructive to note that no one seems to have considered a dual purpose gun at this stage.

In both cases speed was considered unimportant. It was sufficient that such tanks could keep pace with the foot soldiers as they advanced and get across the same kind of coun-

A Medium Tank Mark IA of 4th Battalion, Royal Tank Corps, on a bleak Yorkshire moor.

try. Of course in 1934 no such tank existed and in any case the theory of infantry support by tanks had not been tested in Britain since the end of the First World War. It was therefore agreed that one tank battalion should be singled out for this role and what more natural than to select the poor old Fourth, up there in the wilds of Yorkshire. To add insult to injury Colonel George Lindsay, responding to Hobart wrote;

> *"As obsolescent machines now with the battalions of the Tank Brigade and the battalions in Egypt are replaced by new ones, pass these to the 4th Battalion, Royal Tank Corps, Catterick and let it proceed with the work of experimenting in the close support of infantry. In fact let it specialise in Divisional Tank Battalion work."*

And that is just what did happen.

Of course this period of operational experimenting with a selection of worn out tanks was only to last until the new infantry support tanks were ready and it was now time to decide what should be built. The duty of deciding fell to the Master-General of the Ordnance and at this time the post was held by General Sir Hugh Elles, a complex and contradictory character. On paper he was the ideal person to make such a decision. Elles had commanded the Tank Corps in France right through to the end of the Great War and in doing so had revealed himself to be a gifted leader, personally brave and receptive to original ideas. But these are the qualities of youth. A career soldier, Elles had left the Tank Corps after the war and in those critical years had altered his views. Indeed for some time he became convinced that the tank was already obsolete; believing that power now lay with the infantry, if properly equipped with anti-tank guns. Even the striking success of the 1st Tank Brigade, during tactical exercises on Salisbury Plain in the thirties, did not convince him otherwise although he was now prepared to accept that there was a place for the tank as a supporting weapon for the infantry. He was reverting to what he knew; still seeing the tank in a First World War context. As to what sort of tank, he would be guided in that choice by economic considerations and, that being so it was inevitable that he should decide in favour of the smaller type.

The result was the appearance, in 1936, of the tiny A11 Matilda, or Infantry Tank Mark I. It was followed, two years later, by the larger A12 Matilda, the Infantry Tank Mark II, which in turn was followed in 1939 by the Mark III, the famous Valentine. The next step, the subject of this study, was the Churchill.

However the appearance of the Churchill was not the result of a logical, step by step development. Indeed it evolved in a highly illogical way from something quite different. Whatever else may be said about British tank design at the start of the Second World War nobody has ever suggested that it related in any way to the earlier conflict. Apart from a brief flutter of indecision in 1922, which resulted in the unique *Independent* machine, British tank experts had resolutely ignored all the old concepts of trench warfare and concentrated on designs that suited a more mobile style of conflict. Indeed tactical thinking was already well ahead of what existing tanks

The prototype Infantry Tank Mark I, first of a new generation of heavily armoured infantry support tanks.

Civilian engineers testing an early Matilda, or Infantry Tank Mark II armed with the ubiquitous 2-pounder gun.

could achieve and even the modest work towards the development of infantry tanks could hardly be described as a complete return to old ideas.

But what if the experts were wrong? Once it had been agreed that the next conflict was bound to take place on mainland Europe, with the same old enemies fighting it out once again over the same old ground, might it not be a return to the same old conditions? After all the French had constructed their impregnable Maginot Line, which looked like the ultimate Western Front defensive system and the Germans were known to be working on something called the Seigfreid Line which might well be similar. If that was the case might it not be wiser to design a tank to suit such conditions, even if it was not intended to replace the existing types?

Before we take a closer look at the results of this thinking perhaps it would be sensible to explain the system by which new tanks were designed and built for the British Army. It is quite complicated enough so there is no need to make comparisons with pre-war arrangements since they do not affect this study. Suffice it to say that a Ministry of Supply had been formed within the Government in August 1939. It appears to have been a belated move, especially considering that Winston Churchill had been advocating such an organisation since 1936, but it is worth bearing in mind that it was not until April 1939 that the Chamberlain Government agreed to a suggestion that Britain should raise a field force to participate in a continental war. The Ministry of Supply was responsible for, among other things, the allocation of essential metals and raw materials required for weapons production and for research, design and experiment, production and inspection of all military supplies. Coming down to specifics the first step in the development of any new tank was a proposal by the General Staff of the Army. Within the General Staff the Director of Military Operations and Intelligence pronounced on the type of warfare and the sort of opposition to be anticipated; the Director of Staff Duties worked out the general specifications for new vehicles and advised on the quantities required to suit the current order of battle while the Director of Military Training dealt with troop training of course and the tactical doctrine.

This is all very elevated stuff but it does not suggest detailed technical knowledge. This was supplied by an Inspector, Royal Armoured Corps who brought his particular experience to bear. Even so it soon became apparent that the General Staff would need a wealth of technical knowledge, not just in terms of weapon - in this case tank - design but also on the manufacturing side and it was here that the Ministry of Supply came in, acting as a focal point. Although it takes us ahead a bit it is probably well to mention the Tank Board at this point. It was, in fact, created in May 1940 to advise the Ministry of Supply on all aspects of tank design, production and use. This was recognition of the fact that, of all land weapons with which the Ministry had to deal the tank was, beyond doubt, the most complicated. The tank is an intimate association of matters concerning gunnery, turrets, armour, engines, suspensions and, probably worst of all, the complex subject of transmission as related to the climbing and steering capabilities of tracked vehicles. Knowledge of these subjects generally, and of the mobility element in particular, was sadly lacking in both military and civilian engineering circles, with results that we shall see.

At a meeting of the General Staff, held at the War Office on 1 September 1939, two days before war was declared, the Director of Staff Duties announced that, although the General Staff did not wish to interrupt production of Matilda and Valentine tanks they were concerned about the use of tanks in northern Europe. They anticipated operations resembling siege warfare conditions where extensive shelling was likely to churn up the ground. They therefore offered some very loose specifications which they wished the Director of Mechanization at the Ministry of Supply to consider.

The tank they had in mind should be transportable by rail; it should be armoured to resist the German 37mm antitank gun firing solid shot but they did not insist that the tracks should be protected. They believed a top speed of 10 miles per hour should be sufficient with a radius of action of 50 miles and in terms of armament they proposed a pair of 2 pounder (40mm) guns, three BESA 7.62mm machine-guns and a two inch smoke bomb projector, all operated by a crew of seven men. The mental picture they had of this tank was based upon the Anglo-American Mark VIII type of 1918, the so-called *International* tank. They realised that the proposed armour would present a weight problem but even so required the ground pressure to be as low as possible, bearing in mind the surface such a tank might have to operate upon.

Clearly they were looking at a tank which had tracks running all around the hull because they even suggested that an unditching beam might be carried. It proved to be a useful device for tanks during the Great War but no attempt had been made to revive it since, because it would not work on a turreted tank. Thus, for their new design, the General Staff proposed a return to the old practice of mounting the guns in sponsons on each side so that the final weapon layout would be; a 2-pounder and co-axial BESA in each sponson with the other BESA and a smoke bomb projector in the front of the hull, alongside the driver. They also specified a central command post.

Since it dominates all matters concerning tank design in Britain over this period this is probably a good time to discuss the subject of rail transport. Truck type road transporters hardly existed at this time and those being built were intended for tank recovery on the battlefield, not for long distance road movement. Thus all tank transporting, whether from the factory to the military base, or from there to the docks, or indeed anywhere, had to be done by rail. Now the British railway system is the oldest in the world. The first public steam railway was opened in 1825 and from about 1830 the network expanded in a rapid but disorderly fashion until it virtually covered the land. When these earlier lines were built the majority were expecting to use rolling stock which was little different from existing horse-drawn road vehicles and in order to keep costs as low as possible, bridges across the lines and in particular tunnels, were designed around these wagons with but little clearance. This naturally meant that all future locomotives and wagons had to keep within these restricted dimensions and the measurements that governed their construction were referred to as the Loading Gauge. This had nothing to do with the gauge of the tracks. In Britain this was set at 4 feet 8.5 inches between running rails and this arbitrary figure was adopted by most other countries including those on the Continent and the United States. Thus, in theory, locomotives and rolling stock from all over the world could run on British rails but in practice the restricted British loading gauge severely limited this possibility. Many US servicemen, stationed in Britain during both world wars, remarked on the small size of British trains, often describing them as miniature railroads. Many probably never even realised that the track gauge was the same.

It therefore becomes obvious that any tank built in Britain should be designed within the limits of the British loading gauge. Larger tanks would have to be dismantled in order to travel by rail and that was considered very inconvenient. Thus one may say that most British tanks built up to the end of World War II were limited in their size by civil engineering restrictions ordained by the pioneer railway engineer George Stephenson in the first half of the Nineteenth Century! In basic terms this meant a limit on width and height of about ten feet. It did not affect length of course, at least not in a direct sense, but there were extenuating considerations. Tank design is not really a critical science. Get the design of an aircraft wrong and it will not fly; build a ship which flaunts certain hard and fast laws and it will probably turn over and sink. But you can do more or less what you like with a tank and it will still struggle along. If there is a limiting factor then it is ruled by something called the L over C factor, where L is the <u>L</u>ength of track on the ground and C the distance between track <u>C</u>entres. In other words the longer a tank gets the wider it must be, otherwise it will be very difficult to steer. It was something most designers learned the hard way. So, if width was restricted by the railway loading gauge this indirectly limited length as well.

The Infantry Tank Mark III, or Valentine. One of the most reliable British tanks of World War II. This is the upgunned Mark X with a 6-pounder gun and co-axial BESA machine-gun.

When the General Staff proposed a new tank design they gave it a number, prefixed by the letter A, and the new 'shelled area tank' as one report called it, was designated A20. The Director of Mechanisation, General A E Davidson, showed the General Staff's proposal to two people. One was the president of the Mechanisation Board, Brigadier Hollebone, the other was Leslie Little, chief tank designer at Vickers-Armstrong since the death of Sir John Carden in 1935. Davidson told Hollebone that he reckoned they would find it very difficult to meet the low ground pressure requirement Little's answer was predictable. He told Davidson that the Department of Tank Design should hand the whole project over to Vickers and let them do the design work. This offer was not accepted but Little continued to attend meetings when the tank was discussed and presumably gave whatever advice he could.

It seems that by the end of September 1939 the Department of Tank Design had produced quite detailed drawings which Davidson felt were too restricting. He favoured the idea of planning on a broad basis and had already reached the conclusion that the new tank would have to carry a turret. Indeed he had gone further than that. He specified the new Meadows, horizontally opposed, 12 cylinder petrol engine and combined Merritt-Brown gearbox and steering system then being designed for the new Cruiser tank A13 Mark III, or Covenanter. He also proposed a turret based on that of the A12 Matilda and was working on the basis of 60mm maximum armour thickness which would resist the German 37mm shell, but not the anti-tank round as specified by the General Staff. Even with 60mm armour the use of side sponsons was going to create difficulties and it seems clear that by this time the idea of a turret was being generally accepted.

A model of the proposed tank was shown at a Ministry of Supply meeting held on 25 September at which it was announced that the estimated weight was 32 tons. Someone then raised the question of mine clearing. Anti-tank mines had not been a serious problem during the First World War but their potential was obvious and a good deal of work had already been done to create devices to counter them. The ruling theory at this time was that all infantry tanks, engaged in direct assault operations, should be fitted with a mine plough. However the committee agreed that this was an extra complication that the designers of A20 could do without. At a subsequent meeting it was suggested that the task of dealing with mines should devolve upon the little A11, Infantry Tank Mark I which was now being eclipsed by newer designs. The Assistant Chief of the Imperial General Staff now asked if the new tank was merely an improved version of A12 but was told that it was in fact an entirely new design. He also suggested that, if there were problems finding a suitable petrol engine, he would much prefer a compression ignition, or Diesel type instead. This seems to have been greeted with very little enthusiasm but the committee was relieved to hear that, if the ground pressure problem was solved, the ACIGS would not press for the inclusion of an unditching beam.

Finally the ACIGS asked, in all innocence, whether it would be necessary to build a pilot model. To anyone familiar with engineering practice or weapon development this would appear to be a ridiculous question, but the answer he got was even more astounding. Brigadier Hollebone assured him that they would not need one, that the new tank could go straight into production. He was backed up in this by the Chief Superintendent of Design, Dr. H E Merritt, who claimed that it was quite unnecessary since all the main design features had been tried and tested before; but had they?

A number of tanks built during the First World War had weighed above 30 tons. Indeed the Mark VIII weighed 37 tons, but none of those tanks had been fitted with sprung suspension systems and they had not been expected to move at more than four or five miles per hour. The 1926 Vickers design, known as the *Independent,* weighed around 32 tons but it had not been successful and in any case it was very thinly armoured by 1939 standards while the A12 Matilda, which did have very thick frontal armour, only weighed about 27 tons in battle order. Therefore it seems quite clear that in many respects the A20 design would be breaking new ground. Then there was the engine and transmission. The Meadows Type DAV flat twelve was a 16204cc, overhead valve, liquid cooled, petrol unit rated at 300 horsepower. It could hardly be described as a tried and tested unit. At this stage in its design the Covenanter tank was intended to have a complicated double differential, combined gearbox and steering system developed by Major W G Wilson, the transmission genius and famous designer of the first British tanks. This was also the intended transmission for A20 although Dr Merritt was working along similar lines and his triple differential Merritt-Brown system was shortly to appear and would, for experimental purposes, be fitted into a Covenanter. One could hardly claim that either of these features was fully tested in October 1939. The first Covenanter would not appear until May 1940, and that proved to be an unmitigated disaster. Then again, what of the suspension? Nothing whatever is said about this aspect in contemporary documents concerning A20 but we do know that the system ultimately adopted was quite novel so it is difficult to see how either Hollebone or Merritt could justify their opinion that no prototype was required.

Of course the truth was that they did not believe there would be time to build and test prototypes. It is a problem that has been covered in detail elsewhere,* but no matter how often one considers it the idea of building new tanks for production, straight off the drawing board, seems to be a terrible gamble, and it was one that the British lost every time they played. Expediency, under such circumstances is never justified yet there is little evidence, in the original minutes, of anyone protesting at this foolish practice.

As the meeting drew to a close General Davidson said that he was unable to name a manufacturer, adding that he could see no chance of early production unless assistance could obtained from the United States and he added, rather cryptically, that in such a case the General Staff might have to accept shorter track life for the tank. Quite why this should be is not made clear and no obvious reason suggests itself. Davidson rounded off the meeting by saying that weight would still present problems. He pointed out that none of the floating bridges then in use with the British Army could handle 32 ton tanks.

On 28 October 1929 General Davidson wrote to a Mr. Atholl Blair, of Harland & Wolff in Belfast, Northern Ireland, soliciting his firm's interest in A20. He described it as being based on A12 but with a longer track base in order to cross wide trenches and explained that the General Staff was looking for something with a very low ground pressure which was also simple to build. Harland & Wolff was a general engineering company with branches in other parts of the United Kingdom but they have always been closely associated with the City of Belfast and are best known as a shipbuilding concern. Many of the Light Fleet Aircraft Carriers of the Royal Navy were built by Harland & Wolff.

* Fletcher, *The Great Tank Scandal,* HMSO London 1989.

The company had also built some tanks, having accepted contracts for A9 Cruiser and A12 (Matilda) Infantry Tanks shortly before the war. This was all part of a Government scheme by which many British engineering companies were given what were described as 'educational' contracts to construct small batches of tanks in order to gain experience. It was a sensible move since there were very few firms with any such experience before 1935 and it was now obvious that tanks would be needed in very large numbers, but as usual the authorities left it rather late. Whether the experience gained in building tanks to other people's designs really fitted the firm for design work is debatable. Here we have a new, untried design, which has to go straight into production, built by a firm which had very little experience of making tanks at all. Nevertheless they were told that outline drawings were nearly complete and the firm's task would be to finish the design and prepare working drawings. Construction was to begin early in 1941 and they should be prepared to complete between 50 and 100 tanks with just the first two in mild steel since they would be needed to prove the drawings.

Davidson suggested that if they were happy to accept the challenge, Harland & Wolff should establish a design office at Liverpool or Southampton where they could liaise with the Mechanisation Board staff who would help with the design. He even offered a small incentive by suggesting that if Harland & Wolff were able to complete a 300hp Diesel engine which they had designed it might prove suitable for A20. The Belfast company responded by offering to send three or four draughtsmen to Woolwich and they stated that they also believed it would be wisest to build the mild steel pilot models in England. Obviously they anticipated problems during the design and construction stages with experts shuttling backwards and forwards across the Irish Sea. This, incidentally, is the first time that the tank is referred to by its alternative title of Infantry Tank Mark IV. The difficulties of building tanks in Northern Ireland were also raised by Major Raikes in a report to Colonel Gardiner at the Ministry of Supply in March 1940. Raikes believed that manufacture of the tank and all its components should be concentrated in one area, to avoid the time and expense involved in shipping parts all around the country. If A20 was to be built in quantity then he suggested looking at the industrial area of the Clyde valley around Glasgow, which was within easy reach of Belfast, and other locations in Scotland. "On the face of it," he wrote, "it would appear wrong in principle to transport all the component parts by sea to Belfast, to put them into the hull, and bring them back to this country". However he also appreciated that Harland & Wolff had taken a great deal of trouble in developing the hull construction and, since they had shop floor space to spare, it might make more sense to have them assemble the hulls which would then be shipped across to the Clyde for completion. To be more specific he advised cutting back orders for A12 Matilda with Harland & Wolff, and the North British Locomotive Company in Glasgow and have them concentrate on A20.

By November 1939 it was confirmed that the armour would be on a 60mm basis but by employing a new construction (their words) the hull would be proof against the British two pounder gun and that a test would be conducted at Woolwich to prove this. There is also, for the first time, a hint that the tank would carry a 3 inch close-support howitzer in the front of the hull. It comes in a report from Dr. Merritt who announced that design work on the fighting compartment was nearly complete but that he would like to know if it were possible to cut 12 inches off the howitzer's barrel. Otherwise he feared that it would stick in the ground and become damaged when the tank was crossing a trench. Later that month it was agreed that Harland & Wolff could build up to four mild steel prototypes if they wished and we learn that the engine compartment was being designed around the H&W 300hp Diesel engine but that the Meadows unit should fit as an alternative.

While uncertainty remained over the engine the Chief Superintendent of Design, Merritt, sought to introduce a bit more by discussing alternative transmissions. He announced that he was investigating a Sinclair hydraulic system and David Brown gearbox which was simpler to operate than the Wilson transmission since it employed dog clutches instead of braked epicyclics. Sinclair transmissions were designed to be used in self-contained railcars which required smooth acceleration. However they were not normally subjected to the rough and uncertain treatment that affected tank transmissions and it is probably not surprising that none of the Sinclair systems were ever successfully employed in tanks. It is equally difficult to understand why Merritt was still not recommending his own transmission for A20. It had already been developed in 1938 and was probably more likely to work than Sinclair's untested equipment. All of which again throws into question why it was that Merritt advocated immediate production.

The earlier reference to a new construction for the A20 hull which improved protection is probably explained by a later comment at a meeting on 17 November 1939. This mentions a box girder and is taken to refer to the form in which the hull was constructed, which is probably best understood by referring to one of the Churchill hull diagrams further on in the book. What the report fails to explain is the method of construction. At this time one would expect to find riveting employed for most joints with bolting for certain thicker components. Welding might be used for detailed work but not when it came to joining armour plates. This was partly due to serious reservations about the effectiveness of welding armour but also because firms like Harland & Wolff, traditional shipbuilders primarily, would be experts at riveting with an experienced workforce of riveters.

If there was some confusion over the method of construction it was echoed when the committee came to discuss the matter of what A20 was to be built of. The first mention of armour occurs in a brief note to the effect that CTA plate would be employed upon the outside. CTA, or Cemented Tank Armour, had been developed by the English Steel Corporation in 1935. It was in fact a form of face hardened plate which was calculated to give about 20 per cent better ballistic protection than a similar thickness of the softer, machineable homogeneous armour. Vickers-Armstrong had been using CTA on their light tanks from about 1936 and its good qualities are generally associated with thinner armour plate. This may explain why it is referred to as 'outside'. It suggests that Harland & Wolff was planning to build the tank with an inner structure of mild steel which would have a layer of CTA bolted to the external surfaces. This appears to be confirmed by a report issued in March 1940 which explains that some 14 tons of D.1. steel and 7 tons of CTA plate would be required for each tank. A similar system was certainly used on many subsequent designs. However the committee which governed the future of A20 was not coming down firmly on one side of the fence yet. They mention production problems with CTA and suggest that the design should be capable of being constructed with homogeneous hard plate on the outside as an alternative and round this off by saying that CTA plate could be used on the inside once production difficulties had been overcome. It was also agreed that firing tests should be conducted against sample sections of armour in the meantime.

The hazards of wartime travel were brought out at a meeting held on 20 November. Brigadier Hollebone, in his capacity as President of the Mechanisation Board had visited Harland & Wolff in Belfast. He had flown from Speke Airport at Liverpool in a single engined aircraft which he described as six years old and lacking radio, fire extinguisher and tools. The pilot lost his way and took two and a half hours for a journey that should normally take a fraction of that time. At one point they found themselves approaching Irish Free State territory where Hollebone and his plane would have been interned for the rest of the war. The Brigadier found this all rather unpleasant and felt that the Air Ministry should be told about it. In passing Hollebone also remarked that, while he was in Belfast, Mr. Blair of Harland & Wolff had shown him an eighth scale model of a tank which they were proposing as a rival to A20. It was the same length with much wider tracks but it did not have the prominent front track horns of A20.

Meanwhile new information had been elicited from the Director of Artillery at the War Office. It was learned that the General Staff had discussed the possibility of producing a more powerful tank gun and the design department at Woolwich Arsenal was working on one which could fire a six pound projectile and that this had been going on since 1938. The first question was whether this gun could be fitted in the new tank, but they were told that the new weapon was not yet ready and probably would not be until 1941. Even so it seems hardly credible that the same General Staff should specify a gun which they already believed to be out of date for a tank they had not yet built. Accepting that for the present it was the 2-pounder or nothing, surely the specification should have included reference to the bigger weapon when it became available, and the tank designed to accept it?

This sparked off further discussion on weaponry. The hull mounted three inch howitzer has already been mentioned. It was a low velocity, close support weapon intended mainly to fire smoke rounds and it was designed to fit into the same mounting as the 2-pounder. The A20 team believed that their tank should do more than blow smoke rings and suggested that the gun should be capable of firing high explosive to destroy concrete obstacles. But they had learned that for the present only smoke rounds were being produced for the 3 inch gun. They therefore suggested that A20 should carry the older 3.7 inch close support weapon for which high explosive ammunition would soon be available, but they were told that even this gun lacked the muzzle velocity to do much damage to concrete and one member of the committee remarked gloomily that neither of these weapons looked very promising for the front gun of A20.

It is perhaps in the nature of committees that they should get carried away. Discussion generates discussion until they come to believe that anything is possible, if only one talks about it for long enough. Thus it was that at a weekly meeting held on 24 November the chairman had to remind them that A20 was "a Ford, not a Rolls-Royce and therefore the 1941 production must be to simple designs and new ideas could only be worked out for 1942 production". He was able to show a new set of drawings supplied by the manufacturers and, the committee was told, the turret would not be the same as A12. They were now hoping to create a composite construction of flat plates which would save weight. The original A12 turret being, of course, mainly cast. It was also announced that a contract was being issued to Harland & Wolff for a total of 100 A20 tanks, four of which would be mild steel pilots.

A meeting held on 1 December was something of a progress report. Blair, in Belfast, had agreed to build the four pilots and had already secured the steel plate. The committee also learned that the engine compartment would be able to take the Meadows flat 12 petrol engine or the Harland & Wolff Diesel while the choice of transmissions now lay between a fluid flywheel and clash gearbox, (presumably Sinclair), the Covenanter system (Wilson) and a controlled differential type which would appear to be Henry Merritt's new design. Trials with the armour had shown that A20 should be largely immune to British 2-pounder rounds and that the new turret would save about 80 per cent of the man hours required to make the existing A12 cast turret but that the two types would be interchangeable.

A week later the Director of Mechanisation sounded a note of caution. He told the committee that he had held up the order for 100 tanks with Harland & Wolff but retained the contract for the four pilot models. No explanation was given beyond the view that such an order was inadvisable but the position would be reviewed in the new year. This delay revived the gun issue and the 6-pounder was mentioned again but, after examining drawings of the gun early in 1940, it was considered too long. One statement claims that the muzzle would stick out two feet ahead of the front horns of the tank although this would surely only apply if the gun were mounted at the front, and not in the turret as the committee wished? The statement goes on to say that "on these grounds the gun was ruled out for either turret or hull". Of course the reason why the gun could not go into the turret was that the turret was too small. There seems to have been no obvious reason at all why A20 should not have been designed with a larger turret beyond the fact that it appears to have been decided, more or less off the cuff, to use the A12 design when the original proposal for a turretless tank was rejected. This, of course is the problem. A20, as originally designed, was to be turretless as we have seen. Once it was agreed that a turret was needed the entire design should have been reconsidered to accommodate it, instead the designers appear simply to have sketched a turret onto the existing hull and left it at that. This is no way to design tanks.

Yet if anyone thought that the matter of armament was settled they would have been wrong. A further meeting was arranged with the General Staff and they were offered three options for the main, or turret, weapons. These were; a pair of machine-guns, the 2-pounder or a 75mm gun. The machine-guns would probably be a pair of BESAs; a heavy 15mm and 7.62mm as fitted in some light tanks and armoured cars. The 2-pounder we are familiar with but what about the 75mm? There was no such weapon in the British armoury at this time, assuming one rules out the 3 inch howitzer, and as we shall see, the chances are the General Staff were thinking of the current French weapon. The tank was also going to mount a machine-gun in each side of the hull and something larger at the front. The General Staff were still interested in the 6-pounder and they ordered that the committee should investigate a Royal Naval weapon, a 6-pounder anti-motor torpedo boat gun, which was shorter.

It would be tedious to pursue the armament question very much further. It was never resolved and the endless discussion only serves to point up the utter futility of a system that permitted such committees to go on chewing over a subject *ad infintum*, especially when most of the weapons they proposed could not be fitted anyway, at least not without virtually redesigning the tank. Suffice it to say that among the weapons discussed and rejected were a 40mm Vickers piece, French 47mm and 75mm weapons and the naval 6-pounder. In any case there were other problems.

On 15 January 1940 Brigadier Hollebone wrote to Harland & Wolff expressing doubts about the suspension. He was becoming alarmed at the number of man-hours required to make it and for some reason that is not explained, had asked Vauxhall Motors in England to investigate. This is the first mention of Vauxhall in connection with this tank but they will figure much more prominently soon. There is no reason why they should be expected to know anything of the subject because up to this time they had not had anything to do with tank design, but it seems that their suggestions were well received and they were asked to complete detailed drawings and produce four experimental units for trial purposes. Since we do not know what the original suspension was like we can only assume that it had features in common with the improved Vauxhall system which will be dealt with shortly.

The next day Harland & Wolff received another letter, this time from Colonel Watson, the Vice-president of the Mechanisation Board. In it he told them that, since their 300hp Diesel engine was still in the development stage, and production of the A20 pilots was close, they should go firm on the Meadows engine. A few days later revised weight figures were released, showing that the tank was now expected to be over 40 tons fully laden, which caused uproar amongst the Royal Engineers who pointed out that they had no bridges at all capable of supporting it.

All through the development stage the committee had been considering how production might be expanded. The

The only example of A20 to be fitted with a turret. It was photographed at the Harland & Wolff factory in Belfast. The Matilda turret retains its cupola, which would have placed the tank outside the British railway loading gauge. Slots in the side show where the flank machine-guns would fit.

order for 100 tanks from Belfast had been re-established in February and Mr. Blair had announced that the first pilot should be ready by June. If it was a success there was no doubt that more would be required and Vauxhall Motors was one of the firms that had been considered. The company was told that if they agreed to take on the project the Ministry of Supply would build and equip a shadow factory which Vauxhall would organise and run. Shadow factories were originally an Air Ministry concept, presumably since they were more aware than anyone else of the threat from bombing. The idea was that key plants would have an alternative establishment, not too far away, which they could move to in the event that their original premises were destroyed by bombing. Vauxhall Motors accepted the offer and claimed that they would be able to deliver tanks at the rate of about ten per week.

In March Dr. Merritt visited Belfast and reported back to the committee his view that liaison between Harland & Wolff and the Mechanisation Board was not good enough and that standards were slipping. He also noted that the original A12 turret took the tank outside the railway loading gauge so it was agreed that the new turret would be finished without a cupola to restore the situation. Having sorted that out someone came up with the suggestion that the side machine-guns should be replaced by two small machine-gun turrets, and this at a time when the first pilot model was under construction. Fortunately common sense prevailed and this scheme was dropped. Not only would it have exceeded the loading gauge again, it was estimated to add another 2 tons to the weight.

Major Raikes' report, mentioned earlier, has a number of more specific things to say about progress on the A20, or Mark IV has he terms it. These comments raise some interesting points. Written on 25th March 1940 the report explains that the hull of the first pilot model should be completed by April. It goes on to say that the firm Automotive Ltd., would be in a position to fit the controls by early May, that the engines were nearly ready from Meadows and the final drives from Leyland. On the subject of gearboxes both the Merritt-Brown and Wilson types are mentioned, suggesting that the first two tanks would have different systems for testing. Vauxhall Motors is recorded as supplying the suspension while the London & North Eastern Railway (LNER) company was producing what are referred to as the *main* tracks. This is followed by a reference to 'Auxiliary Tracks and Track Suspension'. No contractor is named and the matter is not explained but it does raise the possibility that A20, like the semi-mythical 'Flying Elephant' of 1916, was to have auxiliary tracks mounted between the main set, beneath the hull, to help reduce ground pressure and maintain traction if the tank started to sink on soft ground.

The term Main Tracks appears again when Raikes reviews progress on the production order for 100 machines. This report is mainly concerned with the availability of steel but it does confirm that no decision had yet been reached on the gearboxes and further explains that Meadows, due to their commitments for supplying engines to the Mark V Cruiser, or Covenanter, would not be able to meet the requirements for A20. A new Vauxhall power unit is mentioned as an alternative. Assuming that other contractors could meet their delivery dates Raikes believed that Harland & Wolff would have five A20 tanks ready by September 1940 with increasing deliveries each subsequent month until the hundredth tank was finished in February 1941.

In April the wooden mock-up was inspected by no less a personage than Sir Albert Stern. As secretary to the Landships Committee in the First World War Stern was always regarded as the driving force who, more or less single-handed, forced the original tanks into production against a lot of official resistance. A hard-headed businessman and born organiser Stern was an individual of very strong character but he never pretended to technical expertise at that time and he failed to gain any specific awards when the invention of the tank was considered in 1919. Now, of course, he was an elderly man who believed he had more to offer than was probably the case but he made the mistake of pretending that he was an authority on contemporary tank design and, for some reason, people were prepared to listen to him. Even so it was made clear to him that A20 was neither an infantry tank, nor a cruiser in the accepted sense and the term *Char de Fortresse* was coined to make the point. Stern appears to have been quite impressed at this stage although he was already at work on a project of his own. In company with many of those he had worked with in 1916 Stern persuaded the Government to allow him to design a tank on even more impressive lines than A20, which ultimately appeared as TOG (for The Old Gang) the nickname of Stern's consortium.

Although it had been agreed that the new Harland & Wolff tank would be shipped by rail there was still the problem of recovery to be considered. This was at a time when armoured recovery vehicles were unknown and, as already explained, wheeled tank transporters were regarded as suitable only for recovering disabled tanks. The type of transporter then in production was equipped to handle tanks up to 30 tons but this could not have coped with A20 so the Director of Ordnance Supplies asked the Ministry to design a suitable tractor and trailer unit for the work. At the time this was regarded as too difficult and the suggestion was made that in the case of A20 the tank would have to be repaired *in situ* so that it could, in effect, recover itself. Later heavy recovery trailers were developed and the hull of an A20 pilot was used as a test load.

On 20 April the mock-up tank was inspected by Monsieur Dupuy, the French Engineer-in-Chief, but his views are not recorded. In any case time was running out. German troops, spearheaded by tightly controlled regiments of tanks were soon pressing deep into French territory and, despite some short lived Allied successes, such as at Arras on 21 May, it would all be over in a few weeks. The war for which A20 had been designed was not going to happen. There would be no enemy fortifications to attack, no shelled area for the

The A20 tank supplied to Vauxhall Motors. A box superstructure is fitted in place of the turret and the tracks have been reversed. Notice also that the hull side door opens the opposite way to the example photographed in Belfast.

tanks to wallow in and no concrete obstacles for it to reduce to dust; at least not yet.

However the project was too far advanced to stop now. A mock-up hull had been built for firing-at trials and it was sent to the gunnery range Shoeburyness on the Thames estuary where they wished to see how it would withstand attack from high explosive projectiles. Harland & Wolff were informed on 27 May that the Meadows/Wilson engine and transmission combination would definitely be installed in the 2-pounder version of A20 and that a Lockheed hydraulic system would be used for throttle and clutch control with compressed air assisted steering. This no doubt means that a 6-pounder version was contemplated after all. Colonel Watson had stated on 20 May that the new turret being designed for A20 would take the 6-pounder but that any hope of installing a 75mm gun in the hull must be forgotten since it could not be done without a complete redesign of the front of the tank. This is probably just as well under the circumstances since the imminent fall of France would have cut off the supply.

Another visitor to Belfast on 1 June was Major John Dodd MP. Besides being a member of parliament Dodd was managing director of his family firm. S Dodd & Son Ltd., of the Lyon Works at Oldham in Lancashire. He was called back from his regiment at the outbreak of war and appointed Honorary Adviser on Tank Production to the Ministry of Supply, which appears to have been a freelance, troubleshooting role. Whether he knew much about tanks is not clear but the impression given is that he was a production expert. In Belfast Blair told him that Harland & Wolff would prefer to complete their full order for A12 Infantry tank and then work on A20, but Dodd was able to tell him that, apart from the four pilot models, the main order for A20 was now in abeyance 'as there was a probability of modification of the A20 to something different although similar in construction'.

On 7 June a meeting of the Tank Board was told that the order for 100 A20 tanks would be cancelled but the pilots must be completed. Indeed pilot number one was already virtually complete at Belfast although early trials had revealed faults in the hydraulic controls, but a Mr. Gates of Harland & Wolff was working on a plan to modify the system. Meanwhile the English Steel Corporation was able to report that it had solved the problem of casting thinner section turrets for A20, which presumably refers to 6-pounder turrets. The first pilot model, now classified A20E1, was photographed in Belfast with a Matilda turret and the same type was earmarked for the second pilot, A20E2. These would employ hydraulic traverse, like the Matilda itself, but the turrets then building for pilots 3 and 4 would feature an electric traverse system.

In July Sir Albert Stern returned to Woolwich to look at the A20 mock-up again. He was now highly critical of the limited internal space but admitted that the tank was a relatively small target. However when he examined Dr. Merritt's new transmission, which presumably was shown independently of the proposed tank, he was scathing. He pointed out that with ordinary clash gears it would always mean the tank getting stuck in the mud - he was thinking in First World War terms again - and he assumed that it would be expensive to produce and never available in sufficient quantities. It is indicative of Stern's poor mechanical acumen that he was criticising one of the most effective tank transmissions ever

Scale plans of A20.

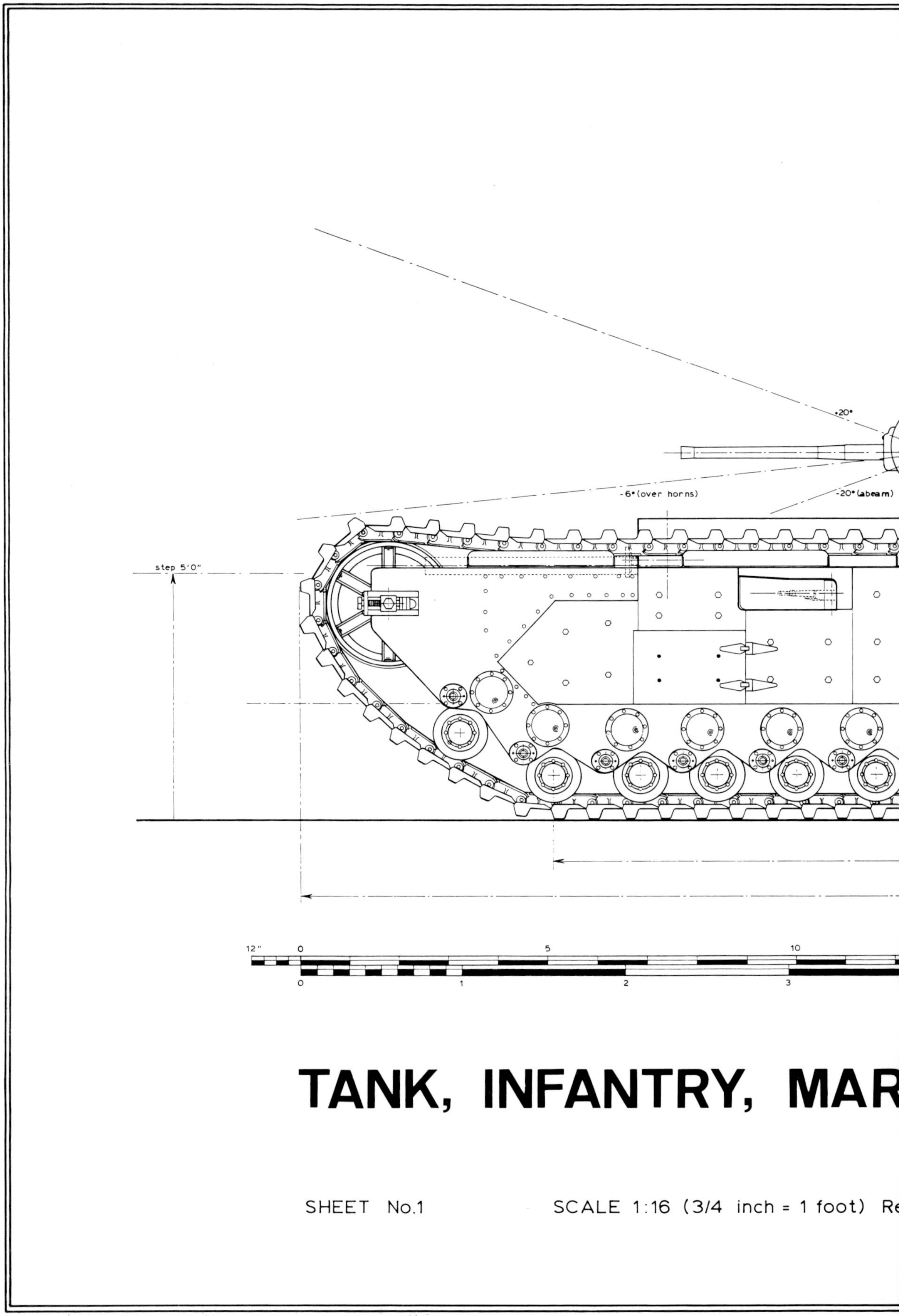

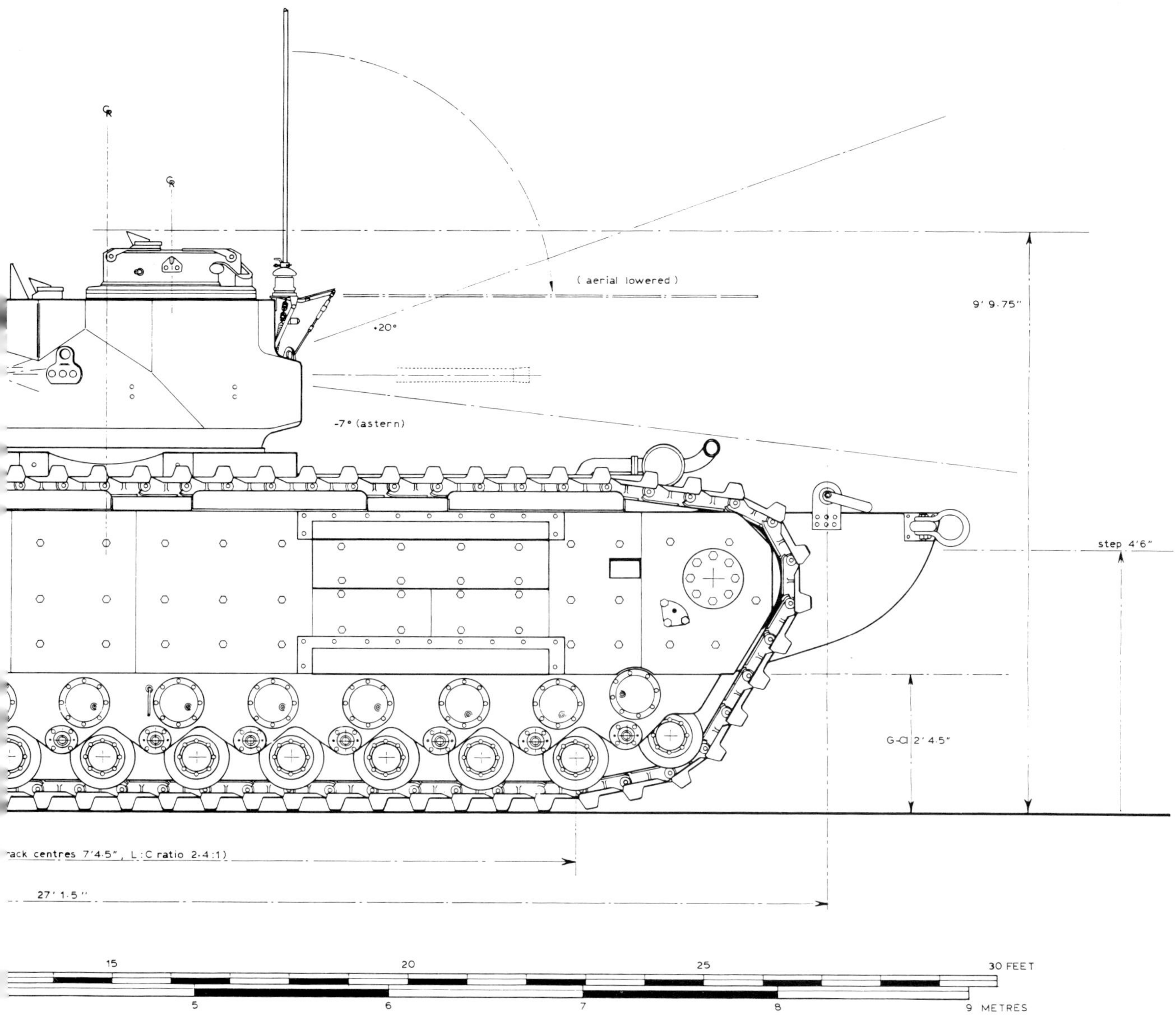

IV (PILOT MODEL, A20E1).

.ruction Drawing by R.C. HARLEY. 16 November 1997. © Richard C. Harley 16/11/1997.

devised. It had its teething troubles, as we shall see, but once these were solved it proved to be ideal and remained in use with British tanks for many years after the war.

The pilot model A20E1 was with Vauxhall Motors by August 1940 because a letter from a Corporal Little at Luton to the officer commanding the Mechanisation Experimental Establishment (MEE) at Farnborough mentions that up to 5 August the tank had completed 110 miles. In that time it had consumed petrol at the rate of about half a mile to the gallon but, more alarmingly, was using oil at the rate of one gallon every twelve miles! Corporal Little explains that the tank was loaded to its full combat weight and that, for some reason, its tracks were taken off and reversed. It is also reported to have taken part in a demonstration at Westbury, in Wiltshire, on 13/14 August. It arrived at MEE from Vauxhalls on 17 April 1941. It was issued with their number 1754 and it is interesting to note that the very next entry in the MEE ledger is for a 40 ton capacity transporter trailer by Cranes of Dereham, which probably explains how it got there. The tank's log at MEE is brief. Eight weeks after delivery it had registered a total distance travelled of just two miles and a month after that it is recorded as moving to the Experimental Bridging Establishment at Christchurch where it stays for just three weeks, no doubt serving as a make-weight for testing new bridges. In July 1942 it was passed to the Wheeled Vehicle Experimental Establishment (WVEE), also at Farnborough, where it was used as a trial load for a new 70 ton transporter trailer, also by Cranes.

The Mechanisation Experimental Establishment had been created, under another name, at Farnborough shortly after the First World War. It was the main testing centre for new military vehicles and the late arrival of A20E1 is a clear indication that it was never considered as a serious contender for service use. Indeed the first A22 pilot model arrived at Farnborough just a few days later and it is that tank which we must study now.

Incidentally there is a reference in the MEE ledger for this tank that some time in June 1941 it sustained damage to the rear towing eyes when it attempted to recover A20 by road, presumably following a breakdown. This may well be the only time the A20 and A22 tanks would have been seen together.

2

Vauxhall Motors and the New Tank

Vauxhall is a district of London, on the south bank of the Thames. In 1857 a Scottish engineer, Alexander Wilson, founded a business there which he called the Vauxhall Iron Works; its main purpose being shipbuilding. The company got into financial difficulties at the turn of the century but was reconstituted in 1903 to manufacture small, 6 horsepower, motor cars. This proved so successful that by 1905 new premises were required so the company moved to Luton, in Bedfordshire, about 30 miles north of London. The company underwent two changes of title, being finally established as Vauxhall Motors Limited in 1907. Throughout the First World War it supplied staff cars to the British Army and after the war established itself as a producer of quality sport and touring cars at the more expensive end of the market; and this at a time when smaller, cheaper cars with mass appeal were starting to dominate the market.

By 1925 the company was not doing so well and it was taken over by General Motors from the United States. Its British commercial vehicle building division, Bedford, was created in 1931 with a factory at Dunstable, close to Luton. Both branches of the firm were soon doing extremely well and Bedfords began building trucks for the British Army. This side of the market expanded rapidly up to the outbreak of war and the company even built a few armoured lorries for the Japanese Army, to be used in Manchuria. This aside they had no experience of constructing combat vehicles and did not include any tracked machines in their catalogue. Indeed none of their commercial lorries were intended for a payload of more than 3 tons and all were fitted with low powered petrol engines to comply with the strict requirements of British road haulage laws of the pre-war period.

From 1939, of course, army lorry production expanded dramatically and one imagines that production facilities at both plants were stretched to maximum. The company enjoyed such a high reputation that it was consulted by the War Office about problems with the cooling of tank engines and then came the request to improve the manufacturing process for A20 suspension. Having authorised construction of a new shadow factory near Luton it was then asked to undertake the design of a new, petrol fuelled tank engine. This engine had to be of the horizontally opposed configuration and capable of delivering no less than 350bhp, so it was clearly intended for A20 which, as its weight increased, was now regarded as underpowered.

At that time the biggest engine Bedfords produced was a straight six, rated at 72bhp. Vauxhall Motors had adopted the overhead valve layout for all their engines back in the twenties but it was soon evident that the new tank engine would have to revert to side-valves in order to fit, so it is quite clear that the company was engaged upon a major engine design project - it was not just a question of sticking a pair of six cylinder truck engines together.

It is a sad fact that in the decade before the outbreak of World War II no serious attempt had been made in Britain to produce a new engine for tanks. Many of the tanks in service in 1936 still employed a 90hp V8 developed in 1922 and efforts to improve it had failed. Lighter tanks, first introduced around 1932, had used commercial truck engines or, in the case of the heavier Matilda, a pair of six cylinder diesels similar to those used in London buses. But the pair of them together did not produce more than 170hp. Vickers-Armstrong had imported a 138hp GM diesel from the United States for their Valentine while the Nuffield Company in Birmingham adapted a tough old V12 aircraft engine, the USA Liberty originally designed in 1917, for their new Cruiser tanks. It could develop 300hp but was too big to fit the A20, even if enough of them could be produced.

In an article about the new Vauxhall engine, published in the journal *Automobile Engineer* of May 1945, an anonymous contributor explained some of the difficulties which faced the design staff at Vauxhalls. Size, as we have already seen, was of prime importance and dictated the horizontally opposed cylinder configuration. Then there was the shortage of quality materials, since hard wearing alloys were only being allocated to priority aircraft production. Enclosed in armour such an engine was difficult to cool and the designers also

His Majesty King George VI in the turret of Mark I T30971. The triangle, at the front of the turret, carries a warning that this tank is unarmoured.

had to consider the layout of fuel tanks and pumps which had to be kept cool to avoid the risks of vapour lock. Power output was an obvious consideration - the writer believed that 350hp was the minimum acceptable for such a heavy tank - and then there was the question of torque. A tracked vehicle, by its very nature, sets up a considerable amount of rolling resistance but when it attempts to turn the demand for extra power increases dramatically. In a vehicle which was often expected to climb or descend steep slopes, dry sump lubrication was desirable and special arrangements had to be made to prevent oil leakage at either end of the crankcase.

Bearing these difficulties in mind, and accepting that there were many others not recorded here, it is impressive to realise that it took less than three months from starting work on the drawing board until the first engine was running on the test bench. The company history gives it more precisely as 89 days. The cylinder and crankcase blocks of the new engine were made from pure cast iron. The crankcase was formed in two halves, each half including a bank of six cylinders. The cylinder heads, in high tensile molybdenum iron, were created for sets of three cylinders. Each cylinder was 5 inch bore by 5.5 inch stroke, giving a total capacity of 21300cc. The crankshaft ran on seven main bearings and the pistons were aluminium alloy. Separate camshafts, driven by roller chain, were provided for each bank of cylinders and early engines employed hydraulic tappets, but these suffered from an ingress of dirt and plain tappets were later substituted.

Dual ignition was employed, with distributors mounted on top of the engine, along with four Solex carburettors, while twin fuel pumps were located beneath the engine where they could be kept relatively cool. The engine had a pair of AC-Sphinx, oil-bath air cleaners which were located in the crew compartment while the oil coolers, situated either side of the engine at the flywheel end, were fed by a large scavenge pump. Separate water pumps served each cylinder block and an induced draught cooling fan was bolted to the rim of the flywheel. A centrifugal governor limited the engine to 2,200 rpm which provided the required 350bhp. Compression ratio was 5.5 to 1 and the torque was measured at 960 ft/lbs. The engine was fitted with a heavy duty, 18 inch Borg & Beck clutch. The belt driven dynamo was located on top of the engine with an axial electric starter motor mounted beneath the left hand block engaging an alloy steel gear ring on the flywheel.

Apart from the valve problems already mentioned there was considerable trouble with cylinder head gaskets, which had to be redesigned twice, and the big ends of the main connecting rods which had to be strengthened. All of this had to be resolved before production could begin and it is worth noting that the company was even obliged to design and build its own dynamometers and testing bays while they were working on the engines because none of their existing equipment was capable of handling it.

The Tank Board, mentioned in the previous chapter, was first suggested by the Minister of Supply, Mr Herbert Morrison, on 29 May 1940, while dramatic events were unfolding at Dunkirk. Morrison announced the new Board in the House of Commons on 11 June 1940. The next two weeks were so hectic that nothing was done so the first meeting did not take place until 24 June, four days after the French government surrendered. The situation had changed drastically. Chairman of the Board was Sir Alexander Roger of BSA; other members were drawn from British industry and the Trades Union Council, along with the Director General of Tanks and Transport, Geoffrey Burton, the Director of Mechanisation and, from the War Office, Major-General Vyvyan Pope and Brigadier Douglas Pratt; both of whom had just returned from France. Item 1 on their agenda was:

> 'To consider (by 27 June 1940) the specification for proposed new Tank A22'.

Which did not give them very much time. These original specifications have not survived but a revised set, resulting from this first meeting, were virtually a description of the tank as built. At the meeting Geoffrey Burton explained that they had been drawn up in order to meet a War Office requirement but bearing in mind most particularly the Prime Minister's wish for a simplified heavy tank to be available in considerable numbers by 31 March 1941. General Pope commented that he had never seen these draft specifications before they were handed to him shortly before the meeting began but at first sight he thought they fell between two stools; being neither an infantry tank, nor a mobile warfare tank.

There is no point in recording the subsequent comments in detail since we do not know what the specifications were that they commented upon, but the following points are of general interest. Clearly the armament was to include a hull-mounted 3 inch howitzer because General Pope criticised this. He believed that two different weapons in one tank could create problems with ammunition stowage and it might slow up production. Pope also wanted more armour, if it could be added without seriously increasing weight and he wanted a cruising speed of 10 miles per hour. It seems that in the original specifications the turret was to be turned by hand and no rotating floor was to be provided for the crew. This was considered a retrograde step and Brigadier Pratt agreed to carry out tests on a Matilda before making further comments.

There was a lot of discussion on methods of transporting the tank, which was quoted as being 12 inches wider than A20. Evidently this was regarded as too wide for general rail travel without making special arrangements and the Board felt that this was unacceptable. The alternative, of designing a trailer for road transportation, was rejected because it would be difficult to find production capacity and in any case the combination of tank and trailer would be too heavy for most British roads.

The Tank Board met again two days later (26 June), which is when the revised specifications were issued. Douglas Pratt reported on his experiments with the turret of an A12. The gunner and commander, he said, would have to be slung. This was not a comment on their usefulness but a suggestion that they be provided with small slings, or seats, suspended from the turret, so that they moved around as it rotated. The loader could manage by shuffling around with the gun. He estimated that this would result in a loss of up to 25 per cent in fighting efficiency but thought it would be acceptable if the alternative was a serious delay in production.

The great man himself, Prime Minister Winston Churchill, cigar in mouth, is assisted aboard a Churchill tank. It should be remembered that he was in his late sixties at this time. The tank is a Mark I, but the 3 inch howitzer has not yet been fitted into the hull.

Colonel Blagden, on behalf of Geoffrey Burton, explained that revised specifications had brought the width down to 9 ft 3 ins., and that it would now be possible to move the tank by rail if the tracks were removed. This seems to be such a ridiculous idea that one wonders why anyone should bother to propose it in the first place. Fortunately the Tank Board rejected it. However the main subject under discussion was that of production. No contractor had been appointed at this stage but it was already clear that manufacture would be by what was known as the parent system.

This involved one company being appointed production parents for the tank but that much of the assembly would be undertaken by sub-contractors. These would be responsible for assembling parts of the tanks supervised by the parent. But the various component parts would be provided by a host of lesser sub-contractors. The parents would also be provided with what were known as 'free issue' items at government expense; things such as weapons, radios, telescopes and so on. The Board was suggesting that each sub-contractor should be made responsible for creating complete tanks so that 'free issue' items would be delivered direct to them. The chairman agreed to give this further consideration but since a similar arrangement already appears to have been operating in the case of the Matilda its discussion here is rather puzzling.

A special meeting of the Tank Board was held on 28 June in order to consider Sir Albert Stern's suggestions for a super-heavy tank, but the A22 was not mentioned. However when they met again, on 6 July, a new problem had been introduced. A few days earlier the Prime Minister issued instructions to proceed, with a demand for a minimum of 500 tanks to be completed by the following March. This set things humming. Geoffrey Burton had already warned Churchill that it was not practicable to guarantee 500 new tanks in such a short time. The Board agreed and admitted that the minutes of their first meeting probably did not go far enough to stress the revolutionary nature of the design and the difficulties of transporting a 35 ton tank. This, they said, was unexplored territory, especially in matters such as weight and engine power.

Vyvyan Pope, who probably did not much like the design anyway, was all for abandoning it. He pointed out that there were already four approved types of tank in production and that was a good enough reason in itself for not adding a fifth. At least not until more experience had been gained. Clearly he preferred to buy more time in the hope of obtaining a better tank later. Geoffrey Burton argued for the longer view, saying that he believed work should go ahead on A22 since economies in production might enable larger numbers to be available by the autumn of 1941. The Board agreed but felt it necessary to add that the military authorities should be advised not to make any plans based on having 500 new tanks ready by next March. Appreciating that, even as they were talking, huge air battles were raging all over southern England, they added that the risks from enemy bombing could easily upset the best laid plans.

According to an official summary, issued in 1942, Vauxhall Motors was instructed to begin detailed design work and manufacture of the new tank on an unspecified date in July 1940. All one can say for sure is that it was some time between the 6th and 22nd. On that last date the Defence Committee met to consider Geoffrey Burton's report on A22. The difficulties of trying to produce 500 tanks by next March were pointed out and it was agreed to request a Priority 1A for them. The priority system was laid out in a printed document, issued by the Ministry of Supply on 14 June 1940. In basic terms it was an instruction to manufacturers on how to grade the work they had in hand. Priority 1A covered virtually anything to do with aircraft, anti-aircraft artillery and small arms. Tanks, at this time, came in category 1B along with field and anti-tank artillery, machine-guns and ammunition. Writing to the Director General of Tanks and Transport in September 1940, John Dodd suggested that this system was a nonsense. He pointed out that any contractor with projects in category 1B found, whenever he made ready to start, that another category 1A job landed on his desk which immediately pushed the 1B order further back. To make matters worse most firms, despite working shifts around the clock, always stopped production when an air raid warning sounded and as an example he cited Henry Meadows Ltd., who had only managed five hours of night shift work in one week owing to this interruption.

By August 1940 most of the main design features had been settled but very few details had been resolved. Even so a wooden mock-up was well on the way to completion and the first Bedford 'Twin Six' engine was on the test bench. This is not conclusive but it points to a date sometime in the middle of May when Vauxhall Motors was first ordered to work on the engine. In September the wooden mock-up was approved by the General Staff and design work was completed, except for the tracks and the gun mountings which had to be modified. Planning for production was said to be fifty per cent complete.

At a meeting of the Tank Board, held at Ivybridge House, London, on 20 September 1940 General Pope sounded another warning about guns for tanks. Having reviewed the situation in detail he announced that all tank weapons, from machine-guns and smoke mortars to 2-pounder guns and 3 inch howitzers, would fail to meet demand by that winter. He pointed out in particular that when orders were placed for the howitzer the A22 infantry tank was not considered, and this had upset all the calculations. With most tanks the 3 inch howitzer was fitted as an alternative to the 2-pounder and only on a very limited scale. In an average regiment of fifty tanks no more than ten, but more likely eight, would require the close support weapon. Yet it had been agreed that every A22 should have one in addition to the 2-pounder, adding an extra 500 to the original order which it would be impossible to meet.

By October Vauxhall had completed the pilot hull and started putting it together but numerous design details remained to be settled and there were problems finding suppliers for suspension units, final drive assemblies and armour castings, presumably turrets. Vauxhall Motors was expressing doubts about their ability to have even 200 tanks ready by March and in response to the shortage of 3 inch howitzers a change in armament was being considered. To aggravate matters further an instruction was received from the Prime Minister to the effect that tank production would go a lot more smoothly if people would simply go ahead and build them, rather than continue endlessly making changes to the design. It was the kind of comment calculated to drive engineers mad and in the case of A22 it was clearly unrealistic. Yet, as usual, Churchill had gone straight to the heart of the problem. Obviously, where something was wrong it would be crazy not to modify it. Yet it was equally foolish to continue making changes for their own sake. It is in the nature of committees to talk every subject to death and, in the process, lose sight of the principle objective. Since the Tank Board was no more than an advisory body its recommendations could be ignored, but they were bound to have some impact and this only served to interfere with production while they were considered.

Most of the information we have concerning Churchill production comes straight from the original contract cards, recorded on a Hollerith type punch card system specially designed for detailing British military vehicle production. The cards are held in the Tank Museum archive. They are not easy documents to interpret, since many have been corrected more than once, but careful study reveals some interesting details, and since these are the first Churchill contracts to be issued they probably deserve closer examination.

Recalling that the initial requirement, authorised by the Prime Minister, was for 500 tanks, the first group of contract cards adds up to 485 vehicles. All cards are dated 5 October 1940 and are mainly in batches of 70 or 75. Vauxhall Motors, as production parent, is not included in this first batch but the Master Card (Contract T301) lists 450 tanks not shown elsewhere so this was probably their initial contribution. Thus the first contract card for Churchill tanks, dated as above, is T747 for 75 'Tanks A22 to be Assembled' by the Austin Motor Company. However this order is then marked 'Cancelled', with the date given as 5 December 1940. This appears to be the only one that is cancelled, all the others are increased considerably. But in their original state they are as follows:

T748 for 70 tanks with the Birmingham Railway Carriage & Wagon Company.
T749 for 70 tanks with Metro-Cammell Carriage & Wagon Company, also of Birmingham.
T750 for 70 tanks with Leyland Motors of Lancashire.
T751 for 75 tanks with Dennis Brothers of Guildford.

A Churchill tank under trial at the Vauxhall Motors testing ground. It is fitted with a superstructure similar to that on the second A20 and this picture shows how easy it was for mud to get all over the hull and into the original pattern air inlets.

T752 for 50 tanks with Broom & Wade Ltd. of High Wycombe.
T753 for 75 tanks with Newton Chambers Ltd. of Sheffield.

The Vauxhall Master Card, Contract T301 dated 9 December 1940 exists in two forms for 1,950 and 2,775 machines but this only reflects the rapid expansion of production and the next set, all dated 21 December 1940, are as follows:

T754 for 70 tanks with Harland & Wolff of Belfast.
T755 for 70 tanks with the Gloucester Railway Carriage & Wagon Company.
T756 for 75 tanks with Beyer, Peacock & Company of Manchester.

A22 Mark I T32246, assembled by Leyland Motors. It has the unarmoured warning triangle on the turret and the howitzer can just be seen at the front.

It must be emphasised again that with the exception of the cancelled order with Austins all the other firms had their order increased subsequently and that, in every case the firm is contracted to assemble the tanks, as distinct from manufacturing them. It is possibly a fine distinction but a significant one where the Churchill tank is concerned.

An undated report, issued at the end of 1940, sounded a warning about weapons. As already explained, the sudden development of A22 seriously upset the production programme for guns and, although attempts were made to meet the new demands, it was now clear that arrears in delivery of the 3 inch howitzer would never catch up. Those tanks for which no howitzer was available would be fitted with a hull mounted BESA machine-gun instead. The first pilot model (T15129) was completed at Luton by Christmas, which by any standards was a remarkable achievement and a second tank was ready by January 1941. Trials, conducted by Vauxhall Motors, began at once. In what will prove to be an impressive piece of understatement a Department of Tank Design monthly report mentioned 'a number of inevitable small defects ... brought to light in the trials' and added that modifications would be introduced in production machines as soon as possible. It went on to say that the 'behaviour of this vehicle under test had been most encouraging'. But in its February report things did not sound so good. Bogie wheel failures were becoming all too common as were bearing failures and problems with the engine cylinder head gaskets.

A War Cabinet Defence Committee meeting held on 26 February announced that two pilot models were being tested and a third was nearly ready. With that kind of optimism only politicians can generate it went on to say the results of these tests were sufficiently favourable to justify starting production without waiting for the results of further trials and that materials for up to 1,500 Churchills were on order. The Prime Minister then said that production should be pushed forward because of the tank's 'superiority over all other models'. The Committee therefore instructed the Ministry of Supply to treat the project as top priority.

Four more pilot models appeared; T15130 from Metro-Cammell, T15131 from Birmingham Railway Carriage and Wagon, T15214 from Harland & Wolff and T65440 from Leyland. Deciding what to do with them caused a slight stir. At a weekly meeting of the Director General of Tanks and Transport's (DGTT) department, on 4 March 1941, it was agreed to issue them as follows; Pilots 1 and 2 should stay with Vauxhalls, number 3 should go to the Mechanisation Experimental Establishment, number 4 to the Gunnery School at Lulworth, 5 and 6 to Bovington Camp while 7 and 8 would be issued to units for user trials. The allocation of eight tanks indicates the use of at least three early production machines as, effectively, pilots. The following week's meeting was told that General Pope did not place MEE high on the priority list. He wished to have the tanks redistributed so that more of them went to units. Geoffrey Burton said that in his view it was up to the DGTT to allocate these tanks, not General Pope, and he wished the General to be reminded of this. The argument rumbled on, mainly focusing on the respective priority of mechanical trials at MEE and gunnery trials at Lulworth. The general view seems to have been that there was little point in checking out the gunnery arrangements until the tank was seen to be working properly and, as Burton pointed out, none of those so far completed could be described as even reasonably fit in a mechanical sense.

Following Churchill's comments on tank modifications, mentioned earlier, the Tank Board met once more in November and then went into hibernation. General Pope was scathing and wrote to Sir Andrew Duncan, the new Minister of Supply, suggesting that he sorted it out. The result was that the Board was reconstituted and met again in January 1941 under its new chairman, Sir James Lithgow. In its new guise the Tank Board was charged with taking decisions on AFV design on behalf of the War Office, which in effect gave it executive powers, while its advisory role to the Ministry of Supply remained as a secondary function. Churchill, wearing his other hat as Minister of Defence, informed the Board that it should concentrate on expanding production of existing types for the present although where the A22 fitted in to this was not made clear.

Not that this changed things very much. At their February meeting the Board spent a good deal of time discussing wider aspects of tank design and in particular the instruction that in future production of cruiser tanks should have priority over infantry tanks. Which was all very well except that the entire production programme, laid down at the outbreak of war and only altered in the light of events in France, was geared the other way. It might have been possible but for the joker in the pack, the A22, which had been wished upon the Army by an impetuous Prime Minister. In a letter to Sir James Lithgow, dated 10 February 1941, Commander E R Micklem of Vickers-Armstrong said; 'it is clear that the Infantry Tank Mark IV must remain an experimental machine for some time to come ... it seems to me quite clear that unless some definite order of priority is given there will be a serious risk of the Infantry Tank Mark IV Group [meaning the companies under the parentage of Vauxhall Motors] absorbing capacity which could be more usefully used on the production of one of the established machines'.

He got his answer two months later when Sir Andrew informed the Board that the Prime Minister had been pressing for the production of the Mark IV and that the Minister of Supply had promised to have 100 ready by the end of May and a further 400 completed by August. It had, therefore, been agreed that the manufacturers should complete 100 tanks to the existing (but flawed) design as soon as possible and that improvements incorporated in the later 400 would be carried out on the first batch later. In other words time, energy and money would be expended on 100 useless tanks to keep the Prime Minister happy rather than holding up production to achieve a better product. Surprisingly General Pope supported this approach and members of another committee minuted the comment that one hundred A22 tanks should be completed 'regardless of what they were like' and the final deci-

An official photograph of a Mark I in which the hull gun can clearly be seen.

sion on their future postponed. Then, like the ghost at the banquet, Sir Albert Stern popped up to say that in his view production of the Mark IV should be halted in favour of the Mark II (Matilda), presumably on the grounds that it was at least good enough. Stern was told firmly by Sir Andrew that whether production of the Mark IV carried on or stopped it would have no effect whatever on the number of Matildas produced.

By March the first pilot A22 had completed 620 miles and there was some improvement in the bogie wheel situation, but there is an odd reference to improved cooling when none of the earlier accounts had mentioned a problem. Yet the report goes on to say that the tank's weight had now reached 38.45 tons, which exceeded the design figure so proposals for weight reduction were being considered. An altogether more bland report, issued to cover the second half of 1940 describes the new tank as weighing 35 tons laden and having a top speed of 15mph. It also explains that the original five-speed gearbox would be replaced by a four-speed unit in due course but it is most interesting on the subject of armament.

To begin with it states that the new heavy tank has a 2-pounder and co-axial 7.92mm BESA in a three-man cast turret, with a 2 inch smoke mortar in the roof, and a 3 inch howitzer or second BESA machine-gun in the hull. It then adds that a plate turret is being developed to take a new triple mounting that is discussed elsewhere in the report. The triple mounting would place a 2-pounder gun, 3 inch howitzer and BESA machine-gun on a common axis mounting in the turret. The scheme was initiated by the Department of Tank Design in September 1940 and a pilot model, expected to be ready by February 1941, was being developed by Vickers-Armstrong in conjunction with the old established engineering firm, Stothert & Pitt of Bath. The logic behind it was inescapable since it provided any tank that could accept the turret with the ability to fire armour-piercing, high explosive or smoke rounds at will. Yet the practical aspects of the project were equally daunting. General Pope had already pointed out the difficulties associated with having two different guns in the same tank, never mind the same turret. The problems for the loader and gunner, having to serve and operated two breech loading weapons and a machine-gun from within a cramped turret will be obvious enough but one must also consider the difficulties that would arise where three different types of ammunition must all be stowed within easy reach of the loader. Fortunately both main weapons could take the same telescopic sight. but that was one of the few positive aspects. In the event this complicated arrangement never got beyond the prototype stage but it is referred to in a number of accounts on A22 development around this time. Interestingly the same report also mentions that a suitable mounting for the new 6-pounder gun was also being considered.

Whatever plans were made concerning armament they could be considered somewhat academic while delivery of guns was falling behind. We have already seen that the BESA machine-gun was being seriously considered when deliveries of the three inch howitzer were in doubt and it now appeared that supplies of BESAs were also in arrears. One school favoured acquiring the American Browning and discussions went ahead with a view to altering the mounting to accept either weapon, bearing in mind that it would also have to be used in the turret. Geoffrey Burton wondered if the Bren gun might not be used instead, but it was pointed out to him that the Bren was too long at the breech end to fit into a tank. As if all this were not enough another report, issued at the end of March, indicated a delay in the supply of gun mountings and it was agreed that tanks for which neither guns nor mountings could be supplied, should be issued with a blanking plate of steel fitted over the opening.

If one aspect dominates all meetings convened to discuss the new tank it is the subject of suspension and tracks. Spring and bogie roller failures were becoming endemic. One solution proposed by Vauxhalls was to fit rubber tyres to the road rollers although new tracks would be required as well and the company was working on a manganese type, but it was not accepted immediately, largely because the supply of rubber was restricted. Yet the rollers, which were of mild steel, seem to have been breaking up with monotonous regularity and a number of alternatives were suggested. One was to harden the original wheels, another was to create wheels with mild steel centres and high tensile rims while cast and forged types were also considered. Other producers came forward with various hardened steel formulations but none of them stood up to the punishment. Vauxhall Motors produced a bump test rig device which simulated the effect of continuous movement over hard ground and this proceeded to destroy virtually every type of wheel tested on it within about two hours. Trials carried out in the field were no less disappointing. One carried out by MEE noted that the roller bearings overheated so much that paint blistered and on one trip eleven gallons of lubricant were used on the suspension in eleven miles, which was ridiculous. A Mr Dean of Vauxhalls came forward with his own modifications, only to discover that entire bogies started to fall off. Tracks were equally troublesome. The Ford Motor Company at Dagenham produced their own steel tracks, and these are said to have been fitted to the first 100 machines. Vauxhalls believed that a shorter pitch track might be easier on the suspension and started to produce something.

The DTD report for April gives an upbeat appraisal of the suspension problems but goes on to explain that the engine situation had not improved. A new type of cylinder head, of stronger construction, was under intensive testing but the problems had not yet been resolved. Indeed it becomes clear in the next report that Vauxhall Motors are having to consider redesigning the heads and block to a more robust standard and that can only mean yet another increase in weight. There is further mention of suspension problems and Vauxhalls are said to be investigating alternative materials for the wheels.

The problem of weight was clearly becoming critical too and, short of reducing the armour protection, there was only one way of keeping it in bounds; that was to weld the tank's hull rather than bolt it together. This matter will be examined in more detail later, when the Mark III version is discussed, but suffice it to say for the moment that by April 1941 a sample of the front end of an A22 hull had been assembled by a welding process and submitted for firing trials. The results had been sufficiently promising for the firm Babcock & Wilcox of Renfrew had been contracted to produce a sample batch of 50 welded hulls; but more of that later.

For the present we are approaching the critical point in our story when the first production tanks were about to appear and we must examine the type in some detail. But before we do it is worth noting some comments which appeared in an anonymous document dated 18 May 1941 since it typifies the sort of thinking that bedevilled British tank development throughout the war. The writer is commenting on a paper on tank design submitted by Mr J Weir who had joined the Tank Board earlier in the year. The precise matters raised in his paper are not known but they mainly concerned gunnery and the ability of tanks to accept the 6-pounder gun. On the subject of infantry tanks our anonymous respondent explains that the Mark II (Matilda) could not accept the 6-pounder gun, that the Mark III (Valentine) might but was it worth it? And that the Mark IV (A22) could but it 'is a heavy tank largely owing to its obstacle and trench crossing capacity. Are the tracks vulnerable as they appear from the front, and can the idler be lowered in future tanks, and the length reduced? If so A22 might become a better all-round infantry tank to carry the 6-pounder'. Perhaps the question should have been; if A22 is not suitable would it be better to mess the design around, or start again?

The earliest completed production Churchill for which we have any evidence is T30971, the first number on the original Vauxhall contract card. It arrived at the Mechanisation Experimental Establishment at Chertsey on 24 May 1941 and took their number 1793. It is described as Tank, Infantry Mark IV but noted against it is the single word *unarmoured*. This is slightly strange since it is not one of the five authorised pilot models, which might well be built from unarmoured plate, but in theory at least a production machine. It was followed by T31347, another of the Vauxhall batch, in July. The full report is not available but from what has survived it is quite clear that suspension and wheel problems remained the most pressing yet the report also mentions other detail defects that have come to light. Given that the new tank would continue to cause trouble for at least another eighteen months this can be interpreted either as a lack of awareness of the seriousness of these problems or the traditional English habit of ignoring unwelcome facts in the unjustified hope that they would go away. The final report for this period, that of June 1941, shows that 14 production tanks had now been completed and blames the delay on modifications. Fourteen in June is certainly not 500 by the end of March no matter how much the Prime Minister may have insisted, and things were not destined to get better for a long time. This same report makes the first mention of another departure from the original programme that we will deal with later, a self-propelled 3 inch gun on the A22 chassis.

Trial reports were, by now, coming in thick and fast and they were uniformly negative. One noted that the side mounted air inlets sucked in mud by the gallon and it seems to have got in everywhere. Mud in the gearbox, mud in the engine valves, mud even in the electrical fittings, so a new design of inlet was devised which drew in the air from above, along with fallen leaves in the autumn. Gearbox trouble was also prominent. It had been agreed that the original five-speed box should only be fitted to the first 100 tanks since complaints were received that it jumped out of top gear, and there was one incident where two gears were engaged at the same time. Unfortunately the habit of jumping out of top gear was also identified in the four-speed box and in time a common fault was identified, not in the gear selectors, which were different, but the drive dogs which were the same pattern. Strangely, reports from users early in 1942, suggested that the five-speed box gave less trouble but this was dismissed by the authorities. The Experimental Wing at the Department of Tank Design pointed out that since four-speed boxes were more common in British service this was a good enough reason to adopt them for A22 but in any case there were more minor faults reported with the five-speed type. Four-speed boxes, it seems, did not produce so many faults, but those that were reported tended to be more serious.

Studying reports from this period one is struck by the lack of complaints about the Merritt-Brown transmission. Since it later became one of the chief sources of trouble this might seem surprising, but it may well be due to the limited mileage achieved in the early trials. One source claims that the original brake drums were too small but there are indications of more serious problems. A meeting of the DGTT's committee suggested employing the American Cletrac system instead but this was rejected on two counts. In the first place it did not provide a neutral turn facility which was believed, in Britain, to be ideal for manoeuvering tanks into tight places, such as the holds of ships. Also it could not easily be adapted to fit into the A22's hull. Oddly, in view of his earlier criticism, Sir Albert Stern reappears in a DGTT report of 16 July 1941. It seems that an A22 tank had been supplied for examination by his Special Vehicle Development Committee. Now that his project for the superheavy tank TOG was coming to a

A Mark I Churchill in service with the 43rd Royal Tank Regiment. The distinctive marking on front and side is a red/white/red flash which was adopted for identification purposes on British tanks at this time.

belated end Sir Albert was looking for other ways to maintain his involvement with tank design. The response to his offer to help improve the new infantry tank is not recorded although it seems clear that nothing ever came of it.

Other reports, filtering back from regiments that had the tanks for trials spoke of a dust problem, when it wasn't muddy. Air, blown from the lower cooling outlet at the back, swept the dust kicked up by the tracks into a cloud which emerged at the front of the tank and blinded the driver. Comments were also made about the tank's ability as a gun platform. The suspension gave such a rough ride that it made firing on the move almost impossible, and firing on the move was regarded as normal practice by the British. And then there was the noise. Another report described it as terrific and forecast that it would damage the hearing of crew members. As it was most men were deaf for 48 hours after a run in the tank and even after that crews were described as looking shaken and incapable of concentration.

While work went on to remedy the problems it was agreed that trouble could be minimised if the tanks were restricted to 10mph and this could be achieved by blocking off top gear. This speed was the maximum that could be achieved without having the tank start to pitch uncomfortably. MEE reported that the tracks and suspension of A22 made it unsuitable for rocky or desert ground and they also registered two gearbox failures. The DGTT, Geoffrey Burton, declared 30 July 1941 as Decision Day for the A22, but it was an empty deadline. The staff at Farnborough stated that their idea of a successful test would be when the tank was able to travel 1,000 miles at 12 mph, and they did not mean cover 12 miles in an hour by varying the pace but maintaining a steady 12 mph for long periods. Burton reminded them that unless A22 could be made battleworthy Britain would be without a 6-pounder armed tank until the Mark VII Cruiser was ready, some time in 1942. And Burton had already promised that a tank with a 6-pounder gun would be ready for service by the end of 1941.

In September 1941, at the suggestion of the Prime Minister, an order was printed which gave all British tanks a name. Until now, with the exception of the Matilda and Valentine, most tanks were known either by their General Staff A number, such as A9 and A10, or their official title such as Light Tank Mark VII or Cruiser Tank Mark V, which became the Tetrarch and Covenanter respectively. The A22, or Infantry Tank Mark IV was to be known as the Churchill. Whether this was intended as an honour or a rebuke is not clear, but it appears to have been taken in the former spirit.

3

The Churchill Described

What follows is a reasonably comprehensive description of the Churchill Mark I and it will serve in general terms for all of the main production models up to Mark VII, which in many respects can be regarded as a new tank. Interim changes of mark will be dealt with as they occur.

Two types of track were provided for the early Churchills, one of which was of First World War style. This consisted of a shaped plate of pressed steel, riveted to internal bars which held the track pins and acted as guides. Each link presented a heavy raised lip to the ground which was supposed to give better traction in mud; designed as a result of experience in France in 1940. It is seen in some of the earliest photographs of Churchill tanks after which it rapidly, and mercifully, disappears. The second type was formed from a single steel casting of more complex shape which offered extra support to the track pin but presented a less aggressive surface to the ground. This, no doubt, as a result of subsequent experience in the desert These cast steel links were shorter than the plate type so 144 links were required to encircle the tank rather than 140. However they were also a good deal heavier, by 11 lbs per link. The tracks were not interchangeable. The difference in pitch demanded that the idler and drive sprockets had to be changed with the tracks. Again, based on earlier unfortunate experience, the track pins were not lubricated.

Simple as they look the suspension units were extremely complicated. The rollers, of which there was a pair on each axle shaft, were flanged like railway wheels and steel rimmed, which probably accounts for a lot of the noise transmitted to

A Churchill Mark I photographed for the frontispiece of the original Infantry Tank Mark IV handbook. This example is fitted with the original style of heavy, built-up tracks.

A new hull on the assembly line showing the flanged rollers and suspension on full stretch. Notice the five humps which carried the top run of the tracks, these were later replaced by full length skid rails.

One bugbear of the Parent system of construction was that parts of each tank had to be moved around the country from one factory to another. Here a basic hull assembly is seen on the back of a civilian lorry, a rigid eight-wheeled AEC Mammoth Major.

the inside of the tank. The shaft ran through white metal bearings (which replaced the original bronze bushes) in a curved steel bracket known as the bogie frame. The axle shaft was hollow and served as a reservoir for lubricant. The bogie frame was pivoted near its centre on a boss mounted on the bogie support bracket. The upper end of the bogie frame pressed against a rebound pad which limited its travel. At the lower end was the roller assembly which was crowned by a steel knife edge. This pushed against a plate at the lower end of the spring assembly. Each bogie included four coil springs fitted one inside the other. The outermost spring, which was also the heaviest, was the main load carrier, assisted by the next inner, which was of lighter construction. Inside that was the bumper spring which only functioned after the outer springs had moved through most of their deflection. As its name implies the bumper spring helped to take the strain of the last three-quarters of an inch when the bogie took a heavy blow. The innermost spring, which was of very light construction, only served to restrain the bumper spring. Each bogie was capable of maximum 3 inch deflection and 2 inch rebound but rebound was restricted on certain bogies to improve the ride. A single bogie unit weighed 336 lbs and there were eleven of them on each side of the tank, all identical except for the thickness of their rebound pads.

The leading bogie on each side normally ran clear of the ground and only came into play when the tank attacked a steep bank or other obstacle. The second bogie had limited rebound and was not load bearing either. On a hard, level surface it stayed clear to shorten the length of track in contact with ground. Bogies three through ten supported the tank at all times. They were arranged, equally spaced, either side of the centre of gravity. The last bogie was regarded mainly as a track tensioner and had limited deflection. Individual bogies were bolted direct to the hull and, apart from the rebound pads, were interchangeable.

The high track profile, inherited from A20, was in keeping with perceived role of the Churchill as an assault tank that might have to deal with substantial trenches, which was another First World War concept. On its return run the track was supported on a series of five curved skids on each side instead of rollers but this feature must have introduced some rolling resistance. To enhance the Great War connection even further both the drive sprocket at the back, and the idler wheel at the front (which was also the main track adjusting wheel) were toothed. This technique had been employed on the very first tanks but proved troublesome because the tracks stretched quickly and failed to mesh with both sprockets. The fact that it could be employed successfully on the Churchill at least suggests that some progress had been made in track materials and design since 1916. The reason for fitting teeth to the idler must have been to prevent the track from working its way off at the front. The Churchill employed a twin drive system, meaning that the drive sprocket had two toothed rims engaging with the edges of the track, which is normal practice although it could be more easily affected by stones or other foreign bodies which were picked up by the tracks.

The armoured hull was arranged in what tank designers called pannier fashion. That is the armoured portion of the hull extended sideways to the outer edges of the tracks. The difference being that in this case the panniers were within the compass of the tracks whereas on tanks such as the American Sherman or German Tiger the hull panniers extended above the tracks. Construction was described as double box girder since each pannier formed a rectangular structure on each side, creating a hull of immense strength. All hull plates were flat and assembled by riveting on early models, the respective thickness of the various plates can be seen from the diagram.

Large, square doors were located in the panniers on each side, just behind the driver's seat, and further back were the air intakes. As already explained these were shaped to draw in air from below but, since the top run of the track was exposed they suffered greatly from mud pouring down from above and water picked up from below, working its way in. A

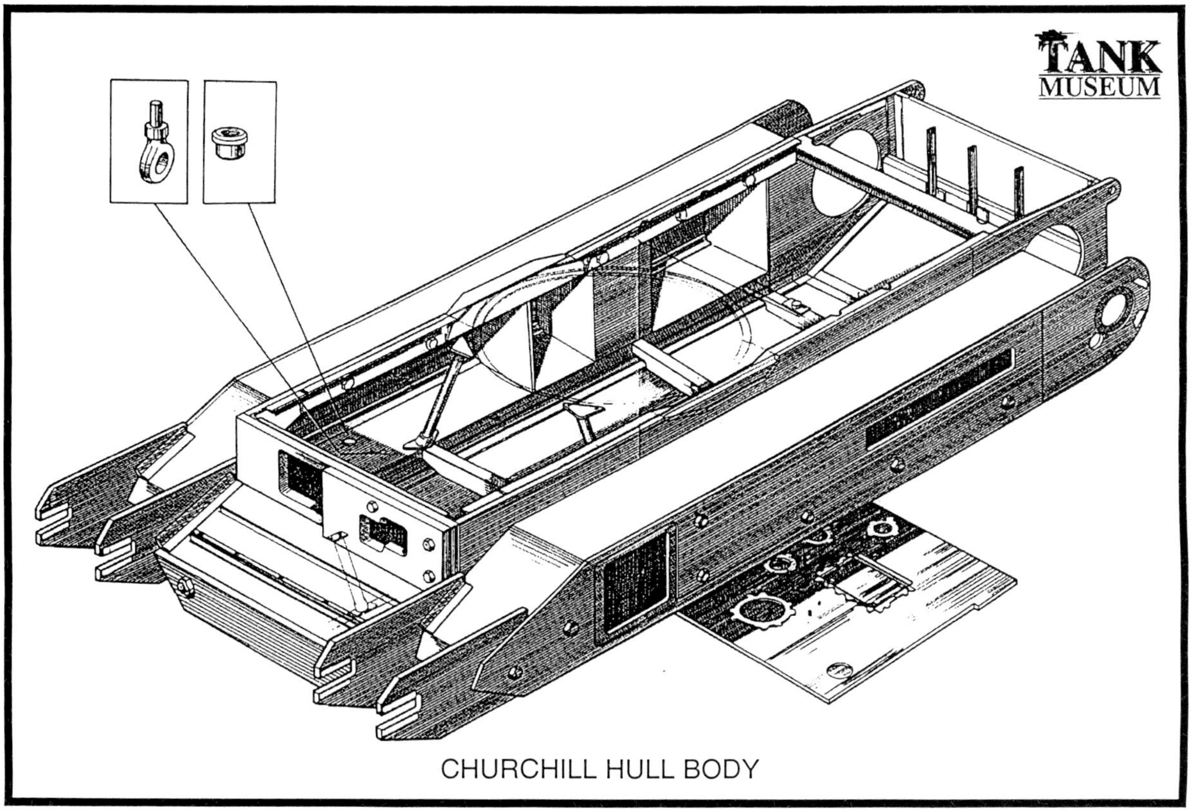

A drawing showing the main hull structure.

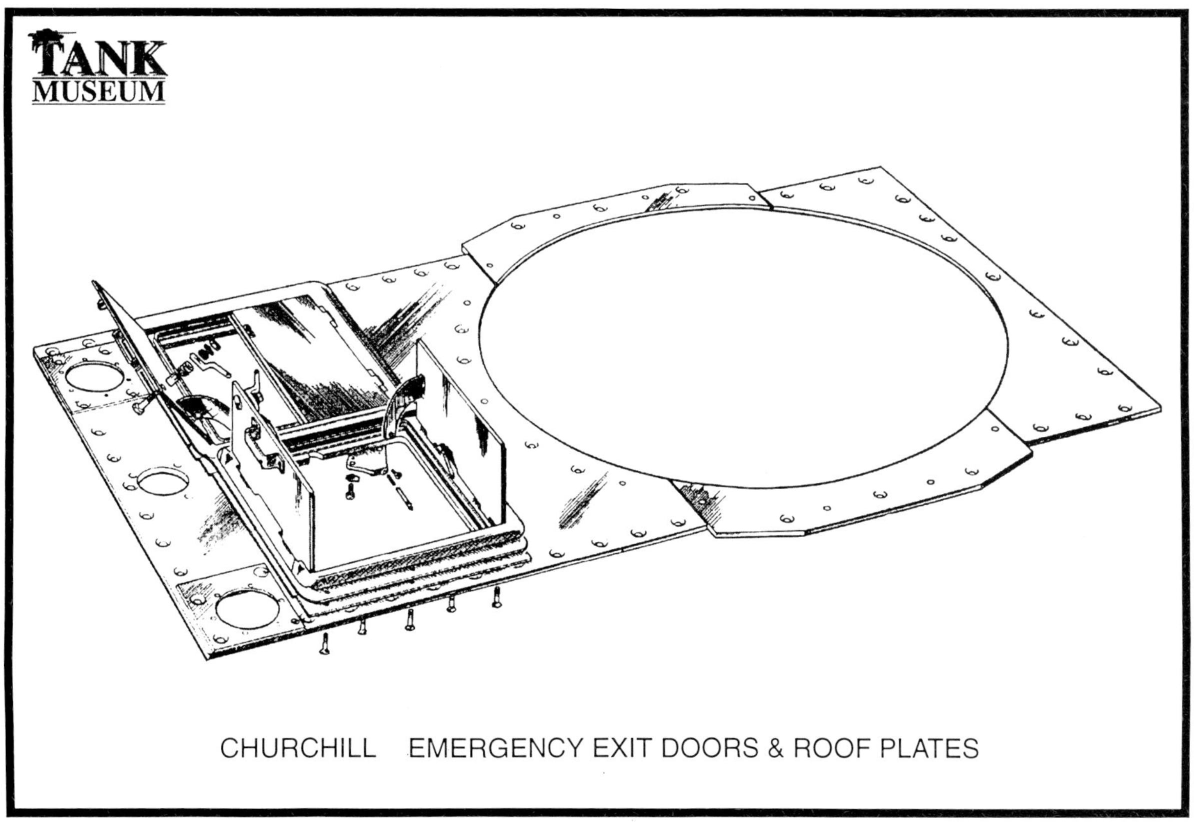

A drawing showing the hull roof including the area around the turret ring and overhead escape hatches for the driver and hull machine-gunner.

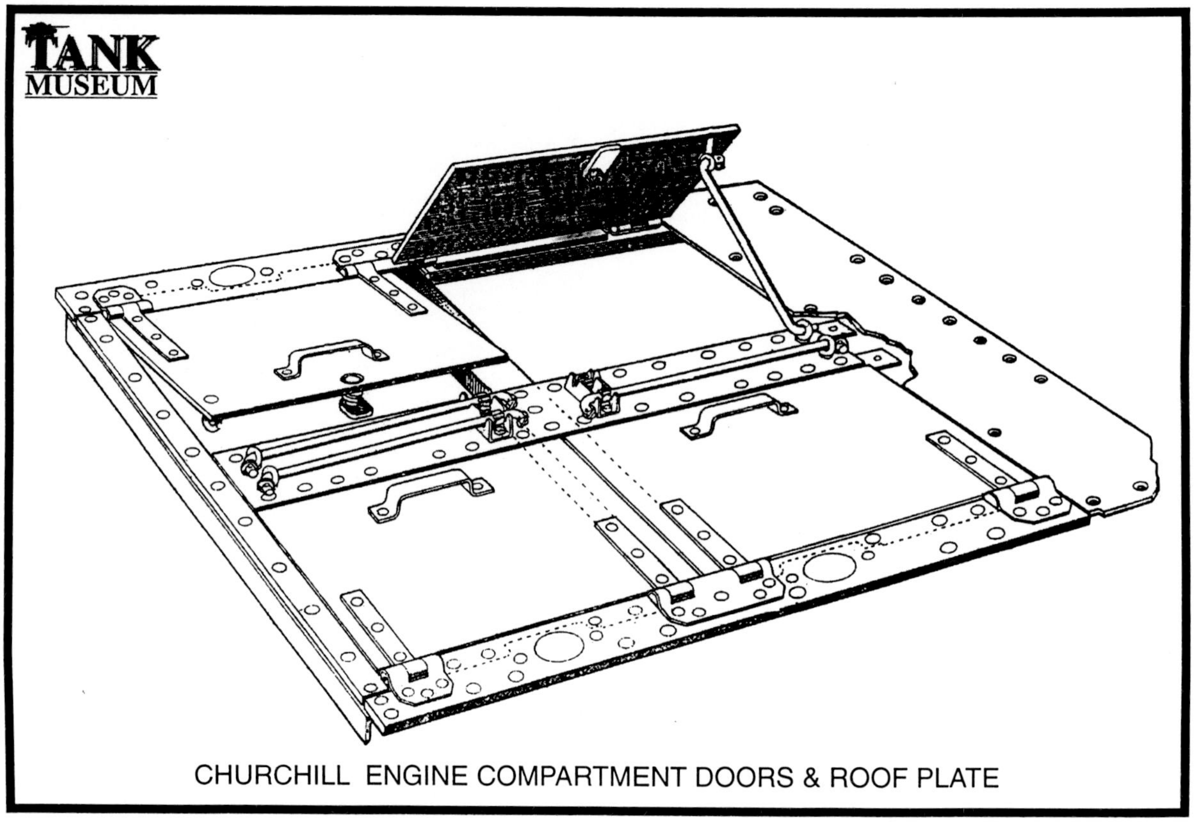

A drawing showing the rear hull roof and hatches above the engine and transmission compartments.

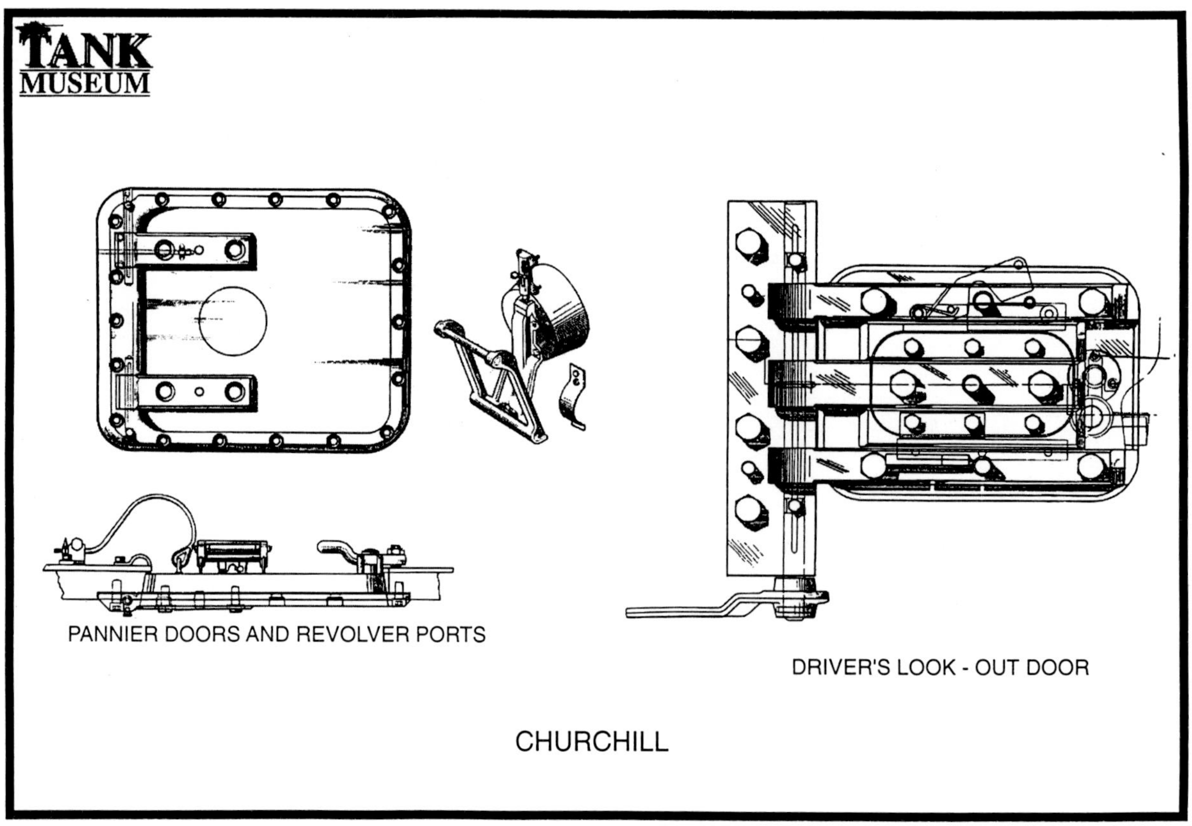

Drawings of the driver's look-out door and hull side door

A scene at the London Docks in July 1941. Prototype Churchill T30972 has just used a special ramp wagon to climb onto a Rectank transporter wagon. High ranking officers from the War Office have arrived by motor coach to watch the demonstration.

Things do not always go according to plan. Railway staff and civilians watch as soldiers prepare to rescue a Churchill tank which appears to have fallen off a railway wagon. The location of this embarrassing event has not been established.

light steel plate was added just above each intake to divert the mud but it did not work very well and in any case so much was thrown up by the tracks that it went everywhere. This problem was only alleviated later when the tanks were fitted with full length mudguards. The intakes extended the width of the tank beyond the railway loading gauge so they were designed to unbolt and were carried on the engine decks when the tank was transported by rail. The rearmost vertical plate was spaced away from the main hull structure to provide outlets, both above and below, for the warm air that had been drawn across the engine. The upper hull, behind the turret, was formed from an arrangement of hatches which provided access to the engine and transmission. This was also the location for the exhaust pipes and silencers. The turret ring itself was just 54 inch diameter and ahead of it was a flat plate containing the overhead hatches for the driver and hull gunner. The former sat on the right, in a part of the hull which was stepped down to provide him with a forward facing door through which he could view the road. To his left was the gunner, serving the 3 inch howitzer on the Mark I. The driver, like his predecessor in the First World War, had a very limited view. Even with the front vision door open his view to left or right was blocked entirely by the high line of the tracks and if he did wish to see to the sides his only means was a pair of rotating periscopes that offered a very limited field of view. In fact he relied to a great extent on instructions from his commander.

The 3 inch howitzer mounting, to the left of the driver, was fitted in such a restricted space that its potential performance was severely limited. The gun was controlled in both elevation and traverse by gears and handwheels but movement was limited. Maximum elevation was 9 degrees and depression 5 degrees while traverse was restricted by the width of the hull between the horns. Nine degrees of elevation, for a howitzer type weapon, was a serious handicap. In ideal conditions the weapon could fire a shot of around 13lbs, at a muzzle velocity of 600 feet per second to a maximum range of 2,500 yards. In trials at the Lulworth Gunnery School the maximum ranges achieved were 2,400 yards firing high explosive and 1,550 with a smoke shell.

Fifty eight rounds of three inch ammunition were stowed, 29 rounds per side, in the hull panniers ahead of the side doors. The official stowage diagram does not distinguish between high explosive and smoke rounds in the Churchill I, although this is normal practice in other close support tanks. In every other case smoke rounds outnumber high explosive by a considerable amount but there is reason to believe that in the case of the Churchill high explosive shells form the main component. This is suggested by a report from Lulworth which noted that the range scale, attached to the gun, was only graduated for high explosive rounds. It would fit with the tank's role as a direct assault weapon. Whether it proved possible in practice, since delivery of high explosive rounds got into serious arrears, is another matter.

The driver's location in a Mark I Churchill. The three foot pedals and gear change lever are obvious but one can also see the handlebar steering control and the bar that links this with another lever, operated by the hull gunner on the left.

The driver's controls are illustrated and the only feature to remark upon is the handlebar style steering lever in place of the normal floor mounted levers. Another unusual feature is the half handlebar provided for the front gunner and the ignition cut out switch. Both would be helpful in an emergency but the steering lever would also be useful to the gunner when laying the howitzer on a target.

The turret for Churchill I is of interest since it was the first attempt by British steelmakers to create a complete cast turret, all in one piece. Infantry tanks Mark I and II had cast turrets but in each case the roof section was made from rolled plate, bolted or welded into place. Infantry tank Mark III, the Valentine, also had a turret combining both types of armour but that designed for the Churchill was entirely cast, and one is justified in asking why? Inch for inch cast armour is not as effective as rolled plate so, in order to achieve the same degree of protection, cast armour has to be approximately five per cent thicker, and consequently heavier. We have already noted that in the case of A20 the designers were planning to create a rolled plate version of the Matilda turret which would not only be lighter, but easier and quicker to assemble. Large castings presented the designer with two difficulties. In the first place it was not always easy to create a turret that was of varied thickness on different planes. Secondly there was a serious risk of distortion when the casting was heated and cooled to achieve the hardening process. That these problems were successfully overcome is regarded as a remarkable achievement for the British steel industry but it still does not explain why it was done. There was no protective mantlet in the normal sense, just three slots in the front of the casting for the 40mm (2-pounder) gun in the centre, the 7.92mm BESA machine-gun on its right and the number 30 sighting telescope on the left.

The turret was rotated by electric power and the controller gave two speeds. Fast had the turret all the way round in 15 seconds while the slow speed completed the 360 degrees in 24 seconds. The control could also be switched to manual so that the turret could be turned by hand. During the design stage there had been talk of providing the tank with geared elevation for the co-axial mounting but this had been abandoned in favour of free elevation, which was standard British practice at this time. This requirement evolved from the British obsession with tanks firing on the move, where most other countries preferred the practice of halting to fire. Firing on the move requires that the gun is easily adjusted to compensate for movements of the tank, and since this was before the days of automatic stabilisers in tanks the Churchill had to rely on its gunner. In action he would stand, or rather stoop, to the left of the gun, his right hand gripping the trigger and his eye pressed to the sight while his left hand grasped the traverse control. As the turret swung round he was carried with it by the rotating floor but in order to counter the pitching of the tank he bent his knees and used his body weight to control the aim of the gun. To operate the gun accurately under these

(text continues on page 42)

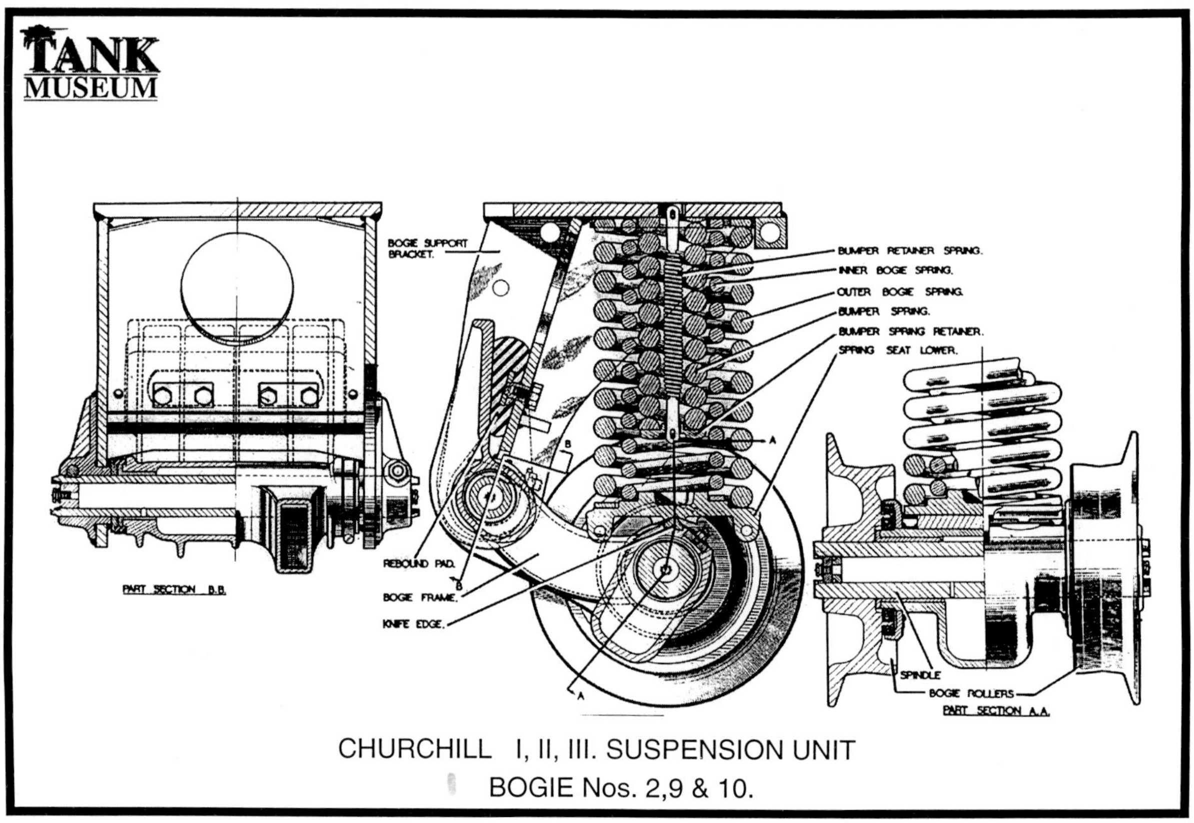

Details of a typical suspension unit.

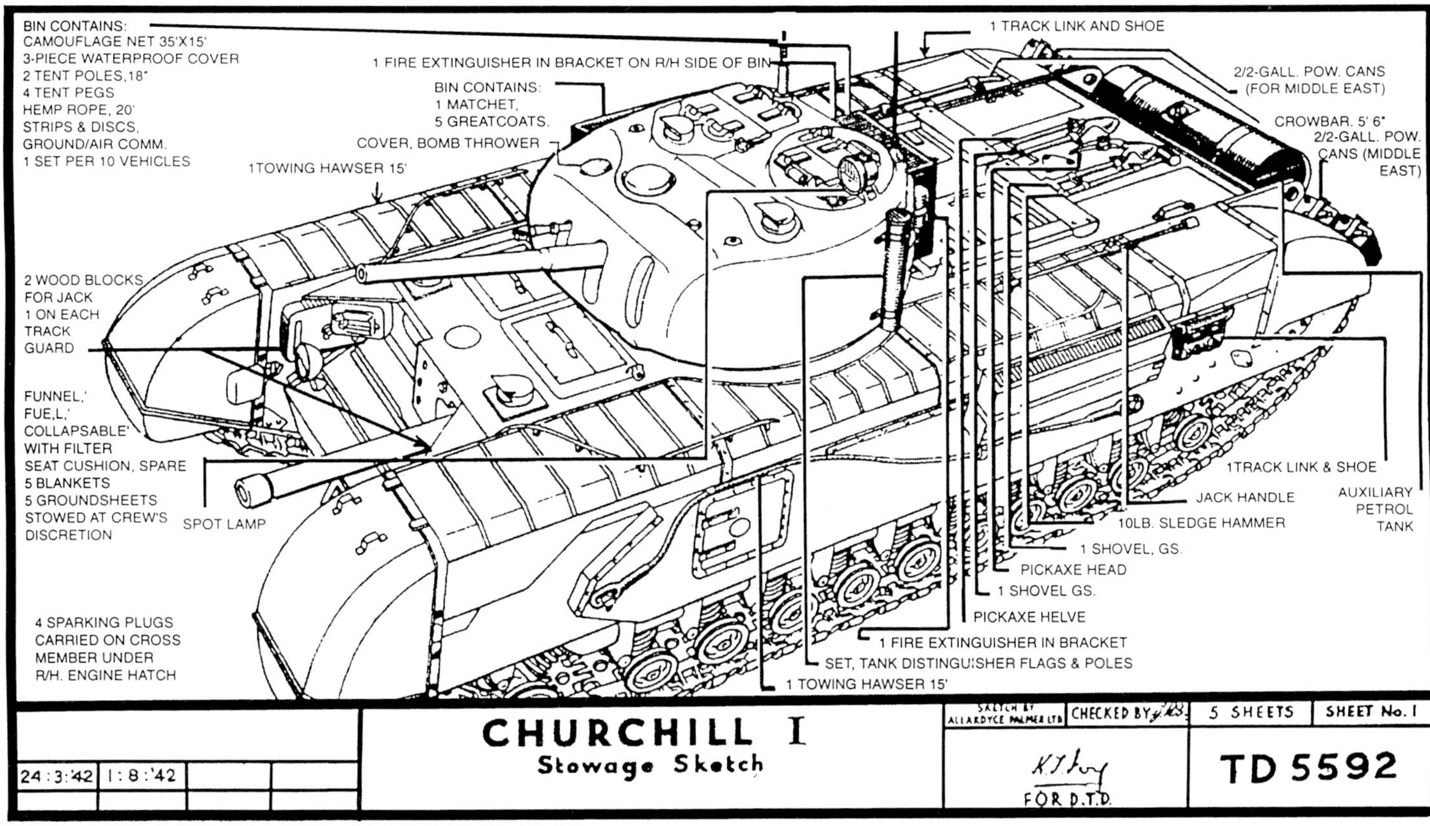

Pages 37-41: A set of drawings showing the stowage arrangements, armour thickness and other details of the Churchill Mark I.

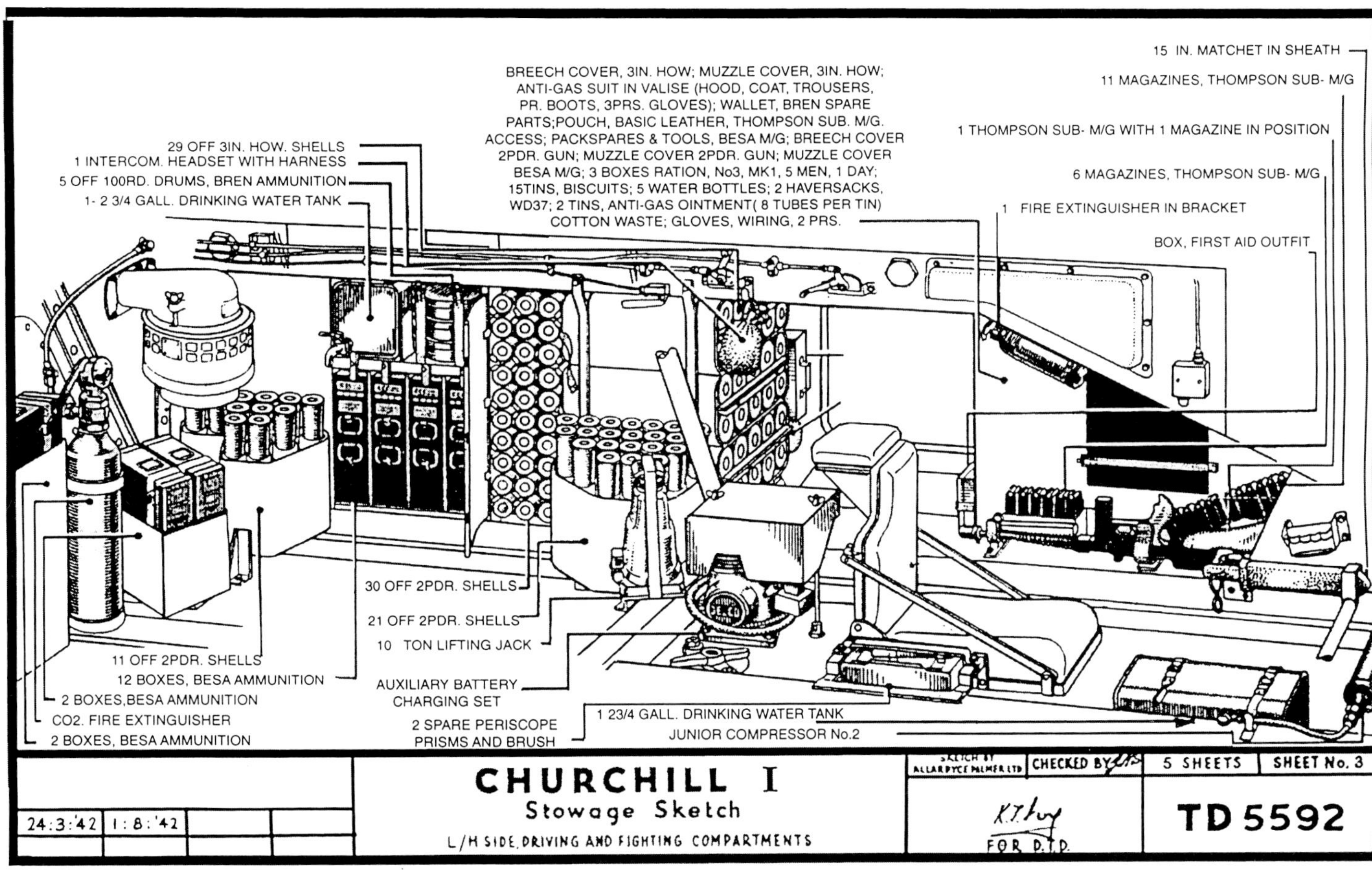

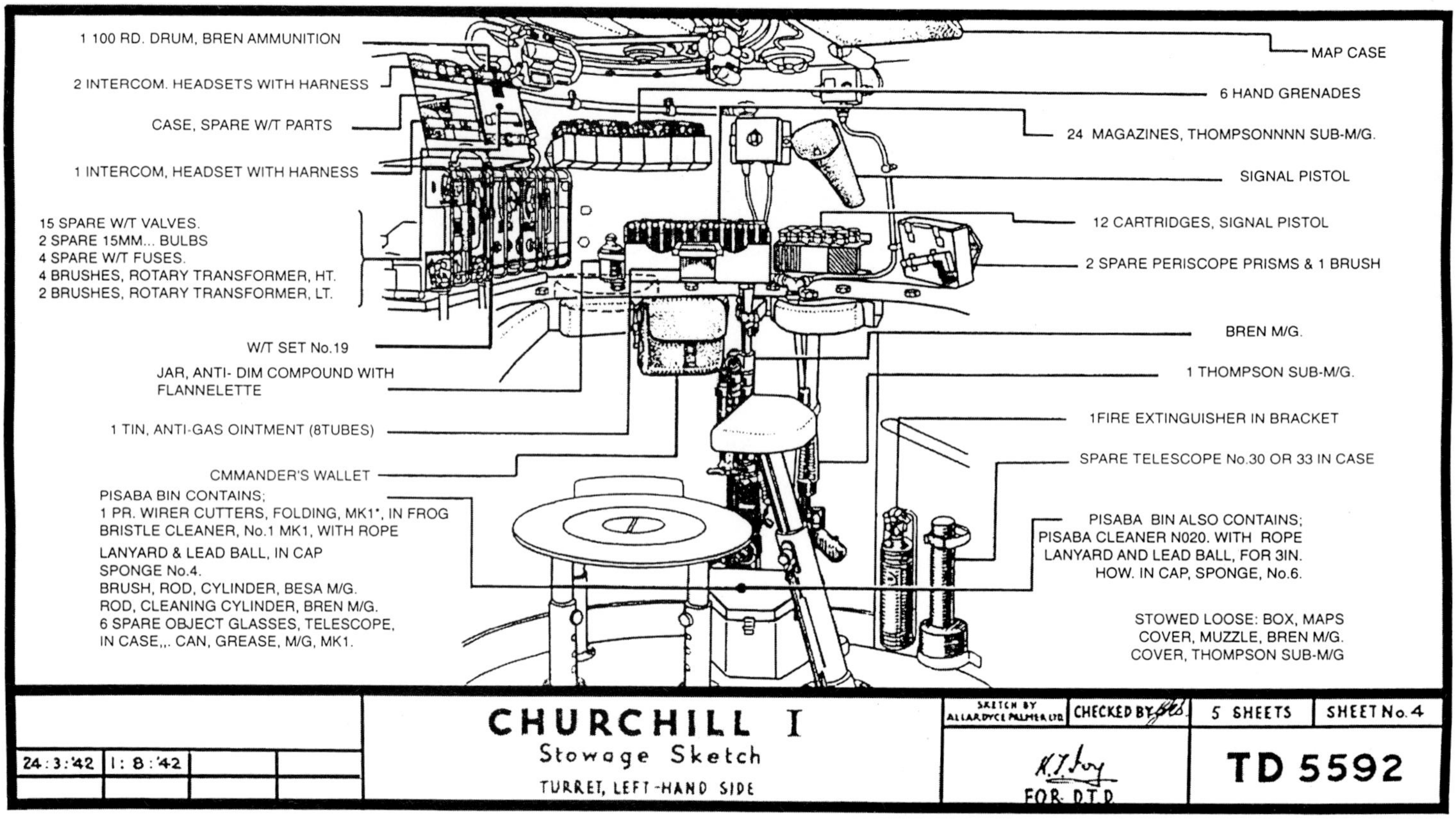
1 100 RD. DRUM, BREN AMMUNITION
2 INTERCOM. HEADSETS WITH HARNESS
CASE, SPARE W/T PARTS
1 INTERCOM, HEADSET WITH HARNESS
15 SPARE W/T VALVES.
2 SPARE 15MM... BULBS
4 SPARE W/T FUSES.
4 BRUSHES, ROTARY TRANSFORMER, HT.
2 BRUSHES, ROTARY TRANSFORMER, LT.
W/T SET No.19
JAR, ANTI- DIM COMPOUND WITH FLANNELETTE
1 TIN, ANTI-GAS OINTMENT (8TUBES)
CMMANDER'S WALLET
PISABA BIN CONTAINS;
1 PR. WIRER CUTTERS, FOLDING, MK1*, IN FROG
BRISTLE CLEANER, No.1 MK1, WITH ROPE
LANYARD & LEAD BALL, IN CAP
SPONGE No.4.
BRUSH, ROD, CYLINDER, BESA M/G.
ROD, CLEANING CYLINDER, BREN M/G.
6 SPARE OBJECT GLASSES, TELESCOPE,
IN CASE,,. CAN, GREASE, M/G, MK1.
MAP CASE
6 HAND GRENADES
24 MAGAZINES, THOMPSONNNN SUB-M/G.
SIGNAL PISTOL
12 CARTRIDGES, SIGNAL PISTOL
2 SPARE PERISCOPE PRISMS & 1 BRUSH
BREN M/G.
1 THOMPSON SUB-M/G.
1FIRE EXTINGUISHER IN BRACKET
SPARE TELESCOPE No.30 OR 33 IN CASE
PISABA BIN ALSO CONTAINS;
PISABA CLEANER N020. WITH ROPE
LANYARD AND LEAD BALL, FOR 3IN.
HOW. IN CAP, SPONGE, No.6.
STOWED LOOSE: BOX, MAPS
COVER, MUZZLE, BREN M/G.
COVER, THOMPSON SUB-M/G
24:3:42
1:8:42
CHURCHILL I
Stowage Sketch
TURRET, LEFT-HAND SIDE
SKETCH BY ALLARDYCE PALMER LTD.
CHECKED BY
FOR D.T.D
5 SHEETS
SHEET No. 4
TD 5592

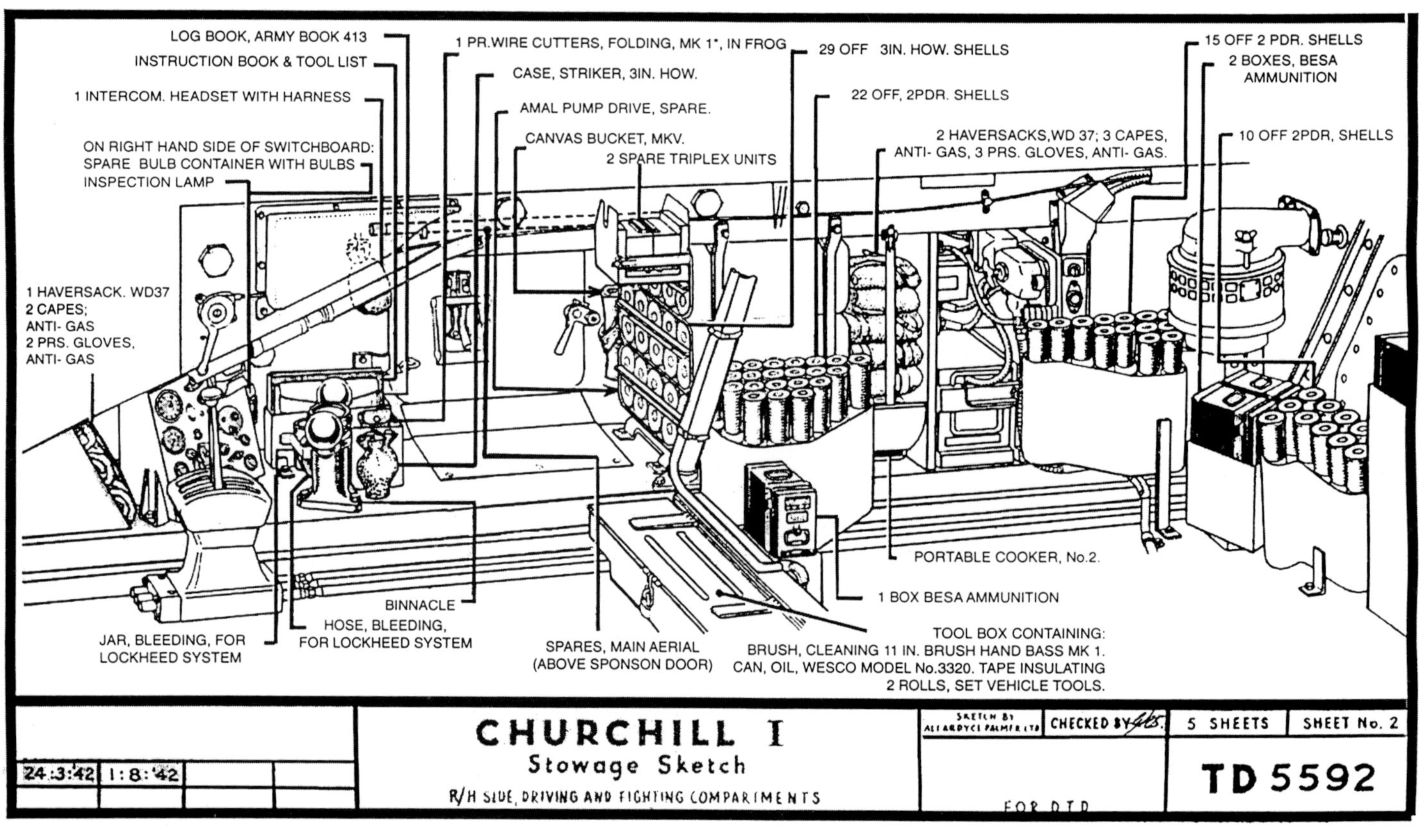
LOG BOOK, ARMY BOOK 413
INSTRUCTION BOOK & TOOL LIST
1 INTERCOM. HEADSET WITH HARNESS
ON RIGHT HAND SIDE OF SWITCHBOARD:
SPARE BULB CONTAINER WITH BULBS
INSPECTION LAMP
1 HAVERSACK. WD37
2 CAPES;
ANTI- GAS
2 PRS. GLOVES,
ANTI- GAS
1 PR.WIRE CUTTERS, FOLDING, MK 1*, IN FROG
CASE, STRIKER, 3IN. HOW.
AMAL PUMP DRIVE, SPARE.
CANVAS BUCKET, MKV.
2 SPARE TRIPLEX UNITS
29 OFF 3IN. HOW. SHELLS
22 OFF, 2PDR. SHELLS
2 HAVERSACKS,WD 37; 3 CAPES,
ANTI- GAS, 3 PRS. GLOVES, ANTI- GAS.
15 OFF 2 PDR. SHELLS
2 BOXES, BESA
AMMUNITION
10 OFF 2PDR, SHELLS
PORTABLE COOKER, No.2.
1 BOX BESA AMMUNITION
BINNACLE
JAR, BLEEDING, FOR
LOCKHEED SYSTEM
HOSE, BLEEDING,
FOR LOCKHEED SYSTEM
SPARES, MAIN AERIAL
(ABOVE SPONSON DOOR)
TOOL BOX CONTAINING:
BRUSH, CLEANING 11 IN. BRUSH HAND BASS MK 1.
CAN, OIL, WESCO MODEL No.3320. TAPE INSULATING
2 ROLLS, SET VEHICLE TOOLS.
24:3:42
1:8:42
CHURCHILL I
Stowage Sketch
R/H SIDE, DRIVING AND FIGHTING COMPARTMENTS
SKETCH BY ALLARDYCE PALMER LTD
CHECKED BY
FOR D.T.D
5 SHEETS
SHEET No. 2
TD 5592

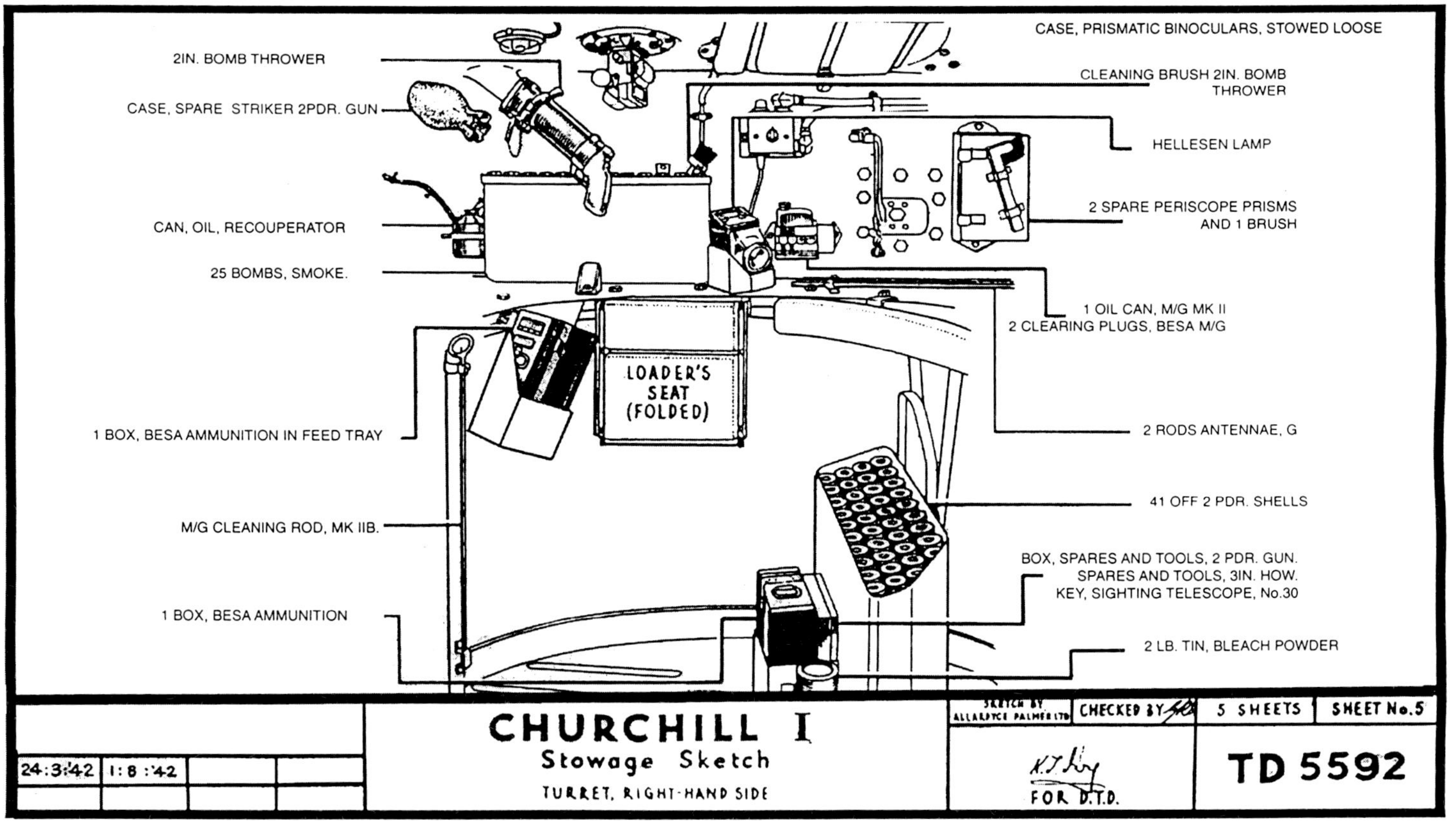
2IN. BOMB THROWER
CASE, SPARE STRIKER 2PDR. GUN
CAN, OIL, RECOUPERATOR
25 BOMBS, SMOKE.
1 BOX, BESA AMMUNITION IN FEED TRAY
M/G CLEANING ROD, MK IIB.
1 BOX, BESA AMMUNITION
LOADER'S SEAT (FOLDED)
CASE, PRISMATIC BINOCULARS, STOWED LOOSE
CLEANING BRUSH 2IN. BOMB THROWER
HELLESEN LAMP
2 SPARE PERISCOPE PRISMS AND 1 BRUSH
1 OIL CAN, M/G MK II
2 CLEARING PLUGS, BESA M/G
2 RODS ANTENNAE, G
41 OFF 2 PDR. SHELLS
BOX, SPARES AND TOOLS, 2 PDR. GUN.
SPARES AND TOOLS, 3IN. HOW.
KEY, SIGHTING TELESCOPE, No.30
2 LB. TIN, BLEACH POWDER
24:3:42
1:8:42
CHURCHILL I
Stowage Sketch
TURRET, RIGHT-HAND SIDE
CHECKED BY
5 SHEETS
SHEET No.5
FOR D.T.D.
TD 5592

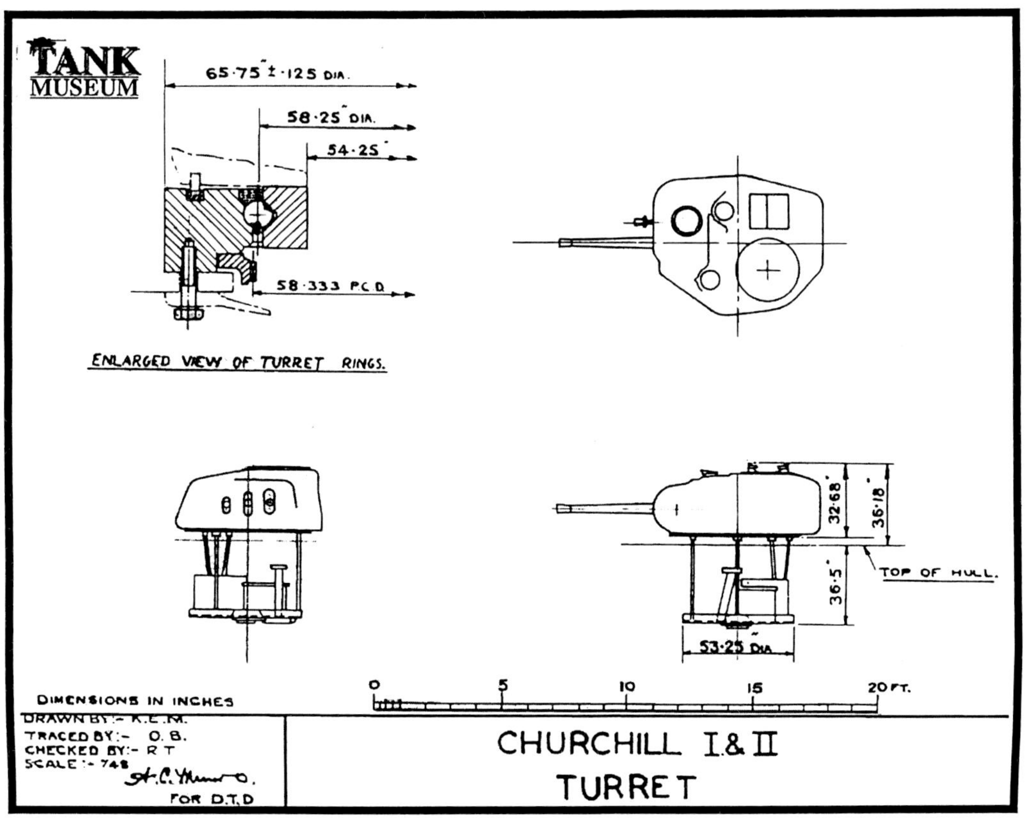
TANK MUSEUM
65·75" ± ·125 DIA.
58·25" DIA.
54·25"
58·333 P.C.D.
ENLARGED VIEW OF TURRET RINGS.
32·68"
36·18"
36·5"
TOP OF HULL.
53·25" DIA.
0
5
10
15
20 FT.
DIMENSIONS IN INCHES
DRAWN BY:- K.E.M.
TRACED BY:- O. B.
CHECKED BY:- R T
SCALE:- 748
FOR D.T.D
CHURCHILL I & II
TURRET

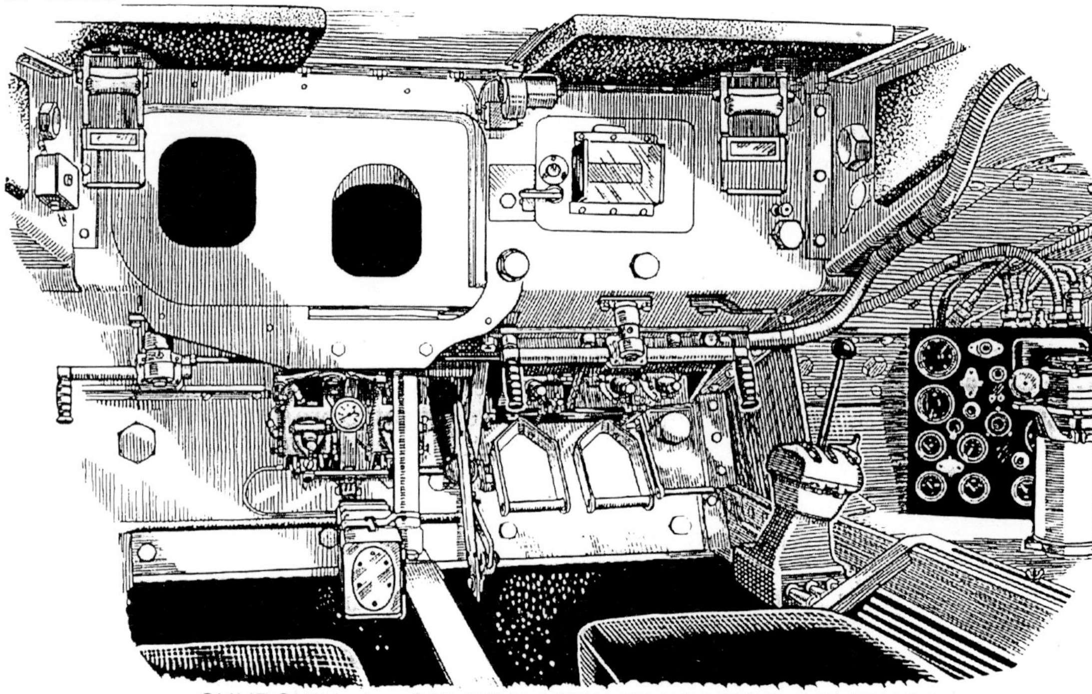

CHURCHILL I GENERAL VIEW OF DRIVER'S CONTROLS

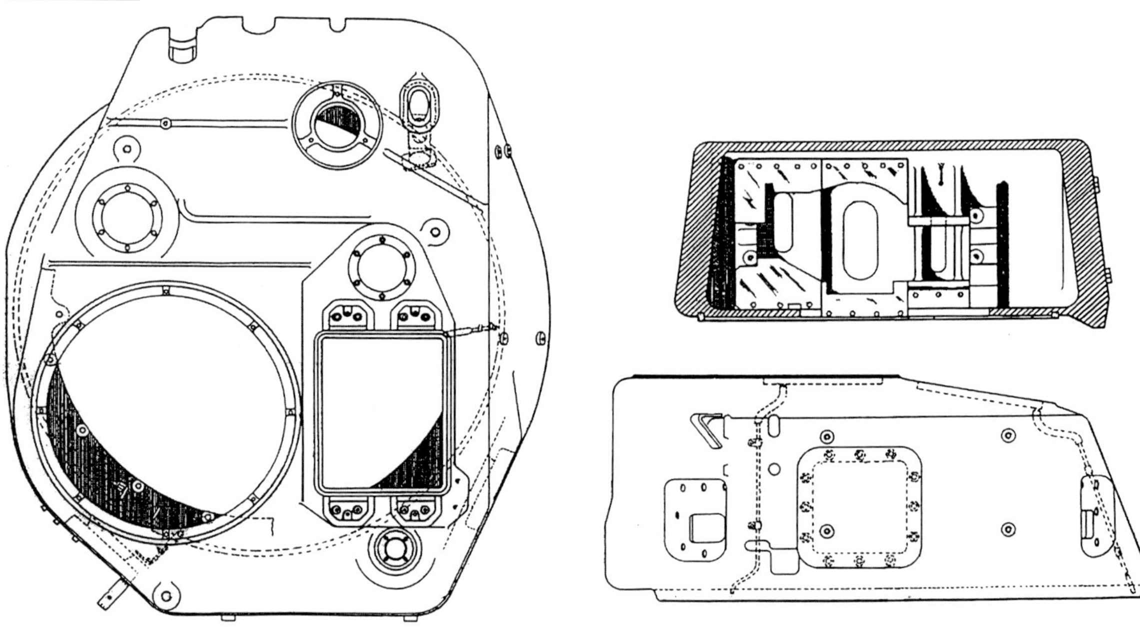

TURRET CHURCHILL I & II

TANK MUSEUM

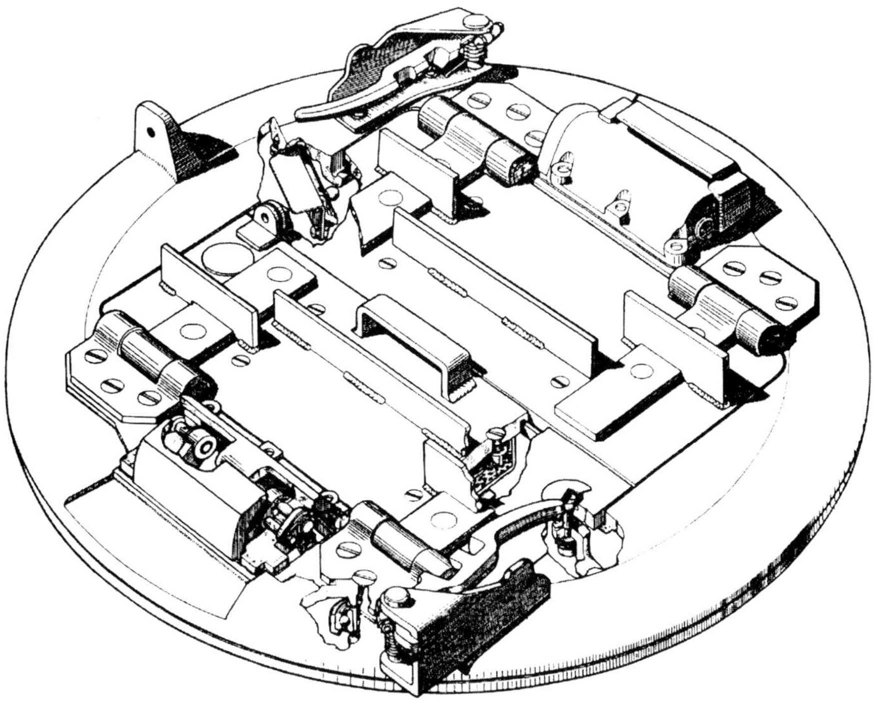

CUPOLA CHURCHILL

TANK MUSEUM

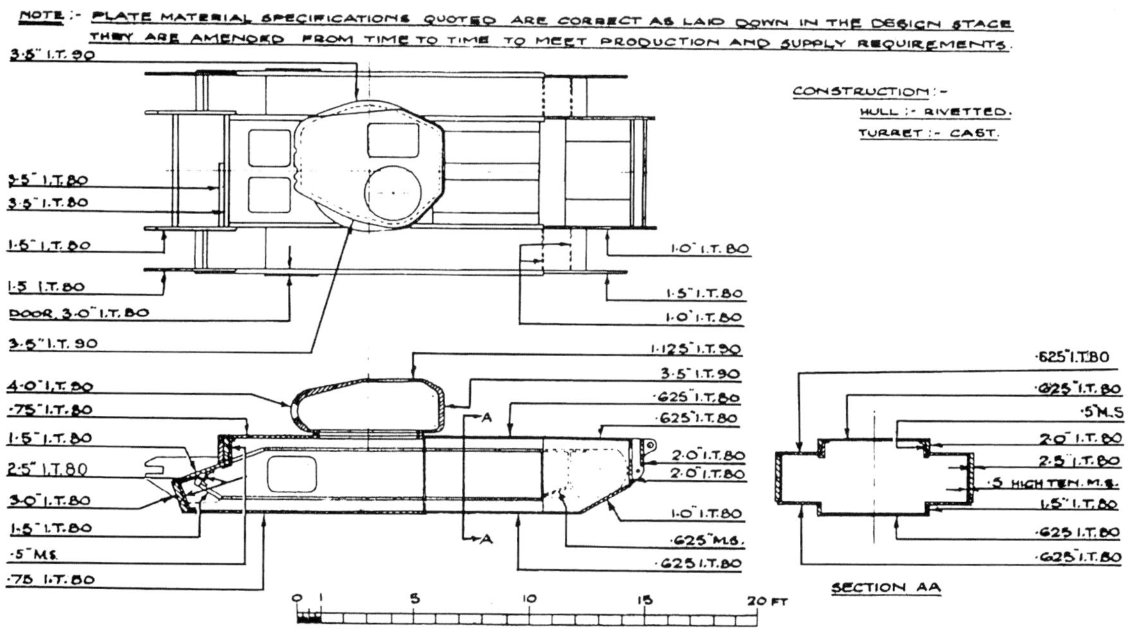

CHURCHILL I & II PLATE THICKNESS

conditions required a great deal of skill and most tank commanders regarded a good gunner as one of his most valuable assets.

The loader, who stood to the right of the gun, also operated the BESA machine-gun, the two-inch smoke bomb thrower in the roof and the radio - or to use the contemporary term - wireless set at the back. The smoke bomb thrower was a small, breech loading weapon which fired on a fixed trajectory through an angled hole in the turret roof. Range was varied by adjusting the gas regulator, a three-way valve which offered the choice of 20, 70 or 110 yards. The radio, at the back of the turret, was the celebrated Number 19 set which provided two-way communication outside the tank and a crew intercom system. The loader also had a rectangular hatch in the turret roof above his head. The third member of the turret party, and fifth member of the tank's crew, was the commander, who was located behind the gunner on the left side. Although described as a cupola the commander's hatch was little more than a large, circular fitting almost flush with the turret top. It was closed by a pair of D shaped hatches and had periscopes front and back which gave a limited field of vision. The entire assembly could be rotated manually by the commander, independently of the turret, but it could be hard work and most commanders preferred to work with their heads out and take a chance for the improved view.

Young soldiers of the Royal Armoured Corps Officer Cadet Training Unit at the Royal Military College, Sandhurst learning to maintain the Bedford engine in an early Churchill. With the top and rear hull plates removed one can also see the gearbox, final drive and brake drums.

The fighting compartment ended in a firewall that separated it from the engine compartment. The engine itself has already been described. It was flanked by four radiators, two serving each bank of cylinders, and outboard of the radiators were six interconnected petrol tanks, three each side, giving

Early Churchill tanks on the production line. Most appear to be of the Mark II type. Why the tank in the foreground should be facing the wrong way is not clear but it provides a good view of the exhaust pipe assembly.

Here a civilian crew is seen taking a newly completed Churchill out of the factory gates for a test run. The exact site has not been identified.

A poor but rare photograph of a new Churchill II being tested by the manufacturer. This tank not only lacks weapons, which one would expect at this stage, but it is also missing its air intakes and the series of little armoured plates that protect the suspension.

A fine close up view of a Mark II built by Dennis Brothers of Guildford which clearly shows the hull BESA mounting and the co-axial machine-gun alongside the 2-pounder in the turret.

A dead Churchill being winched onto a transporter. This close view of the idler enables us to see the teeth and the rubber tyre which were peculiar features of the Churchill. The presence of Jeeps in the field suggests a date no earlier than 1942.

a total capacity of 150 gallons; outside the hull at the rear was an auxiliary fuel tank, containing another 32.5 gallons. This was also connected to the main petrol system but could be jettisoned in an emergency. Directly behind the engine, and running on the main crankshaft, was the flywheel that incorporated a large Sirocco fan which provided air circulation and stirred up most of the dust already described. The rear face of the flywheel was also the driving member of a Borg & Beck single, dry plate clutch. Beyond the clutch was the gearbox, transmission and final drive.

Again, as already described, the first 100 tanks (one source claims 115), which were not necessarily all of the Mark I type, employed a Merritt-Brown type 301c, five-speed combined gearbox and steering system. From the 101st (or 116th) machine the type H4, four-speed transmission was used instead but in most respects they were similar. Drive from the

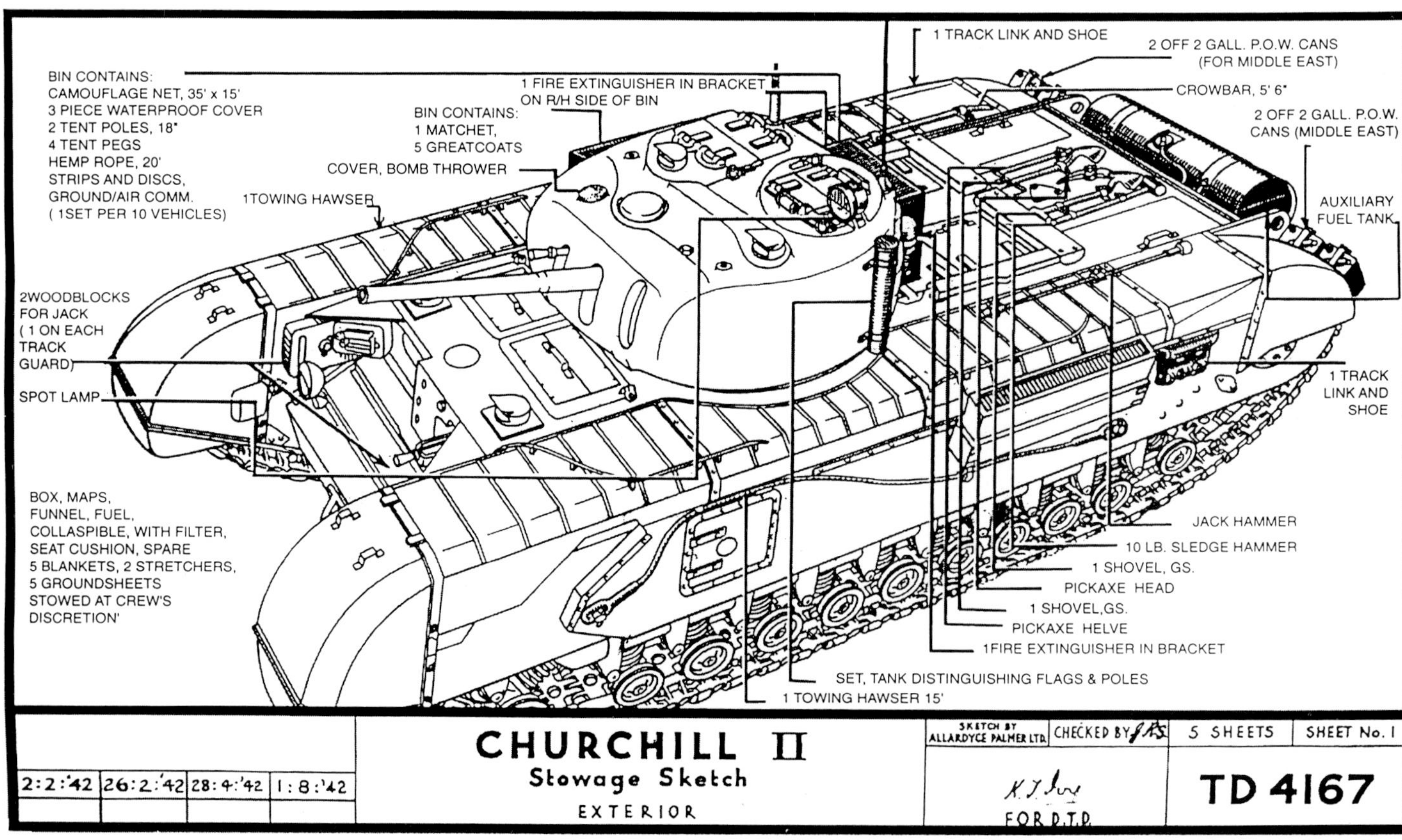

Stowage diagrams for the Churchill Mark II.

An overhead view of a Mark II in original condition. From this angle one can clearly see the layout of hatches, periscopes and ventilators.

clutch passed into a large casing, situated crosswise at the back of the tank, which contained the gearbox, differential and steering epicyclics all in a remarkably compact shell. Where the drive enters the box it effectively splits so that power is diverted in two ways. One goes direct through the gearbox to a layshaft linked to the epicyclics while the other diverts through a differential to the epicyclics by a different route. The transmission is fully regenerative so that the radius of turn depends on the gear selected but with the gearbox in neutral, drive still passes direct to the epicyclics and enables the tank to perform a neutral turn, with one track running in either direction. According to the first edition of the Instruction Book the five-speed gearbox moved the tank at 2.04 mph in first gear and 15.95 mph in fifth while the respective turning circles were 12.3 ft in first up to 95.9 ft in fifth. In the four-speed box the ratios were further apart. In first gear the speed was just 1.78 mph with a turning circle of 10.85 ft while fourth gear in the H4 box was identical to fifth in the 301c. Dr Merritt's transmission was a complicated piece of equipment which is reputed, by imaginative British Army instructors, to have driven its designer mad, but his sanity was never in doubt and the transmission forms the basis of many modern systems world wide.

Steering brakes were attached to the output shafts from the epicyclics and the main tank brakes were mounted outboard again. The final drive assemblies, located in the track frames, carried the drive to the sprockets and reversed the direction of rotation at the same time. Brakes were hydraulically operated using Lockheed equipment with separate systems for the steering and stopping . The main clutch and steering epicyclics used air assisted controls fed by a Clayton Dewandre compressor and 1,050 cubic inch capacity reservoir. The air system was activated by the normal controls so that, for instance, when the clutch pedal was depressed it opened an air line which activated a servo at the clutch. On the five-speed tanks the air compressor was situated in a corner of the gearbox compartment while on later models it was fitted directly on to the gearbox casing.

The Churchill operated on a 12 volt electrical system from two 6 volt batteries located with the master switch in the right side hull pannier. The main generator was attached to the engine but there was another in the gearbox compartment

A frustrated and very dirty soldier works to replace the nearside track on an early Churchill. The censor has not helped by obliterating the unit sign but T68504 was built by Broom & Wade.

providing power for the turret. However the tank also carried an auxiliary generator. It was stowed behind the front gunner's seat and bolted to the roof of the gearbox compartment when in use. This was a Delco-Remy single cylinder, four-stroke petrol engine which could generate sufficient power to start the main engine if required.

The Churchill II, originally referred to as the Infantry Tank Mark IVa, mounted the 7.92mm BESA air-cooled machine-gun in place of the 3 inch howitzer. The weapon was fitted into a gimbal mounting which also incorporated a small telescopic sight.

4

Early Trials

When a new tank has been built in haste, without the benefit of a long period of prototype development and testing, there is really little point in postponing the evil day. Thus as soon as the pilot and early production models left the factories they were despatched to various trials establishments around the country and tank regiments then stationed in Britain. At the former locations they would be subjected to arduous and thorough tests of every aspect of their performance in the hands of experienced staff. Those delivered direct to regiments were simply handed over to the tender mercies of soldiers who, as is the way of soldiers, probably did their level best to wreck them in order to show the manufacturers what was really expected of their products.

Starting with the professionals we can observe the results of both driving and gunnery tests carried out in the summer of 1941. Most of the early defects have already been recorded in Chapter 2 and it would be labouring the point to repeat them. On the other hand it is important to emphasise the scale of the problem. Remember that we are studying a period when Britain's fortunes were at their lowest ebb. Evicted ignominiously from France and now under constant attack from the air the very spirit of the nation was in danger of being sapped. True the mood was not defeatist but there was little occasion for optimism. Everyone, from the farmer to the bus driver, felt that they were in the front line and major inconveniences such as food rationing and the blackout made everyday life bleak, even for those who were not subjected to air raids. Designing and building a new tank under such conditions must have been perceived as a positive indication of defiance so the fact that this tank produced nothing but a catalogue of failures would have been hard to bear. Even its most ardent supporter, the Prime Minister, knew despondency and is recorded, on one occasion, as saying that even if the tanks did not work, at least they could be parked at strategic points and used as pill boxes in the event of invasion!

Not that it was all doom and gloom. The engine was described as very satisfactory on the whole. There was a list of faults but the report is extremely optimistic. Cooling, both of water and oil, was regarded as more serious but it is difficult to find any British tank which did not suffer in this respect when it first appeared, although the Churchill is the only one that stands accused of sucking in leaves through the air inlets which then plastered themselves all over the radiators.

In wet weather the tank leaked badly and, since it was without mudguards, it covered itself in mud as we have already been informed. A potentially more dangerous problem was revealed by the statement that ammunition would fall out of the stowage racks and roll around on the floor. However all of these things could be put right.

In the light of all the later troubles that beset the Churchill the two final paragraphs of this report, dated 19 August 1941, might well be quoted in full;

> *'Although this vehicle has never completed a run, however short or however long, without need of at least half a day maintenance, it has always got home under its own power. This cannot be said of any other make of tank.*
>
> *The vehicle has no major trouble, the cure of which is not in hand, but has a mass of small defects which make its maintenance excessively arduous. These small points are all capable of being cured once the development work is completed.'*

It is a remarkably confident report, which was echoed by one issued by the Gunnery School at Lulworth.

The reporting officer, Major FDW O'Rorke was commenting on Mark I tank T32246 which they had been testing since it arrived on 3 July. He noted that instructions were received not to run the tank any faster than 10 mph but then immediately goes on to say that at 15 mph the Churchill pitched worse than any other tank in the service, including light tanks which really is saying something. At 8 mph or less it was fine; after that it pitched and it was impossible to fire the turret gun at any speed above 12 mph, but at 15 mph it was dreadful. Otherwise O'Rorke felt, the tank performed well. He approved

A fine view of a Churchill Mark II storming along a country road in England on a bright winter's day. Notice that the two leading rollers on each side are not carrying any load when the tank is on level ground.

A summer shot of a Churchill II on a route march. The diablo sign on the lower hull front, driver's side, indicates an Army Tank Brigade. The figure 175 suggests the second regiment in the brigade and the white bar, below this figure, shows that the brigade is serving under an army control.

of the 2-pounder gun mounting and turret traverse but he was very unhappy with the 3 inch howitzer which, he believed, should be exchanged for another 2-pounder for dealing with anti-tank obstacles - presumably by chipping them down to size - a bit at a time.

Dust, or rain and mud, were again the biggest complaints. Firing the 2-pounder with the turret reversed threw up such a cloud from the engine deck that the gunner could neither see the trace in the shell nor whether he struck the target or not. Even when the tank was fully closed down, O'Rorke claimed, dust came in everywhere. The turret hatches were criticised for not being secure. One man is reported to have lost a finger and there were other injuries when catches failed and hatches slammed shut prematurely.

This summing up also bears repeating;

> *'Considering the number of new features incorporated in the design and the short time taken in its production, the tank is very free of defects which cannot be overcome on later models. This tank is a great improvement on any vehicle yet received for trial at Lulworth.'*

So much for the professionals; what about the users?

It may well prove impossible to discover which regiment first received the Churchill tank. Surviving regimental war diaries should make it possible to decide but unfortunately they were compiled in such varied ways that there is no guarantee of certainty at all. Some, for example, virtually ignore tanks altogether. They prefer to concentrate on such mundane things as postings of officers to and from the regiment, or the results of courts martial, that one wonders whether the adjutant ever left his office at all, let alone ventured down to the tank park? Others will admit to having tanks but are vague about which sort. They record the arrival and despatch of infantry, or I Tanks, as they are often called, but fail to specify the type. At the other end of the scale are those who list not only every tank and truck which passes through the regiment's hands but give the precise type and its War Department (WD) number.

One possible contender for the first Churchill is 12th Royal Tank Regiment which, in 1941, was stationed at Rufford Abbey in Lancashire. Its war diary records the arrival of one Churchill tank on 14 June 1941, and it does refer to it as a Churchill. This is some months before the name became official and proves that it was common currency a good deal earlier. The regiment was operating Matilda tanks at the time but in April an officer is reported as going on a Mark IV course and in early June other officers and NCOs are shown as away at Vauxhall Motors and Metro-Cammell for the same purpose. By the end of July the regiment has thirteen Churchills and the war diary remarks; 'they are coming in frequently now'. In August the adjutant notes that everything is 'quiet on the surface but intensive training on the Mark IV continues'.

Ninth Royal Tank Regiment was based at Otley in Yorkshire in June 1941 with a sorry collection of tanks. The weekly progress report for the first week in June shows that they had

A Churchill II in the woods, another winter scene. This tank carries the figure 40 in a circle, an indication of its weight for bridging purposes; the figure 68 shows the second regiment (probably 144 RAC in 33rd Armoured Brigade) and the red/white/red Allied marking. The Wyvern badge alongside the hull machine-gun suggests that the brigade was serving with 43rd (Wessex) Infantry Division.

An excellent view of the wireless operator/loader of an early Churchill in his turret. This view gives some indication of the complexity of the turret casting. The four discs on the side show where a stowage bin would be fitted and just to the rear is the hinged pistol port.

two infantry tanks out of an establishment of 50, which were probably Valentines. It also had two old medium tanks of about 1925 vintage, seven light tanks and a motley selection of other tracked vehicles for driver training. Six close support infantry tanks, which were probably Mark I Churchills, followed the next week and then seven more in July, one of which was a Mark II. By the end of September they had 24 Churchills and still the two Valentines but the war diary itself makes no comment upon them at all.

Tenth Royal Tanks was also at Otley but it is even more coy about its tanks. They do record a talk entitled *Repair and*

A Churchill II named *Babe* of the Three Rivers Regiment, Canadian Army, fording a shallow river during exercise Spartan in March 1943. It is clear that in deeper water, with these early style air intakes, there would be considerable risk of drowning the engine.

Recovery of Mark IV Tanks in the Field, given by a Royal Army Ordnance Corps Major in August 1941 but even that does not induce them to reveal whether they have any or not. In November they moved down to Eastbourne, on the south coast, and a month later a party is sent to study combined operations at Inverary in Scotland, about which more later. In January 1942 they were visited by Geoffrey Burton, chairman of the Tank Board, who discussed Churchill tanks with the commanding officer, Lieutenant Colonel K C Johnston-Jones, which suggests that they were having problems. It is even more revealing to note that when the Prime Minister himself visited Otley in April he was treated to a demonstration of Churchill tanks given by a special composite battalion created from the three regiments of 31st Army Tank Brigade, which strongly suggests that this was the only way in which the brigade could muster a full battalion of working Churchills. The military habit of putting on its best face when authority comes to call is typical behaviour, but one wonders if it might not have done more good to let the great man witness a little more of the truth.

Another regiment stationed at Eastbourne at this time, and in the same brigade as 10th RTR, was 141 Regiment Royal Armoured Corps, or 141 RAC as it was known. The regiment had been created in November 1941 by converting an infantry battalion, in this case 7th Battalion the Royal East Kent Regiment, popularly known as the Buffs. The need to expand the tank arm of the British Army had already been agreed but it could only be done as sufficient tanks became available. Nothing could have been more soul destroying for these proud regiments, who probably regarded tanks as nasty, dirty things at the best of times, than to take away their rifles and give them nothing in return. Even so the Buffs behaved true to form and did their level best to avoid mentioning anything to do with machinery in their war diary. Indeed their preoccupation with trivia is almost breathtaking. On 15 January 1942 the war diary records that a Court of Enquiry was convened, headed by a Major with two subalterns in support, to enquire into how a certain Corporal had managed to wear out two pairs of boots to the extent that they were beyond repair in a matter of weeks. Surely the good corporal could not have been marching that much, now that his regiment had tanks? But the outcome is not recorded.

On 24th January they also had a visit from Geoffrey Burton, presumably on the same mission as his visit to 10th RTR, but they do not actually say so. On 6 February 1942 the war diary does mention Universal Carriers and on the 18th they go to Linney Head in South Wales for gunnery trials, but they still do not mention tanks. Eight days later they are forced to reveal the truth. Second Lieutenant Neal and four men take a Churchill to the Royal Air Force station at Biggin Hill in Kent to take part in a demonstration to show how armoured formations should co-operate with aircraft.

The 43rd RTR was one of twelve Royal Tank Regiments that had been created from infantry battalions of the Territorial Army; in this case the 6th Battalion the Northumberland Fusiliers from Newcastle-upon-Tyne. This transition took place in 1938 but on 19 October 1941 they had just moved from Brighton to Welbeck Abbey. They had three Churchills at this time, 18 by the end of the month and 43 at the end of November. By the end of the year they were able to report that the changeover from Matildas to Churchills was almost complete. On 7 February nine Churchills from the 43rd, along with three from 144 RAC and three from 148 RAC, undertook a 32 mile road march to Midhope Ranges near Sheffield for a live firing exercise. They reported various minor troubles on the way but only one tank broke down completely, and that was taken back to the tank park. They came back on the 14th, and the war diary refers to 'a number of mechanical breakdowns'. Then, on the 25th, comes their big test. For the first time since it was equipped with Churchills, the entire battalion took to the road for a 32 mile march with 36 tanks. The war diary is brief but positive, saying that they had little mechanical trouble but eight wireless sets failed, which is not bad going. On 3 April they record the arrival of their first Churchill Mark III.

At Crowborough in Sussex 51st Royal Tank Regiment (RTR), another Matilda unit, records the arrival of its first A22 on 10 July 1941 and its second two days later. This is one regiment that goes into explicit detail and shows that both tanks are of the Mark IVA, or Churchill II type, although it never mentions the name at all. Further the tanks are shown as T31496 and T 31497 which would have been the first two production Churchills completed by Metro-Cammell. More Mark IVA tanks arrive in a steady trickle from late July and on 29 September the regiment participates in Exercise Bumper, of which more anon. In November and December the war diary records the arrival of some Mark IV (close support) tanks which must mean Mark Is with the 3 inch howitzer. This reveals an interesting detail.

As we have already seen the original plan was to equip every Churchill tank with the howitzer so that the close sup-

Familiarisation; young soldiers from an infantry regiment get to know Churchill tanks of 43rd RTR during an exercise near Brockenhurst in the New Forest in 1942.

Two Churchills, the leading one with the unlikely name of *Casket* or *Gasket*, are directed off a country lane into woods during an exercise.

port distinction would not exist, but this entry seems to indicate that regiments were not aware of this and, being used to the idea of having a small proportion of close support Matildas with their regular gun tanks, they make the same assumption about the Churchills. What they fail to explain is whether, by this time, the priority on smoke and high explosive ammunition has changed. On 16 December 1941 a Lt Col Disney Barlow arrives from the War Office 'to carry out special tests on the Mark IV' but there is no explanation. Presumably the regiment had been complaining but if so not a hint of this appears in the war diary and of all those inspected for this period none makes any comment whatever about the new tanks, good, bad or indifferent.

Quite different again was the case of the North Irish Horse. This regiment was a Yeomanry, or volunteer cavalry force raised before the First World War. It subsequently converted to armoured cars but then changed again to become an infantry tank regiment, equipped with Valentines. Stationed at Ballykinlar in Northern Ireland it suffered from being at the end of the supply line, and the delivery of tanks was painfully

A partially reworked Churchill Mark I swings out of a side road under the direction of a military policeman. The turret crew have a Bren light machine-gun standing between them.

slow. Yet one wonders why, being on the doorstep of Harland & Wolff so to speak, they did not receive Churchills a lot earlier - unless the Belfast firm was late with its deliveries.

As it was the regiment moved to England in October 1941 and the war diary, which seems obsessed with disciplinary matters, only mentions the movement of B vehicles, the regimental transport. Nothing is said about tanks at all. The North Irish Horse is now based at Westbury in Wiltshire but on 26 February 1942 the regiment goes briefly to the firing ranges at Castlemartin in South Wales where it does firing practice on Churchills. Nowhere else in the war diary around this time does one find tanks mentioned at all. It must be assumed that they took over Churchills on arrival in England, or shortly afterwards, but official confirmation is lacking.

This survey is by no means comprehensive. At least five more British regiments converted to Churchills over the period up to April 1942, and others would soon follow, but it does serve to provide a flavour of how the tanks were distributed, the extent to which regiments reported on them and, to some extent, the ways in which they were used. There are no overt signs of trouble, although the tour of Churchill regiments by Geoffrey Burton provides a hint. Apart from those with a penchant for concentrating on the crimes of their personnel the war diaries give a consistently up-beat impression. The extent to which this was all part of a positive, war-winning outlook is impossible to judge, but there is another, less creditable explanation. The war diary is normally compiled by the regimental adjutant and then signed, on a day to day basis, by the commanding officer. Now although it is by no means true of the entire British Army; the Royal Artillery and Royal Engineers being among the notable exceptions, the average British officer was not really expected to know how things work, or even why. It was his job to command, and set an example for those that do. Again, with due allowance for individual exceptions, it has to be said that this tendency was honoured, almost to the point of becoming *de rigueur,* in the Royal Armoured Corps. Perhaps, then, the attitude enshrined in some of these war diaries was primarily the result of studied indifference, or ignorance. Quite possibly those that compiled the diary, who must have been aware that all was not well on the tank park, could not explain why, and probably did not wish to know, or display their ignorance.

This attitude contrasts sharply with that of the manufacturers. In May 1941 Vauxhall Motors printed a small leaflet on cheap, yellow wartime paper which they caused to be inserted into early editions of the vehicle handbook. The six page document was headlined POINTS TO WATCH and what followed was a list of known faults along with tips on how to minimise them and an explanation of what was being done to correct them. Tucked inside the front cover of each handbook it was available to be read by all who had anything to do with the Churchill tank and as such is probably unparalleled, not just in the tank world but in the defence industry generally. It must have had official sanction to appear there in the first place but it would have been exceptional in any army, never mind the British.

A Canadian trooper clutching an early pattern Thompson sub-machine-gun, shelters at the back of his tank. Tank crews carried revolvers as their personal weapons but the Tommy Gun was also available and can be seen on the stowage diagrams.

Of course there was an ulterior motive in such transparent honesty. It was not simply a matter of laying their cards on the table, Vauxhall Motors management were recruiting the rank and file to their side. They explained that these faults were not endemic; they were the result of hasty development under urgent, wartime conditions, but they would be put right. By treating the men as adults, and taking them into their confidence, the company deflected a lot of unhelpful criticism. This was further improved by sending teams of engineers from the factory to work with the men on their tanks and it was a refreshing approach. Other companies, which like Vauxhall Motors were household names, were responsible for supplying equally dreadful tanks but nobody ever heard them admit it.

Because it was organised along British lines the Canadian Army was also persuaded to create one army tank brigade in addition to its more conventional armoured brigades. First Canadian Army Tank Brigade was created at Camp Borden, Ontario, in February 1941. It comprised the Ontario Regiment, *Trois Rivieres* (Three Rivers) Regiment and the Calgary Regiment which joined in March. They were otherwise known more prosaically as the 11th, 12th and 14th Army Tank Battalions respectively.

The Canadian Official History is a little confusing in its treatment of them, claiming that most of the brigade had arrived in Britain by July and was virtually complete by September 1941. At first the tank battalions appear to have been equipped with Matildas although Canadian industry was gearing up to build Valentines. A photograph in the official history shows an early Churchill, apparently belonging to the Calgary Regiment, driving through a bombed house on the Isle of Wight and this is said to have taken place in June 1942. The

1st Canadian Tank Brigade (less the Calgary Regiment which was preparing for the Dieppe Raid) had two brushes with English history at around this time. It took part in a mock action near the site of the Battle of Lewes, in Sussex, which was fought in May 1264 between the forces of King Henry III and the pretender Simon de Montfort. During Exercise Tiger it found itself under the steely eye of a certain General Bernard Montgomery, who would have further dealings with Churchill tanks before his military career was over.

If the exercises proved anything, as far as the Canadians were concerned, it was the clear superiority of the Calgary Regiment over the other two battalions in the brigade. This was attributed to the fact that the men of the Calgarys came from prairie farming stock. This made most of them familiar with heavy machinery, meaning that the transition from a tractor or combine harvester to the driver's seat in a Churchill tank did not present great difficulties. The other regiments were presumably composed mainly of city boys who took longer to adapt. By January 1942, according to the Official History, 1st Canadian Army Tank Brigade had 157 Churchill tanks which should have been sufficient to equip all three battalions and brigade headquarters.

A summary of faults, detected during official tests and regimental field trials, was issued in July 1941 and by comparison with the comments recorded above it is scathing. The officer who compiled it, whose signature cannot be deciphered, began with a sly dig at the manufacturers. He pointed out that the faults listed had in fact been reported to the chief engineer at Vauxhall Motors and then goes on to explain that the firm does not inform the service of troubles that show up during their own tests. The summary deals with reports on a dozen tanks. It is too long to cover in detail but the comments are short and to the point.

Incidence of wheels and bumper retaining springs breaking are described as 'too many to enumerate ... practically every vehicle made is concerned'. Loss of track pin retainers 'looks like reaching the epidemic state' while gearbox selector troubles and main brake drum bolts working loose are said to be 'general'. The same remark is used to describe the problem of sucking up leaves. The twelve tanks involved certainly include T31496 and T31497, which we know went to 51st RTR on the 10th and 12th of July, yet this report was compiled and issued by the 25th so whatever conclusions this officer reached would appear to have been on the basis of very limited trials. In fact T31496 is only mentioned once, in connection with suspension damage, while T31497 just suffers from an engine gasket failure. The Churchill from Hell is undoubtedly T30971, which appears in the list no less than 72 times! This, of course, is the first Churchill to go to MEE at Farnborough and they might be expected to make a thorough job of testing it; almost to destruction by the sound of it.

Yet note that the tank had been with MEE just eight weeks when the summary was compiled which, no matter how one calculates it, means that T30971 was developing faults at the rate of 1.28 per day, assuming that it was being run every day. But we know, from MEE's own account, that no matter what went wrong the tank always managed to drag itself back somehow at the end of each trial and they do not appear to have been too concerned about it. Seen in that light the summary might be regarded as rather exaggerated but it was an early step on the path towards giving the Churchill tank a very bad name.

A Tank Board meeting on 1 August heard a report which stated that it would take six months to redesign the Churchill sufficiently to eliminate all fundamental defects. In response the General Staff wanted to know what would happen to tank production during that time while the War Office, clearly getting desperate, said that it was prepared to accept any tank that could run for 10 miles without breaking down! Losing patience with the soldiers Sir William Rootes explained that you could not simply switch off production and then switch it on again six months later. The Board was then told that the Prime Minister had suggested, through the CIGS, that some Churchills ought to go out to the Middle East for desert experience, but the Board thought this was inadvisable. Commenting on this the DAFV, in a letter to the Tank Board, pointed out that all Churchills delivered up to the end of October 1941 were not only unfit for service in the Middle East but were not even fit for sustained operations in Great Britain. Even so, as we shall see, two Churchills, modified to suit conditions in the Middle East, were at Farnborough by the second week in December and in Egypt by the end of the month.

Returning to September 1941 we find the War Office notifying the Tank Board that the Ministry of Supply wished to take out contracts for a further 1960 Mark IV tanks, bring the total up to 4300. These figures, clearly stated in an official document, are not confirmed by any other source. Yet by November the General Staff is saying that a definite limit should be put on the number of tanks built, unless they can be delivered in a battleworthy state. They go on to say that they would be prepared to accept a reduction in output if 600 serviceable tanks could be ready for the army by the end of March 1942 and 1,400 by the end of June. Conflicting signals seem to be flying all over the place.

Meanwhile the entire future of the tank hung in the balance. At a meeting of the Tank Board held on 9 December 1941 it was clearly stated that:

i. The War Office were discussing specifications for a new infantry tank. And.
ii. It was not intended to develop the Mark IV.

Yet at the meeting of 8 January 1942:

'The General Staff are clear that there is a requirement for an infantry tank. For this they are prepared to accept the Churchill carrying a 6-pounder gun and made reliable'.

On 21 January Geoffrey Burton received a strongly worded letter from General R M Weeks at the War Office. He

Shipley, **a Mark II Churchill of 43rd RTR, leads one column of tanks across the heather covered ground of the New Forest in Hampshire. Markings include that of 3rd Infantry Division with which 33rd Armoured Brigade was serving in 1942.**

The Maple Leaf emblem on the front of the nearest tank reveals that these two early Churchills belong to 1st Canadian Army Tank Brigade. This is another view of Exercise Spartan.

stated that as of 15 January 1942, 42 per cent of Churchills were off the road.

> *'This is an appalling number inasmuch as ... the majority of the tanks are new.*
> *The casualties are due almost entirely to defects in design.*
> *We view with the gravest concern the apparent inability of the Vauxhall Company to remedy design defects.*
> *Previous promises given to the Tank Board were to the effect that by the end of March, a Churchill battleworthy in all respects would be produced. We have no confidence in these promises being fulfilled.*
> *We have already stated that from April onwards we do not want any more Churchills unless they mount a 6-pounder.*
> *The Ministry of Supply has ordered 3,000 Churchills and are asking us to place orders for more. We are not prepared to order more until and unless we can have assurances that battleworthy tanks are going to be produced.*
> *We state most emphatically that reliability is vital, and we are not satisfied that attempts to achieve this reliability are taking priority over the attempts to produce more numbers.*
> *The War Office Maintenance Department cannot be expected to (nor are they designed to) deal with a fleet of tanks which break down for design reasons almost as soon as we receive them.'*

The next day Geoffrey Burton passed these comments on, in somewhat milder terms, to Mr C J Bartlett of Vauxhall. On the 23rd a meeting of the Tank Board announced that no further orders would be accepted for Churchill tanks until a satisfactory reply had been received from Mr Bartlett. One may scan the reports of subsequent meetings in vain for reference to any such statement. Yet production of Churchills went on as usual.

Although it takes us a little ahead in time this is probably the logical place to review another experimental modification which reveals that even quite fundamental matters concerning the Churchill design had not been entirely resolved.

On 13 February 1942 the Leyland built Mark II Churchill T32256 arrived at the Fighting Vehicle Proving Establishment (FVPE) Chertsey, from the manufacturers. It was in fact the 10th Churchill that Leylands completed to their original contract. It was powered by a 12 cylinder, horizontally opposed, supercharged diesel engine. According to the company's wartime history *And then - Tanks - And still more Tanks* this engine was designed specifically for use in tanks and is illustrated on the same page as a picture of T32256 in the Lancashire factory, linked up to a dynamometer. Evidently the engine had been designed with tanks like the Churchill or Covenanter in mind, to judge from its configuration, but it appears to have fallen at the first hurdle.

Specifications in the FVPE report show it as developing 352bhp at 1800 rpm, against 340bhp at 2200 rpm for the petrol fuelled Bedford equivalent, with torque of 1146 lbs/ft at 1120 rpm against 908 lbs/ft at 1580 rpm for the Bedford. Equipped with the new engine, and a five-speed gearbox with revised ratios to suit, the Churchill put up an excellent performance, especially on climbing trials, but there was a failure of the Lanchester vibration damper so it had to be removed. Further inspection revealed that the supercharger gear train on the left hand cylinder bank was damaged so in March it was returned to the manufacturers. Unfortunately it remained there until October and when the tank was reassembled it was given a comparative trial with the temperamental T30971.

On the flying half mile there was hardly anything to choose. Both tanks rumbled by at a steady 16mph but on steep slopes the diesel Churchill romped away with power to spare. This is undoubtedly where the extra torque came in and the testing authorities described it as "an appreciable improvement in performance". Yet it was not adopted and one can guess that this had something to do with the uncertain future of the Churchill tank. The trial ended on 6 December 1942 when the tank was returned to Leyland for disposal. The report ends by saying that further development would be required to improve mechanical reliability before it was suitable for service overseas. This would seem to be a remarkably harsh verdict on an engine which only had about four months trials in a tank so there may well be other, unrecorded, reasons for its demise.

5

Great Guns

Blessed with the invaluable faculty of hindsight some modern writers, and this includes many veteran tank men, find it difficult to mention the 2-pounder gun without employing the epithet pea-shooter. To anyone writing of the middle years of the war this might be justified, but it would be unfair to apply the comparison universally. In 1940, when the Churchill was designed, the 2-pounder was one of the best anti-tank weapons in the world; marginally better than the American 37mm gun and vastly superior to Japanese and German guns of the same calibre.

Even so, as we have already shown, larger weapons had been considered from the outset and it was only a matter of having a new gun available, for the Churchill to be modified to accommodate it. That new gun was the 57mm (six-pounder) anti-tank gun. It was simply a matter of coincidence that this was the same calibre and shot weight as the weapon provided for British tanks in the First World War, and that is as far as the coincidence went. The new 6-pounder was first proposed by the Director of Artillery early in 1938 and outline design work started that April. It was a low priority scheme. Funds were limited and effort was being concentrated on the 2-pounder which was in demand as an anti-tank gun and for tanks. A mock-up of the 6-pounder, along with the design of a field carriage, was ready by September 1939 but the war intervened and once again it became a matter of providing existing guns, not experimenting with new ones.

It was the 2-pounder which started the trend of employing the same gun in tank and anti-tank mountings. It was a practice favoured by most countries and although the field and tank weapons were not exactly interchangeable they had the same characteristics and fired the same ammunition. A similar policy would be adopted for the 6-pounder. Precisely how things developed from there was the subject of some disagreement between the War Office and Ministry of Supply but the basic facts are not in dispute. At this stage consideration was being given to the development of the gun purely as an anti-tank weapon, although as we have noted a tank version was considered during discussion of the A20 design back in December 1939.

The early part of 1940 was given over to consideration of the kind of field carriage to be designed and in August a statement of requirement was placed before the Ministry of Supply. This, coming shortly after the fall of France, did not get very far. For one thing the design of the new weapon had not yet been approved and matters were further complicated by General Pope, fresh from his experiences in France, declaring that while the 2-pounder was good enough, for now, a more powerful tank weapon would soon be required. This was no time to be asking for such things. The British Expeditionary Force had lost virtually all of its 2-pounder anti-tank guns in France and while these were being replaced even more guns would be needed for the extra tanks that the Prime Minister was now demanding.

The design of the 6-pounder was approved on 9 September 1940 and two months later the War Office asked the Ministry of Supply if the new gun could be fitted into Churchill tanks. However this was all academic while the demand for 2-pounders claimed priority. A new factory, at which it was proposed to manufacture the new gun, was turned over instead to increased 2-pounder production on the grounds that 600 of the smaller weapon could be produced while only 100 of the new weapon might be expected, and things did not change until February 1941. At that time the War Office felt confident that 2-pounder production was satisfactory but realised that unless something was done soon to produce the new weapon British troops would be at a severe disadvantage.

Early estimates suggested that 600 six-pounder guns could be ready by the end of the year although more pessimistic pundits thought 100 pieces more likely. Production facilities were also being sought in the United States and in due course a version of the gun was produced in America and used by the U.S. Army. The other problem facing the British was whether to build them as tank or anti-tank guns.

A fine view of an early Mark III Churchill at Lulworth Camp. Notice that on this, and subsequent models, the co-axial BESA machine-gun is mounted to the left of the main armament. T31184 was built by Vauxhall Motors.

This overhead view of a Mark III makes an interesting comparison with the similar picture of a Mark II in chapter 3. The increased size of the 6-pounder turret being the most obvious difference.

In its original form the barrel of the 6-pounder Mark 1 was 50 calibres long. This was a standard method of describing the length of a gun's barrel as a multiple of its calibre - in this case 57mm by 50 or 112 inches. In January 1941 the Director of Artillery announced that a weapon of this length could not be fitted into the turret of a tank where the turret ring diameter was only 54 inches, as was the case with the Churchill. This is one explanation given for the development of a shorter version of the gun but other sources claim that the main reason why Britain could not mass produce a gun with this length of barrel was because no suitable lathes were available; but the Director of Artillery merely referred in rather vague terms to production difficulties in Britain and the United States. In fact there is no reason to believe that the difficulty with lathes existed in the USA so this may well be no more than an excuse.

In Britain, therefore, the gun was redesigned with a 43 calibre (96 inches) barrel. This would mean a reduction in muzzle velocity but at the time there was no obvious alternative and preparations went ahead to produce the gun in two forms. The Mark 2, the first production version, was designed to sit on an anti-tank field carriage but it could not be fitted into a tank turret. The Mark 3 design was intended for tanks but could, at a pinch, be mounted on a field carriage, yet there appeared to be no urgency to produce it until a suitable turret was available.

The question of a turret to take the 6-pounder on Churchill raised enough problems of its own. There were two schools of thought. One faction wished to use a cast form, relying no doubt on the success of 2-pounder turret, but others advocated a plate type. What they could not decide was how this should be built. Welding seemed to be the obvious choice but nobody in Britain had ever tried to weld a heavy turret before and the general view was that it would break up under fire. Welded turrets had been developed for armoured cars, and they appeared to be satisfactory, but unless they are very unlucky armoured cars do not normally take strikes from tank guns so presumably they did not count. Those who doubted the wisdom of welding favoured bolting, or as some contemporary reports put it, screwing the plate turrets together.

It is not altogether clear what was meant by this. Assuming that the turret was to be of single plate thickness the obvious method would be to create a skeletal frame to which individual panels could be bolted. However certain cruiser tanks being developed at this time had a double thickness turret; an inner structure welded from mild steel plate with an armoured skin bolted to the outside. Since no bolted turrets were ever produced for the Churchill this is only an academic question at best.

Two matters indirectly connected with Churchill turrets might be noted briefly at this stage. Initial design work had already begun on a new cruiser tank, the Mark VII, which

A pair of Mark III Churchills leading a column of tanks on a cross-country driving course on a summer evening.

Firing trials of the 6-pounder on a Mark III Churchill at Lulworth Camp. Bigger guns and the need for more space caused the firing range to be enlarged and many homes, like the house seen here, had to be evacuated. Despite promises made at the time, none of these homes have ever been reoccupied.

was intended to take the 6-pounder gun. But when this looked like being delayed an alternative plan was introduced to fit the gun into the Cruiser Mark VI, or Crusader. This was not a popular move because the Crusader's turret was already rather cramped and the DGTT reckoned he had enough work on without expecting his people to tackle something else. But the project went ahead anyway. In the meantime another proposal, to fit 2-pounder Churchill turrets to Matildas was being considered, but this came to nothing.

Another matter raised during March 1941 concerned gun depression. Some commentators calculated that the maximum depression that could be expected with the 6-pounder turret was five degrees, which was less than the General Staff would normally consider sufficient. There were some pessimists who even suggested that no depression would be achieved at all, and that was quite unacceptable. The proposed solution was to insert a spacer ring between the turret and hull, in effect raising the turret. This was rejected as being too complicated and likely to increase weight beyond the acceptable level. It should also be noted that the 6-pounder mounting in the Churchill was described as having geared elevation, manually operated, whereas the same gun fitted into the Cruiser Mark VII employed a system of free elevation. This may well have been due to the perceived difference in operation between a cruiser and infantry tank. The former would be expected to exploit its greater speed and mobility in action, firing on the move and taking snap shots at targets of opportunity. In such circumstances a skilled gunner could achieve good results with this technique. But for a slow moving tank, making a deliberate attack upon a fixed target it probably helped to have a gun which could be held at particular levels. However, as we shall see, this subject is not as simple as that. It is also worth pointing out that the 6-pounder mounting had the BESA machine-gun to the left of the main weapon, that is on the gunner's side and here it remained on all subsequent versions of Churchill.

Weight looked like being a problem in any case. Reporting in April Mr Durrant of the Tank Board stated that Vauxhall Motors was conducting a weight analysis to see if it could be trimmed anywhere. The Churchill was already well over the 35 tons of its original specification and with the new turret and gun would probably touch 40 tons. Pending the result of this study Colonel Bouchier was asked to approach the War Office and see what sort of reductions they would consider. Mr Little of Vickers suggested reducing armour thickness around the engine compartment but this was rejected because the General Staff had always insisted on all-round protection which, it was believed, the Germans did not have. The question of a cast turret was also raised again but discarded on the grounds of weight.

On 9 April 1941 it was reported that a mock-up of the 6-pounder turret was now complete and General Pope was invited to go down to Farnborough to inspect it. The mock-up, of course, would be made of wood which presented no problems. Armour plate was another matter. Supplies were administered by the Iron and Steel Control Board and it was suggested that Beardmores, one of the firms engaged to produce plate for the new turrets, was being a bit too optimistic in its promises. Yet this was not the only problem. There were good reasons to believe that the various armoured castings required to complete the gun mountings would be in short supply and this was seen as likely to create a serious bottleneck in production. This shortfall would not be resolved until deliveries began to appear from the United States.

A week later, on 16 April, the matter of cast turrets was raised again. Reports received from the USA, following tests

(text continues on page 65)

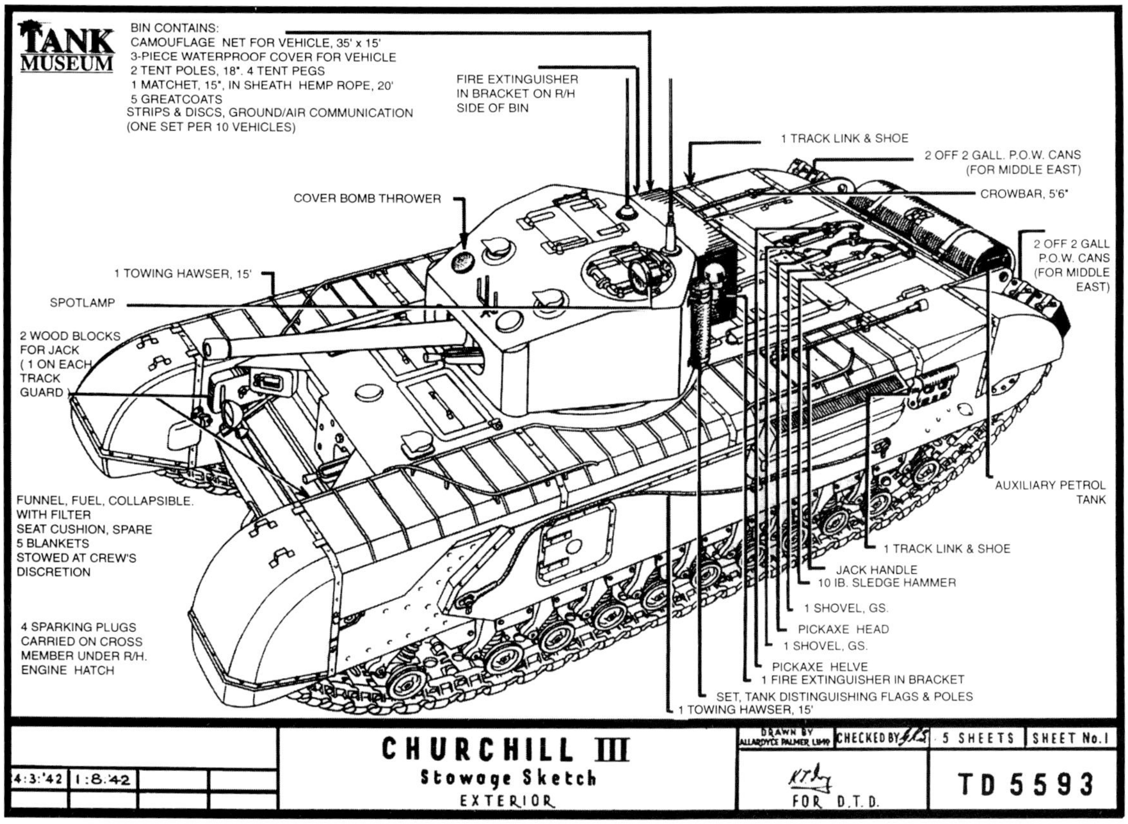

Pages 61-64: Stowage, armour thickness and other diagrams relating to the Churchill Mark III.

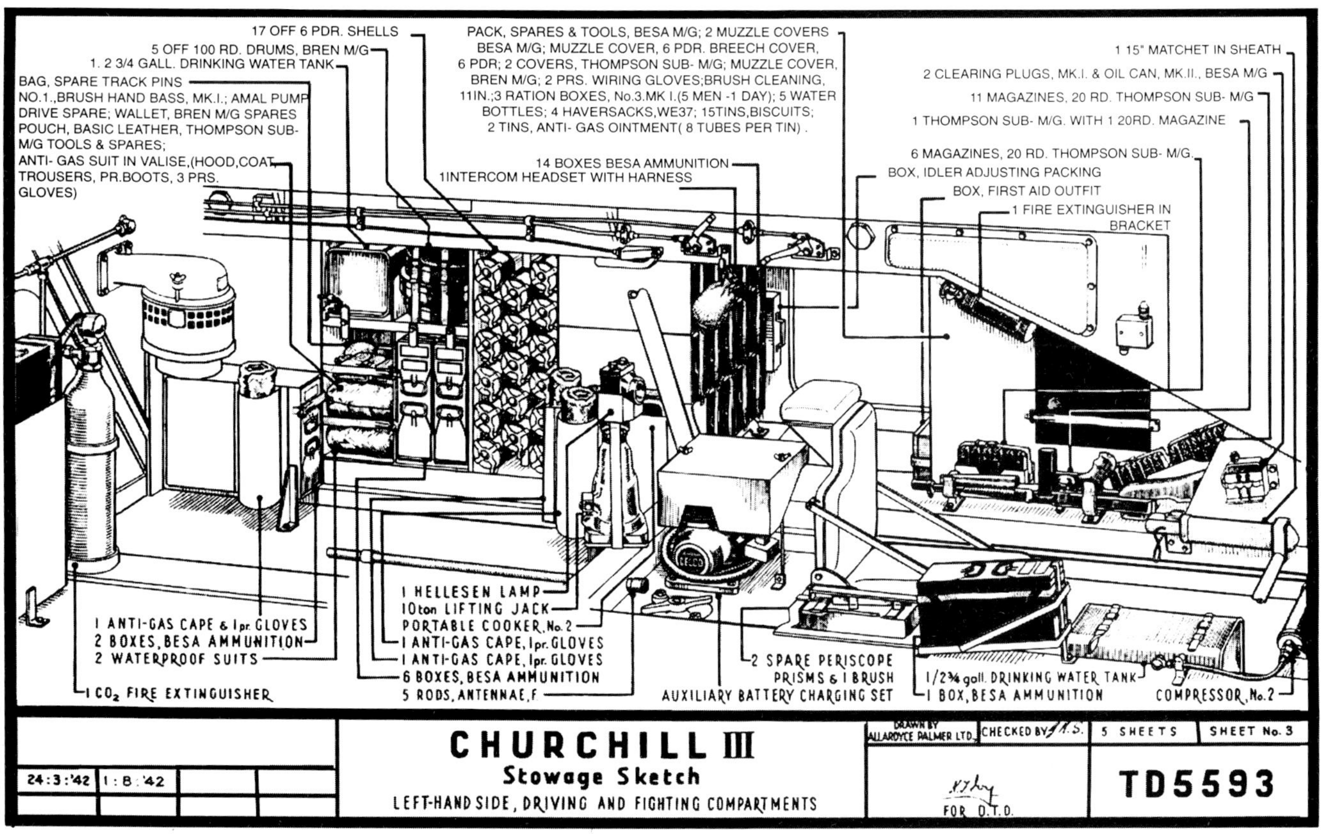

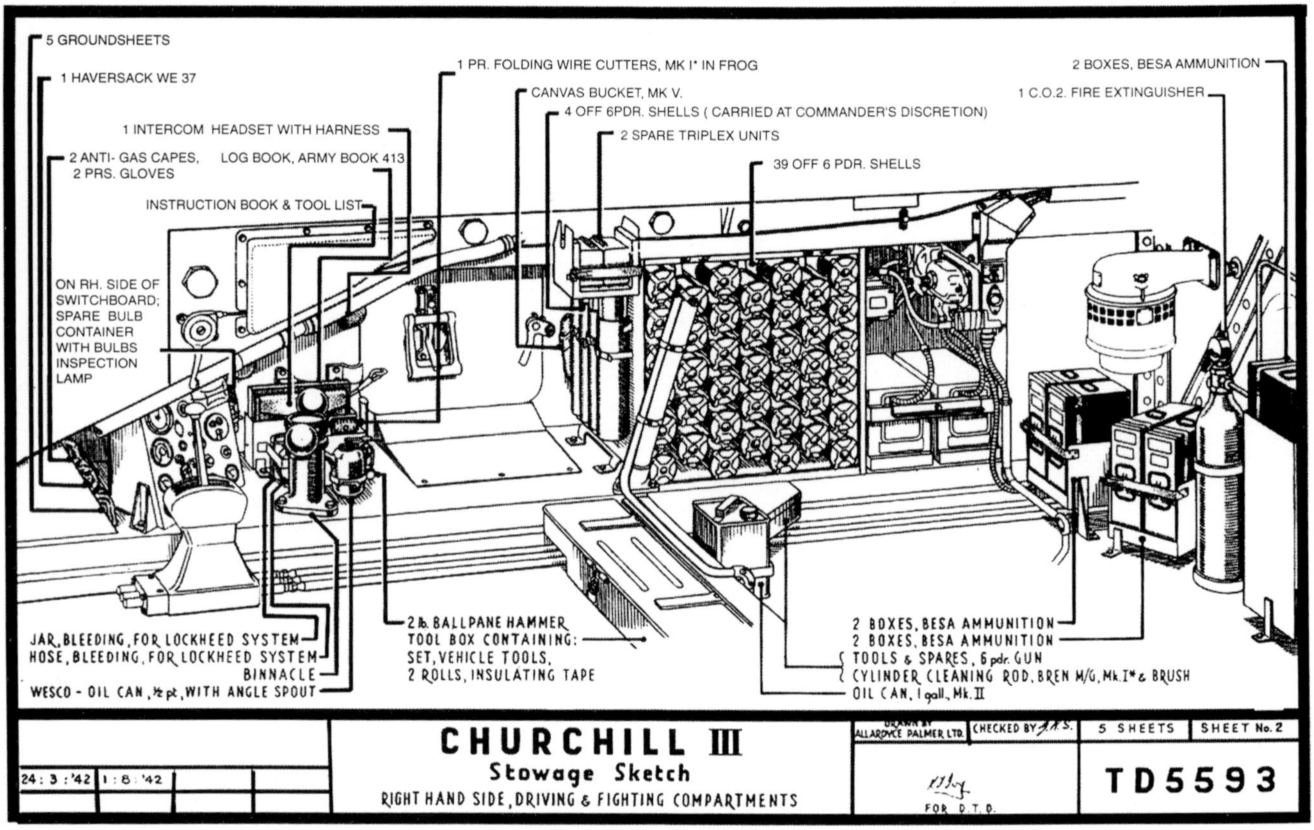
5 GROUNDSHEETS
1 HAVERSACK WE 37
1 INTERCOM HEADSET WITH HARNESS
2 ANTI- GAS CAPES, 2 PRS. GLOVES
LOG BOOK, ARMY BOOK 413
INSTRUCTION BOOK & TOOL LIST
ON RH. SIDE OF SWITCHBOARD; SPARE BULB CONTAINER WITH BULBS INSPECTION LAMP
1 PR. FOLDING WIRE CUTTERS, MK I* IN FROG
CANVAS BUCKET, MK V.
4 OFF 6PDR. SHELLS (CARRIED AT COMMANDER'S DISCRETION)
2 SPARE TRIPLEX UNITS
39 OFF 6 PDR. SHELLS
2 BOXES, BESA AMMUNITION
1 C.O.2. FIRE EXTINGUISHER
JAR, BLEEDING, FOR LOCKHEED SYSTEM
HOSE, BLEEDING, FOR LOCKHEED SYSTEM
BINNACLE
WESCO - OIL CAN, ½ pt, WITH ANGLE SPOUT
2 lb. BALLPANE HAMMER
TOOL BOX CONTAINING: SET, VEHICLE TOOLS, 2 ROLLS, INSULATING TAPE
2 BOXES, BESA AMMUNITION
2 BOXES, BESA AMMUNITION
TOOLS & SPARES, 6 pdr. GUN
CYLINDER CLEANING ROD, BREN M/G, Mk.I* & BRUSH
OIL CAN, 1 gall., Mk.II
24:3:'42
1:8:'42
CHURCHILL III
Stowage Sketch
RIGHT HAND SIDE, DRIVING & FIGHTING COMPARTMENTS
DRAWN BY ALLARDYCE PALMER LTD.
CHECKED BY
FOR D.T.D.
5 SHEETS
SHEET No. 2
TD5593

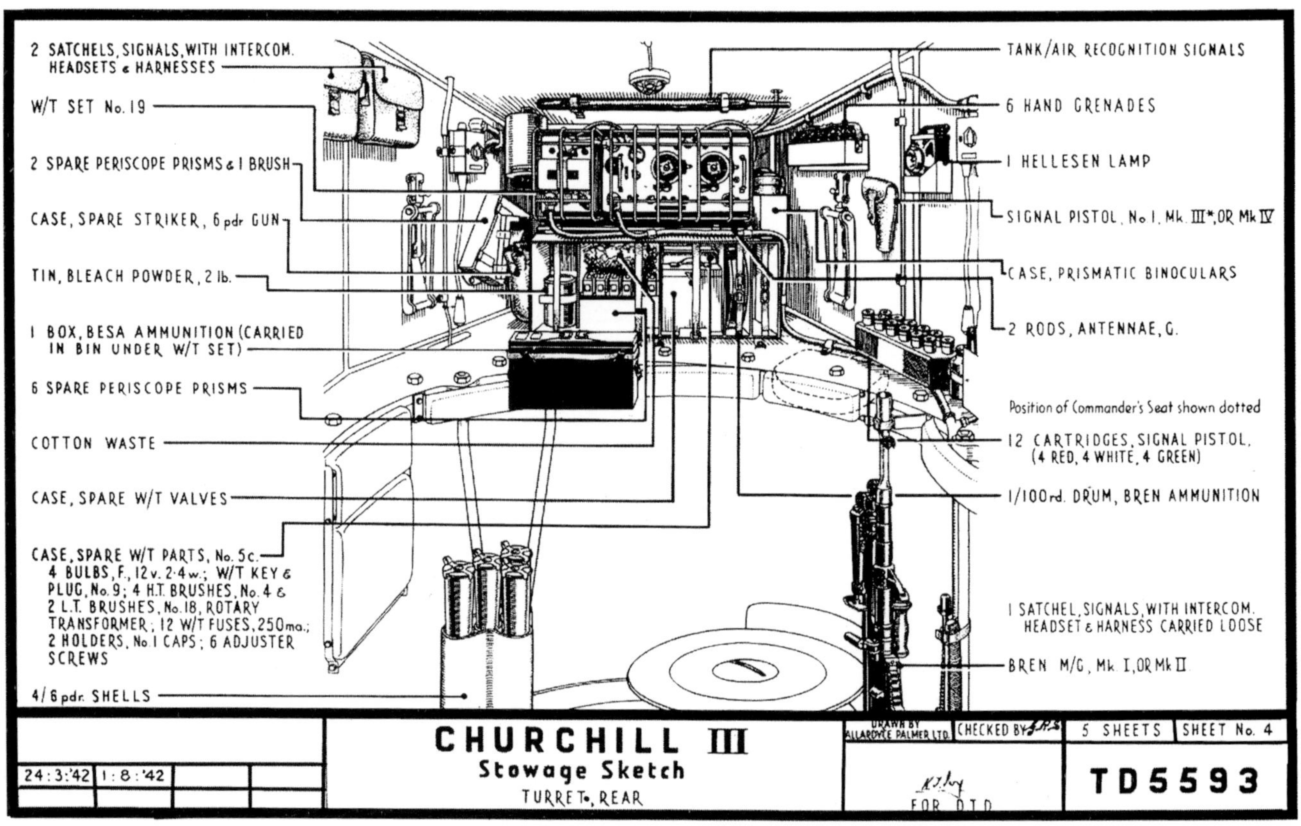
2 SATCHELS, SIGNALS, WITH INTERCOM. HEADSETS & HARNESSES
W/T SET No. 19
2 SPARE PERISCOPE PRISMS & 1 BRUSH
CASE, SPARE STRIKER, 6 pdr GUN
TIN, BLEACH POWDER, 2 lb.
1 BOX, BESA AMMUNITION (CARRIED IN BIN UNDER W/T SET)
6 SPARE PERISCOPE PRISMS
COTTON WASTE
CASE, SPARE W/T VALVES
CASE, SPARE W/T PARTS, No. 5c. 4 BULBS, F., 12v. 2·4w.; W/T KEY & PLUG, No. 9; 4 H.T. BRUSHES, No. 4 & 2 L.T. BRUSHES, No. 18, ROTARY TRANSFORMER; 12 W/T FUSES, 250ma.; 2 HOLDERS, No. 1 CAPS; 6 ADJUSTER SCREWS
4/6 pdr. SHELLS
TANK/AIR RECOGNITION SIGNALS
6 HAND GRENADES
1 HELLESEN LAMP
SIGNAL PISTOL, No. 1, Mk. III*, OR Mk IV
CASE, PRISMATIC BINOCULARS
2 RODS, ANTENNAE, G.
Position of Commander's Seat shown dotted
12 CARTRIDGES, SIGNAL PISTOL. (4 RED, 4 WHITE, 4 GREEN)
1/100 rd. DRUM, BREN AMMUNITION
1 SATCHEL, SIGNALS, WITH INTERCOM. HEADSET & HARNESS CARRIED LOOSE
BREN M/G, Mk. I, OR Mk II
24:3:'42
1:8:'42
CHURCHILL III
Stowage Sketch
TURRET, REAR
DRAWN BY ALLARDYCE PALMER LTD.
CHECKED BY
FOR D.T.D.
5 SHEETS
SHEET No. 4
TD5593

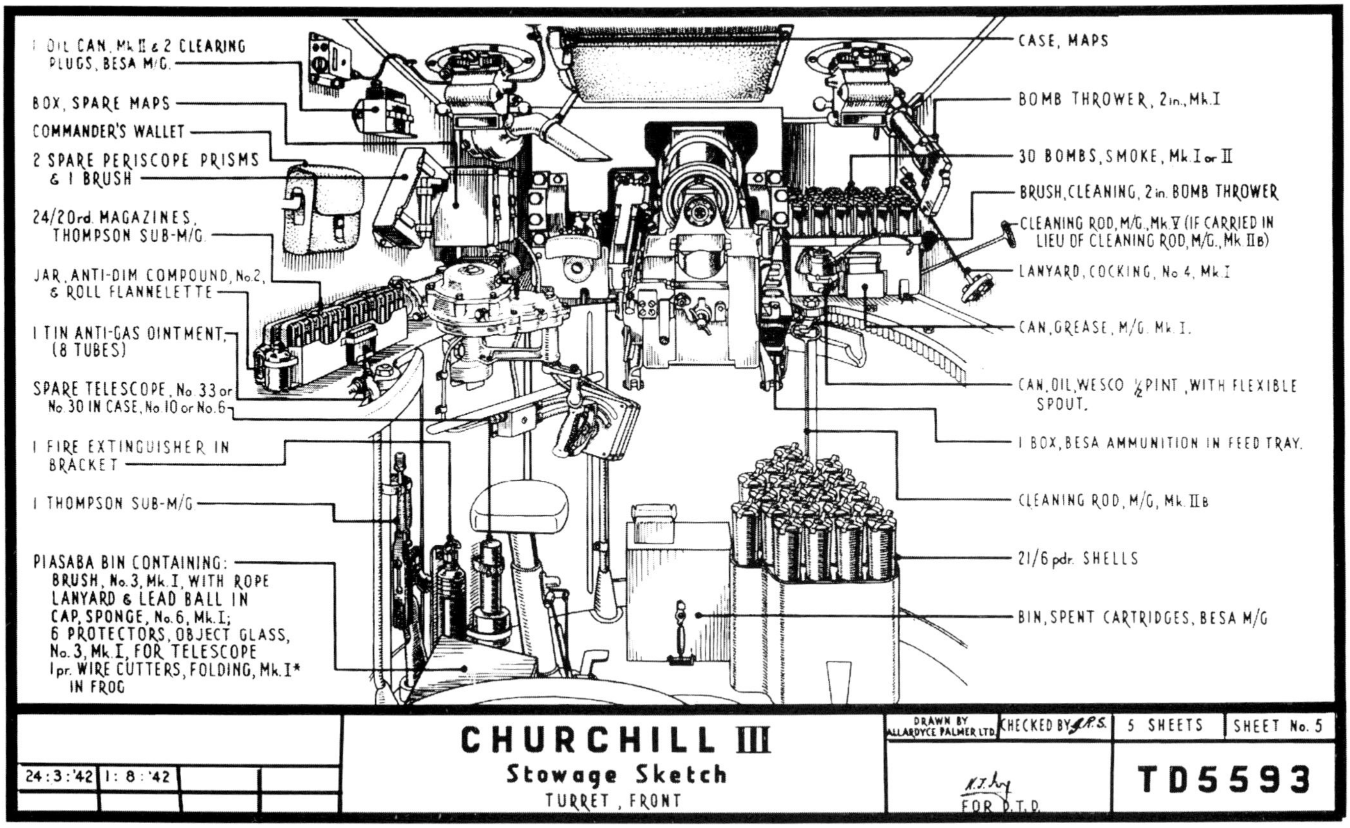
1 OIL CAN, Mk II & 2 CLEARING PLUGS, BESA M/G.
BOX, SPARE MAPS
COMMANDER'S WALLET
2 SPARE PERISCOPE PRISMS & 1 BRUSH
24/20rd. MAGAZINES, THOMPSON SUB-M/G
JAR, ANTI-DIM COMPOUND, No.2, & ROLL FLANNELETTE
1 TIN ANTI-GAS OINTMENT (8 TUBES)
SPARE TELESCOPE, No. 33 or No. 30 IN CASE, No. 10 or No. 6
1 FIRE EXTINGUISHER IN BRACKET
1 THOMPSON SUB-M/G
PIASABA BIN CONTAINING: BRUSH, No.3, Mk.I, WITH ROPE LANYARD & LEAD BALL IN CAP, SPONGE, No.6, Mk.I; 6 PROTECTORS, OBJECT GLASS, No.3, Mk.I, FOR TELESCOPE 1pr. WIRE CUTTERS, FOLDING, Mk.I* IN FROG
CASE, MAPS
BOMB THROWER, 2in., Mk.I
30 BOMBS, SMOKE, Mk.I or II
BRUSH, CLEANING, 2in. BOMB THROWER
CLEANING ROD, M/G, Mk.V (IF CARRIED IN LIEU OF CLEANING ROD, M/G., Mk IIB)
LANYARD, COCKING, No. 4, Mk.I
CAN, GREASE, M/G. Mk.I.
CAN, OIL, WESCO ½ PINT, WITH FLEXIBLE SPOUT.
1 BOX, BESA AMMUNITION IN FEED TRAY.
CLEANING ROD, M/G, Mk. IIB
21/6 pdr. SHELLS
BIN, SPENT CARTRIDGES, BESA M/G
CHURCHILL III
Stowage Sketch
TURRET, FRONT
24:3:'42
1:8:'42
DRAWN BY ALLARDYCE PALMER LTD.
CHECKED BY
5 SHEETS
SHEET No. 5
FOR D.T.D.
TD5593

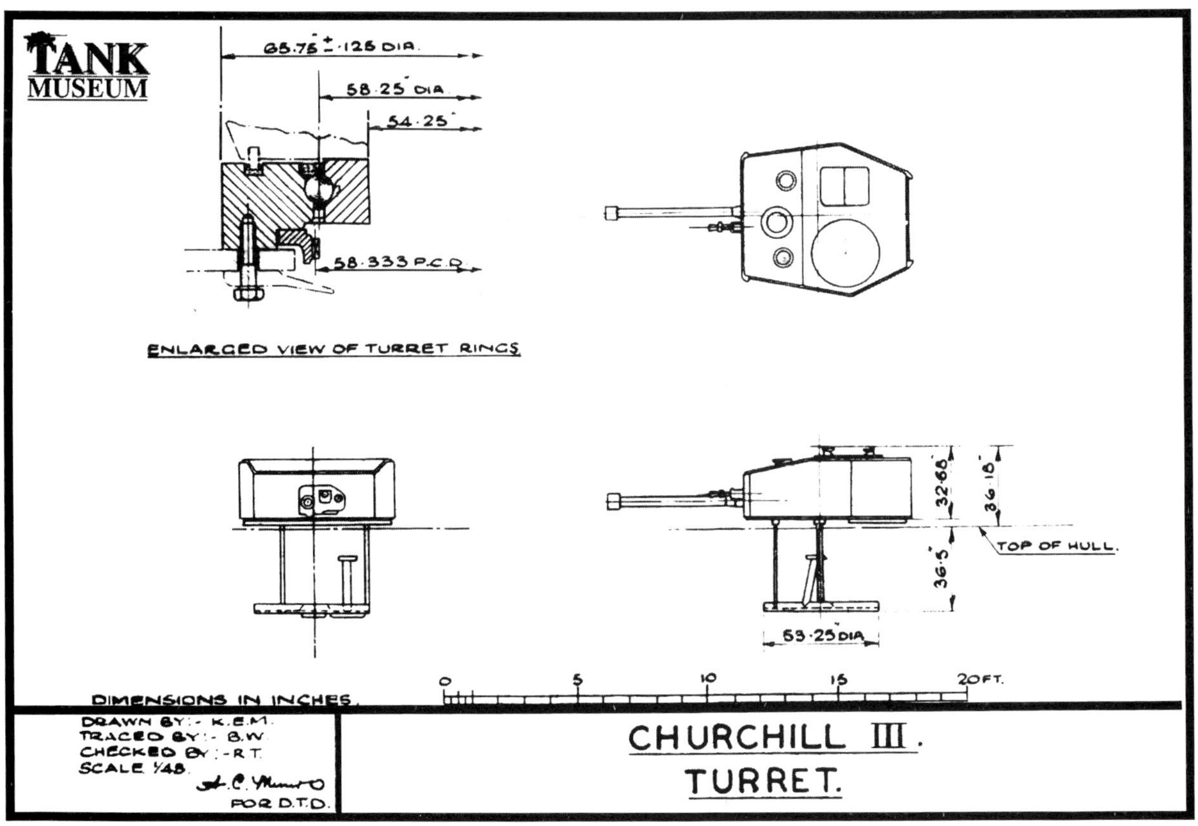
TANK MUSEUM
65.75 ± .125 DIA.
58.25 DIA.
54.25
58.333 P.C.D.
ENLARGED VIEW OF TURRET RINGS
32.68
36.18
36.5
TOP OF HULL.
53.25 DIA.
0
5
10
15
20 FT.
DIMENSIONS IN INCHES.
DRAWN BY:- K.E.M.
TRACED BY:- B.W.
CHECKED BY:- R.T.
SCALE 1/48.
FOR D.T.D.
CHURCHILL III.
TURRET.

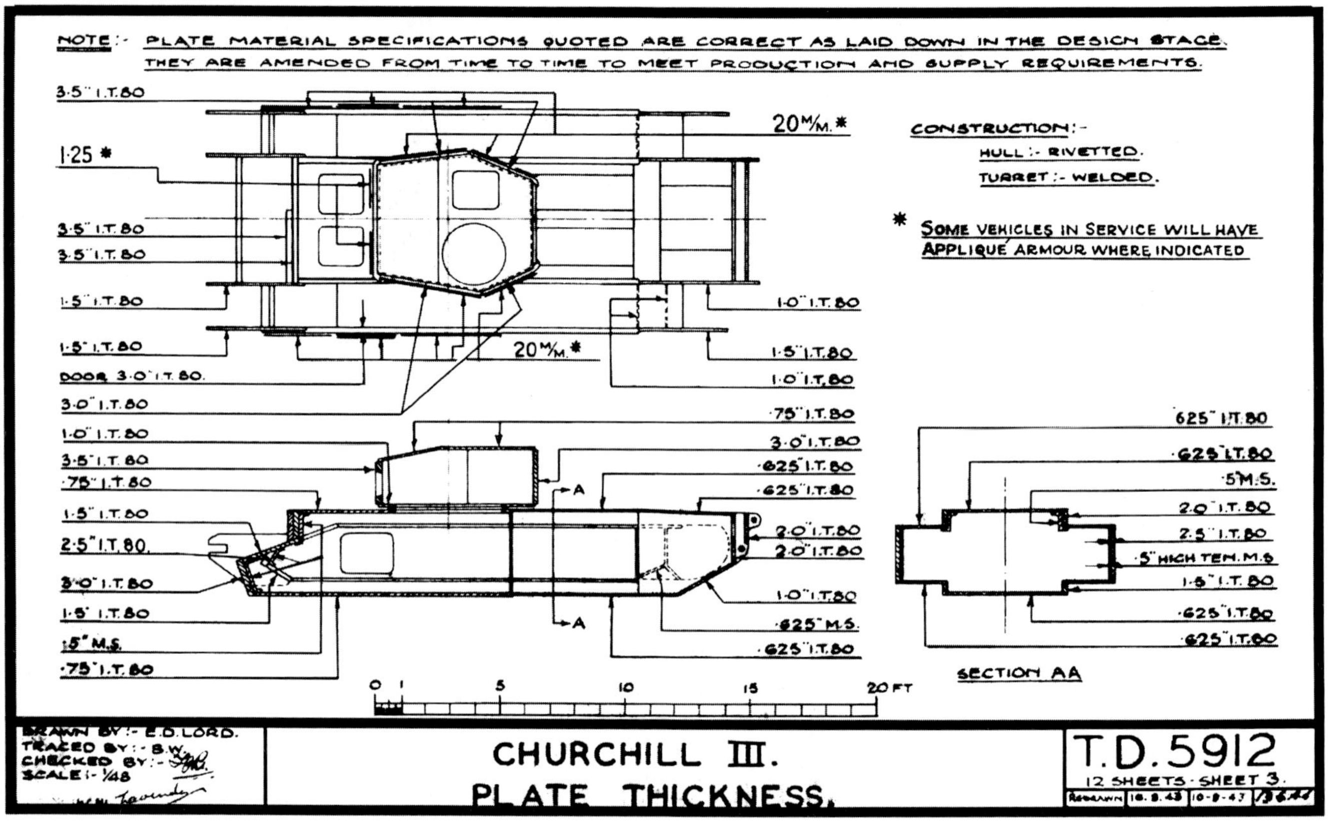

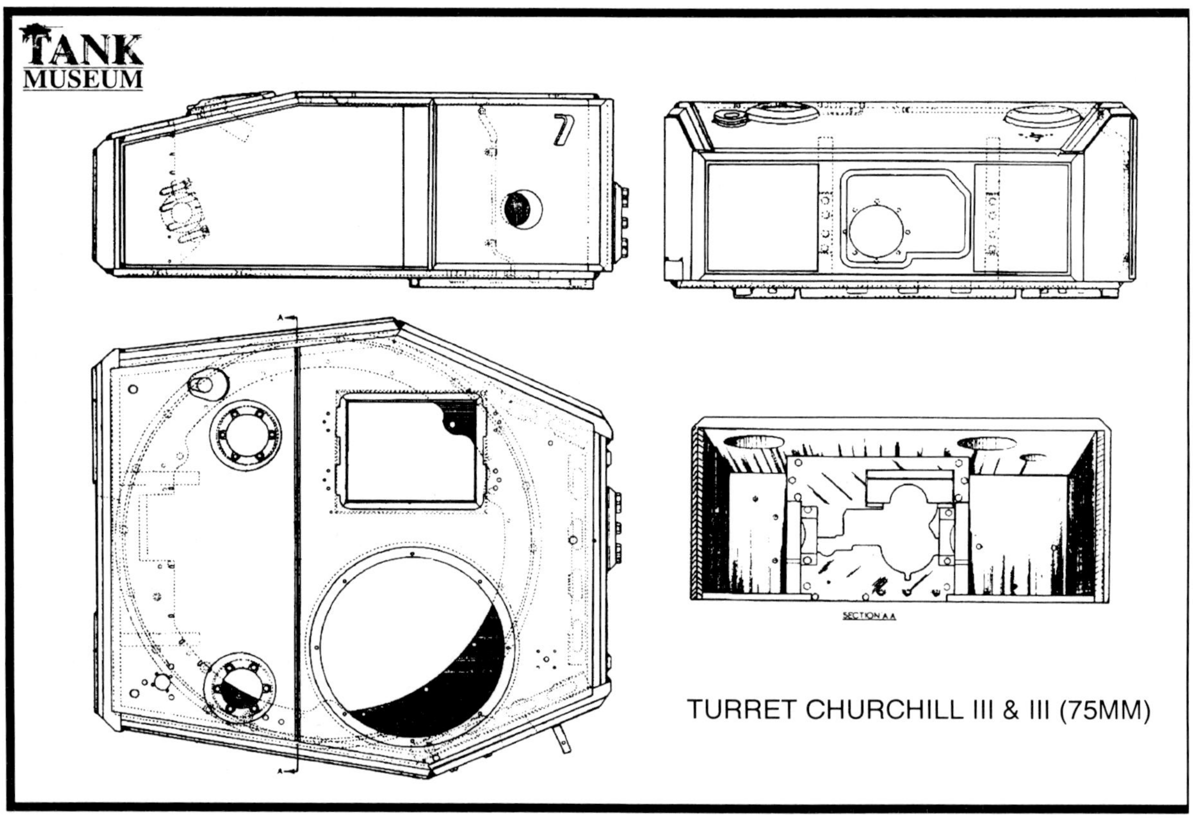

TURRET CHURCHILL III & III (75MM)

conducted with the 37mm gun, suggested that cast armour could be produced of equal quality to rolled plate. Some British experts found this difficult to believe but they could not agree on the ideal type of rolled plate either. Some advocated face hardened armour while others preferred what was described as homogenous, or machineable quality (MQ) plate which was better known. After a long discussion it was agreed to order the pilot turret in MQ plate and then place a contract for six turrets of each type for tests.

By May revised figures for the 6-pounder gun had appeared. Now, it seemed, at least 300 could be expected by the end of the year and the new chairman of the Tank Board, Sir James Lithgow, was anxious to speed up the production of turrets. Vauxhall Motors announced that it was eager to see the welded turret put into production but it seems that the bolted type still had its supporters. The Iron and Steel Control Board, having examined the drawings, calculated that they could release enough material for 500 turrets, of which 400 would be bolted. The Scottish company Babcock and Wilcox of Renfrew stated that they had spare welding capacity and estimated that, once production was under way, they could deliver turrets at the rate of six a day.

A fine study of a Churchill Mark III with its crew. The tank commander is festooned with his communication harness but one questions the standard of maintenance, the track seems incredibly slack.

Members of the DGTT learned in June that the 6-pounder gun mounting for A22 had undergone a successful firing trial and that Stothert & Pitt of Bath should be made the parent firm for production of this item. Vauxhall Motors meanwhile complained that they had not been consulted on the turret order and they did not have much faith in the bolted design. Geoffrey Burton, for the DGTT, admitted that they should have been informed but stated that it was up to his department, and not Vauxhalls, to say which type should be built. However he qualified this by adding that if the welded type did prove satisfactory it would be possible to assemble the plates for the bolted type in the same way. A week later it was learned that a Mr Laskey of Vauxhalls was refusing to sanction an order for bolted turrets on the grounds that plans for production had not been approved.

This suggests that Vauxhalls were prepared to be awkward in the hope that, by delaying matters, they would get their way. This clearly annoyed the authorities but Vauxhall

***Sprinter IV* is an early Mark III in full operational trim, probably in service with 43rd RTR in the New Forest, a favourite hunting ground of King William (the Conqueror) and his sons 900 years earlier.**

stuck to their guns. They were convinced that production of the bolted turret would present major problems and intended to by-pass the Board and write direct to Sir James Lithgow and tell him. A Board member, General Crawford, was told that if Vauxhalls refused the work he should start negotiating with Vickers but the former firm was urged to press on with the welded design as quickly as possible.

The subject of depression was raised again early in July. General Pope, the least compromising individual at the War Office where tanks were concerned, had apparently agreed to accept a figure of 2.5 degrees as the absolute last resort but he discussed with Geoffrey Burton the rather bizarre idea of painting a white line around the inside of the turret to show the gunner where he could depress to 7 degrees and where 2.5 was the limit. Bearing in mind that elevation and depression of the gun was a straight physical operation by the gunner and considering that this extra factor would have to occupy the gunner's mind in the heat of an action, it seems to have been expecting a lot.

The fate of the bolted turret was finally decided by default. Three manufacturers had been approached and all stated that they would not be able to undertake the work. A fourth firm said that it could produce the plates but not assemble them, and this was the final stumbling block. The authorities agreed that unless production capacity could be found within one week the bolted turret would have to be abandoned. Meanwhile Babcock & Wilcox announced that they could increase production of welded turrets if they were relieved of the responsibility for designing welded hulls for the A22, of which more later. This was accepted and it is the last one hears of the bolted version.

Babcock & Wilcox Ltd. completed an experimental welded turret and offered it for firing trials in July 1941. The turret was the correct shape and size but lacked any sort of openings such as hatches or gun apertures. Once assembled the turret was heated up to 600 degrees centigrade for 3 hours to restore its armoured integrity. It was then subjected to 18 rounds of 2-pounder shot, and one of 25 pounder which it survived, more or less. Dire prophesies of disaster preceded the trial. Many forecast that when a weld was struck it would fracture but in fact the effect was invariably local. There were instances of rounds striking the edge of a plate and prising it away from the weld, but all things considered the turret stood up to the punishment very well. Even so two recommendations were adopted on the production turret. Extra fillets of weld were to be added inside the turret to strengthen weak spots and the vertical joints were bevelled to reduce the risk of rounds lodging in the weld.

As we have already seen the first Mark III Churchill tanks started to enter regimental service around April and May of 1942. It is not possible to give a precise figure for the number of Mark III Churchills built since almost every contract was altered, some more than once, and none of the figures add up, but a rough guide may be quoted from notes found on one of the Vauxhall Mastercards for Contract T301. This gives 303 Mark I, 1120 Mark II and 692 Mark III. A later source,

This Mark III, named *Betty*, belonged to B Squadron, 14th Canadian Army Tank Regiment (the Calgary Regiment). It is seen here being recorded by an Army Film Unit but the tangle of ropes at the front suggests that there is something wrong with the locking device for the gun.

issued by the Royal Armoured Corps in June 1944, quotes 675 Mark III. Even if it is not precise it does appear to give a fair representation of the relative quantities of each type.

Although the British Army had developed self-propelled artillery equipment during the First World War, and used it to some extent, the concept did not prove popular. A few experimental machines, known as Birch Guns, were produced after the war but by 1939 there was no perceptible interest at all and no prototypes to provide inspiration. By 1940 the subject was up for consideration again, but with two quite different objects in view. The Royal Artillery was interested in obtaining self-propelled mounts for two of its weapons, the 25-pounder field gun/howitzer and the six inch howitzer. The former ultimately emerged as the Bishop, based on the Valentine tank chassis, but the bigger weapon never developed beyond the discussion stage. All we learn from surviving pa-

A Churchill 3 inch Gun Carrier with the main armament on full elevation, as it would be when the vehicle was in travelling mode.

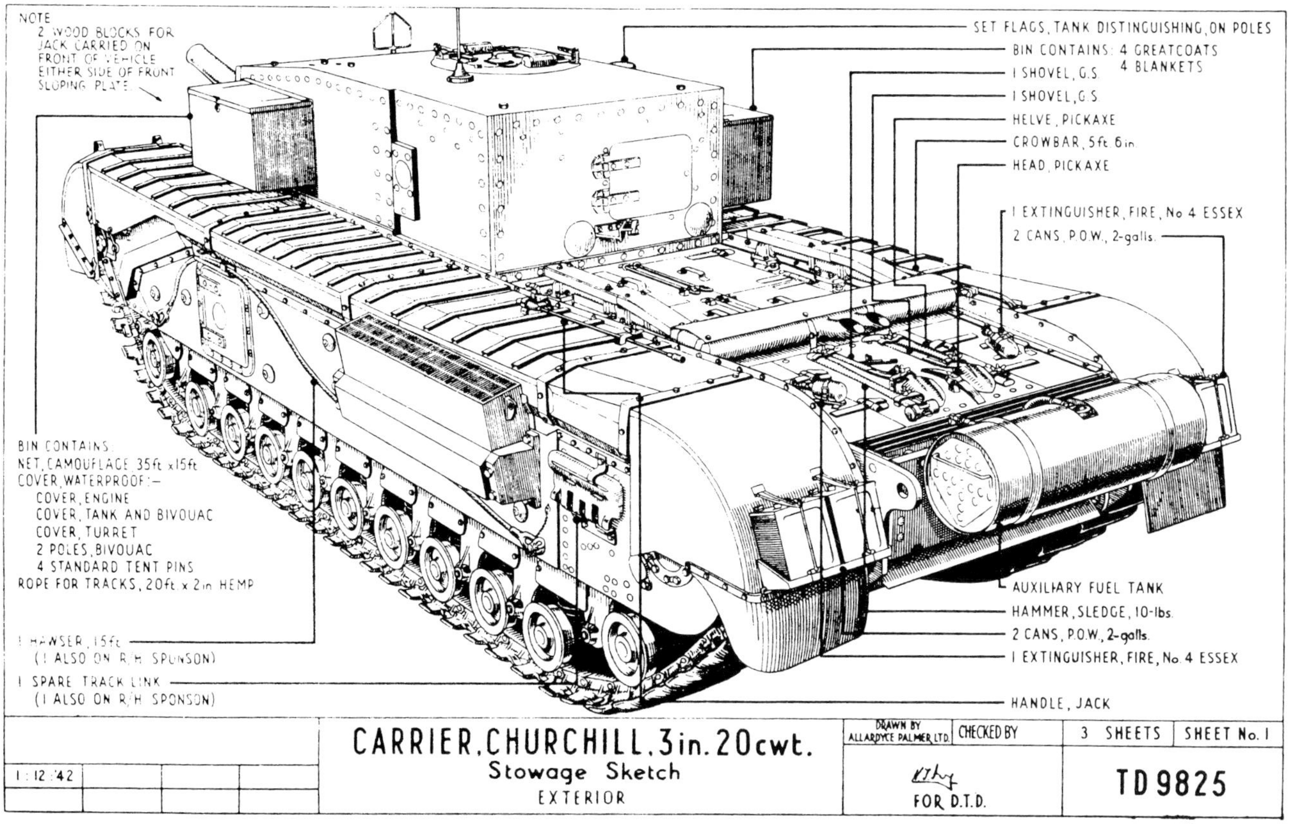

This page and top of page 68: Stowage diagrams for Churchill Gun Carrier.

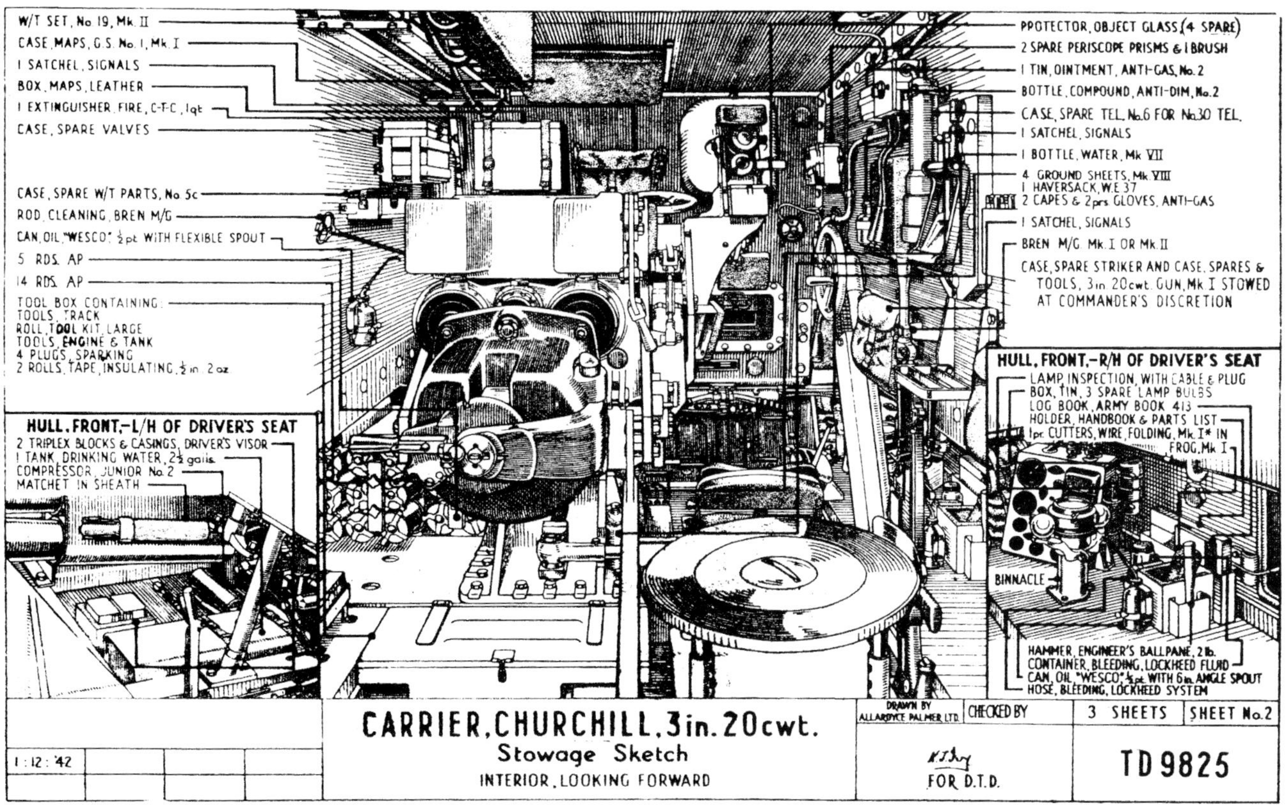

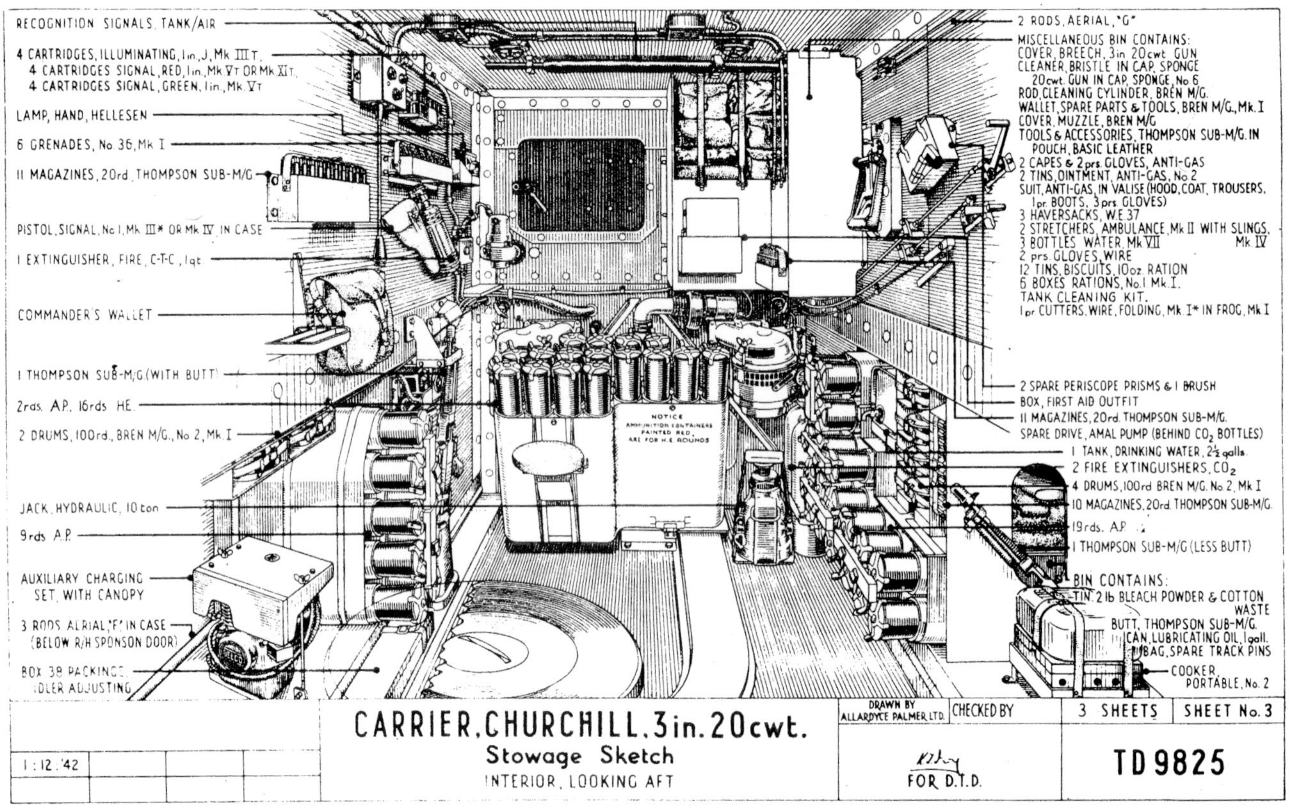

pers is that it would have been fitted onto the Churchill chassis and one curious hint that it might have been mounted in a turret. Exactly how a turret, large enough to accommodate a six inch gun, could have been fitted to a Churchill is not explained, probably for a very good reason. However it may have taken the form of some sort of open top barbette or even a poorly chosen term to indicate a fixed superstructure. This project was finally cancelled in December 1941.

The other self-propelled gun project was quite different. It was born of the desperate need to find a weapon powerful enough to defeat German armour in the event of invasion. No one was expecting a sophisticated weapon system, which might take years to develop; all they wanted was a quick and effective means of transporting the heaviest possible anti-tank gun on some sort of tank chassis. But which gun?

As we already know 2-pounder production was fully committed and the 6-pounder only at the design stage in 1941, while a 17-pounder was being considered. Experts were consulted and they came up with the novel idea of adapting an anti-aircraft gun and the weapon they selected was the 3 inch, 20 cwt. Three inch is clearly the calibre; 20 cwt the weight of the barrel and breech mechanism. In fact it was an odd expression. Twenty hundredweight (cwt) is an Imperial ton but presumably someone thought that the former term sounded better. In any case the parts in question actually weighed 2250 lbs, which was 10 lbs over 20 cwt, but such inconsistencies never seem to have bothered anyone when describing weapons.

The 3 inch AA gun dates from 1914 and it was common to both the Army and Royal Navy, so stocks were good. By 1940 it was slowly being replaced on land by a superb new 3.7 inch weapon and there was no problem in providing 100 or more barrels for this project. Quite who came up with the idea of converting an anti-aircraft gun into an anti-tank weapon is not clear. We know now that German troops had used their formidable 88mm anti-aircraft weapon against tanks both in Spain in 1936, and more recently in France, but it is unlikely that anyone in Britain was aware of that in 1941. However the performance of the British gun speaks for itself, and availability must have been another key factor. Firing horizontally the gun had a range of some 12,000 yards and, firing a 12.5lb armour piercing round, could penetrate 84mm of armour, at 30 degree angle, at 1,000 yards. The German 88mm was better, but it was larger and fired a heavier shell at a higher muzzle velocity (2660fps against 2000fps for the 3"). Compared with the American or British 75mm tank guns of 1942/43 the 3 inch, 20 cwt was very impressive indeed.

The matter is first mentioned at a meeting of the DGTT committee in March 1941 when Geoffrey Burton reported that the Defence Committee had requested that the 3 inch AA gun should be carried in some tanks until sufficient 6-pounders were available, as an emergency defence measure. The Defence Committee was a Churchillian creation dating from August 1940. In his other role, as Minister of Defence, the Prime Minister effectively kidnapped the Joint Planning Committee, which had hitherto been an independent tri-service body, and made it his personal advisory committee. It is, therefore, not beyond the bounds of possibility to hint that it was Churchill himself who suggested this emergency conversion.

The project was authorised by the War Cabinet two weeks later, and the Prime Minister also dominated the War Cabinet. Certainly Geoffrey Burton was impressed by the urgency of the scheme and suggested that as a first step a gun should be obtained. Then the committee could seek around for something to carry it. General Crawford mentioned the Birch Gun and Burton asked if they might see one, but a phone call to General Sir Campbell Clarke, the Director of Artillery, indicated that there were none left. As an alternative Colonel Blagden arranged to produce drawings and photographs.

A week later Geoffrey Burton told the committee that a meeting at Farnborough had agreed that A22 was the best choice for a chassis but, should it not prove suitable, the A12 Matilda could be examined. Another committee member, Mr J Weir, pointed out that the project would absorb a lot of valuable time and technical expertise. A week later, when it was established that A22 was the only tank suitable for the scheme, Weir again voiced his doubts. He explained that it would upset Churchill production and that there was no proof that the 3 inch gun was superior to the 6-pounder. Colonel Bouchier said that there was plenty of evidence that the 3 inch was a much better anti-tank gun and went on to say that the Germans were considering tanks with 150mm armour. How he knew this is not explained but he was prepared to state that the 6-pounder could not deal with such heavy armour, whereas the 3 inch gun could.

By the end of April 1941 it was learned that general arrangement drawings for the new mounting were more or less complete. A turret, it seems, was out of the question but things were not going well. Vauxhall Motors had been tasked to develop the design but they were not happy. They could see no future in the work and questioned the sense of it, but Geoffrey Burton explained that they must keep on with it and do so as quickly as possible 'otherwise the use for which it was designed would disappear'. Unless this is taken to mean that the enemy would conquer the country and put an end to all weapon production this phrase would appear to be self contradictory.

Early in July a wooden mock-up was nearly complete although arguments had broken out between the Director of Artillery and the Department of Tank Design over whose responsibility it was. On 25 July 1941 a contract for 100 'Tanks A22 "Special Type" (Churchill 3" 20 cwt Self Propelled Mounting)' was issued in favour of Vauxhall Motors. But it was agreed that Vauxhalls, having built the pilot model, would sub-contract the assembly work. Investigations by the Tank Board in December revealed that work on the project had almost ceased and Vauxhalls claimed that they had been told to drop the scheme. No such order being traced Sir George Usher, the Director General of Tank Supply, announced that 24 of the guns should be completed as quickly as possible and within days the order had been official reduced to 24 machines. In January 1942 it was revised again to fifty; that is one pilot model and 49 production machines to be assembled

A rear view of the 3 inch Gun Carrier gives a good idea of the size of the superstructure. Although it does not show up clearly here there is a hatch in the rear plate which has to be opened when the gun fires but closed when the tank moves.

by Beyer Peacock & Company of the Gorton Foundry, Manchester, a firm famous for constructing the huge Garratt articulated steam locomotives. All fifty were completed by November 1942 and issued with the War Department numbers S31273 to S31321.

The first edition Instruction Book for this vehicle, published in August 1942, gives it the official title of Carrier, Churchill 3 in. Gun Mark I. The book makes no attempt to explain its role and contents itself with the phrase that it 'is a heavily armoured vehicle' which is fair to the extent that armour thickness was the same as the Churchill I or II. The appearance may be judged from the illustrations which show that it was little more than a standard Churchill hull with a large, box-like superstructure and the gun mounted low down at the front, to the left of the driver. The frontal plate, which is a direct extension of the vertical driver's visor plate was 89mm thick, with 76mm at the sides and 15mm on the roof. Overall weight is given as 39 tons; length 26 ft 1 in. (including gun) and height 9 ft 1 in. This was nearly a foot taller than the Churchill I but it was exactly the same width.

The Carrier had a crew of four. The commander was stationed on the right side, about midway along the superstructure and he was the only one with a roof hatch, or cupola. In front of him was the gun layer with the loader on his left and the driver, in his normal location, in front of the gun layer. Indeed it is difficult to imagine how the gunner was able to function without virtually standing in the driver's lap. He would have to lean forwards to use the sighting telescope, with his left hand resting on the elevating wheel and his right groping for the traversing wheel. The gun had maximum elevation of 15 degrees, depression of 10 degrees and fractional traverse just 5 degrees either side of the centre line. Thus it was essential that the driver be in his place with the engine running whenever the Carrier was in action because lateral movement of the gun beyond 5 degrees was his responsibility, achieved by use of the neutral turn facility. Stowage was provided for 49 rounds of armour piercing ammunition and 16 rounds of high explosive. There was no permanently mounted secondary armament but a Bren gun was stowed inside the crew compartment along with two Thompson sub-machine guns.

During an engagement the commander first selected his target by directing the driver, lining the tank up, using his external sighting vane. To assist this part of the process the driver had his own traverse indicator alongside his visor. Final target acquisition being taken care of by the gun layer. Firing on the move was impossible since the gun itself was too heavy. When the Carrier was travelling the gun had to be locked in full elevation to prevent damage. Once the Carrier halted to fire the Instruction Book stressed that the rear bulkhead door in the superstructure should be opened, and the engine held at about 1,500 rpm. This was to provide adequate ventilation for the crew. When the machine was moving, on the other hand, the rear door had to be shut. Otherwise the airflow through the radiators would be disrupted.

If, as the contract cards appear to show, the order for these Carriers was completed at the beginning of November 1942, they appear to have been emerging at the rate of about one a week, starting in May. One explanation of this might well be production problems but it is also reasonable to assume that they were, by then, a very low priority. Geoffrey Burton's concern appears to have been justified; the purpose for which they were designed had disappeared. The threat of an invasion of Britain had passed and, indeed, a substantial attack on occupied France was in the offing. The question of what to do with them must have been exercising some minds but documentary evidence is almost entirely lacking. Only one vague clue can be found, and it takes the form of an

Seen in profile it becomes clear that the fixed superstructure of the Gun Carrier brought the centre of gravity too far forward. It must have placed enormous strain on the suspension and made the machine difficult to drive, especially across country.

A Churchill 3 inch Gun Carrier in full service trim and classic British brown and black shaded camouflage. The beautifully presented WD number is odd. T31996 should have been the first of a batch of Churchill gun tanks built by Newton Chambers. In any case, as a self-propelled gun, the number should have been prefixed S, rather than T for tank.

entry in the War Diary of 14th Canadian Army Tank Regiment, otherwise the Calgary Regiment of 1st Canadian Army Tank Brigade. The entry records a call for volunteers from the regiment to form a Tank Brigade Heavy Support Company and it goes on to say that a company "of Churchill tanks mounting heavy guns is being formed within the Brigade. The role of the Company is to provide mobile self-propelled anti-tank artillery to be employed in a manner like that in which the enemy used his anti-aircraft guns in Libya". There can be little doubt that the Churchills referred to here are the 3 inch Gun Carriers.

Twenty four men were selected from the volunteers and duly posted. The term Company, where tanks are concerned, is relatively rare. Most battalions or regiments were divided into Squadrons, and the Squadrons into Troops. But in British Army Tank Brigades up to 1942 Battalions were divided into three Companies, and each Company into four Sections (each of three tanks) and a Company Headquarters. A brief contemporary report claims that 'certain Army Tank Brigades' would be equipped with 3 inch Gun Carriers on the scale of three troops of three tanks each per battalion. The report also explains that the 3 inch gun Churchill had only been produced pending the development of something better, mounting a 17-pounder gun.

In saying that they would be used as the Germans used anti-aircraft guns in Libya the Canadian war diary is not very helpful. The German tactical technique, as practised by General Rommel, involved mixing tanks and anti-tank guns in a kind of leapfrog battle, either in attack or defence. The tanks would move up, halt and fire; anti-tank guns would pass through and repeat the performance, then the tanks and so on. But the fearsome 88mm anti-aircraft gun was too big to handle in this kind of situation and would normally remain dug in as a rear screen. In the attack its range enabled it to provide support over a considerable distance and in defence it was formidable. But the idea of using the Churchill Carriers in this way seems improbable. More likely they would function as mobile anti-tank support in a brigade attack. Moving up with the rest of the armour and then using their firepower to cover the conquered position and beat off armoured counter attacks.

However all of this is little more than hypothesis and there is no evidence to show that these Heavy Support Companies were ever formed. Those volunteers from the Calgary Regiment may have gone off to train with the new weapons but they might well have been sent straight back. Their regiment was destined, within a few weeks, to take their Churchill tanks into action and it would be a historic debut.

As a footnote to this machine there is a note in the RAC Six Monthly Report of December 1943 which claims that a plan to convert them to carry the 3.7 inch gun had been dropped, and that they would now be converted into armoured recovery vehicles. However one, at least, retained its superstructure and was used as a test vehicle to carry and launch demolition equipment.

6

With the Canadians

In the House of Commons, on 4 June 1940, Winston Churchill gave one of his most stirring speeches; 'we will fight on the beaches and landing grounds, in the fields and streets and we shall never surrender'. It was entirely defensive in spirit yet it did not reflect the attitude of the speaker in every sense. On the same day Churchill wrote to Lord Ismay, his military chief on the War Cabinet 'But if it is so easy for the Germans to invade us ... why should it be thought impossible for us to do anything of the same kind to them?'

At this stage Churchill was considering small scale raids, intended to keep the enemy off balance, and it was this thinking that led to the appearance of the Commandos. In his personal history of the Second World War Churchill explains that he was always fascinated by amphibious warfare and had proposed a number of such schemes, involving tanks, in 1917. In another letter to Ismay dated, appropriately, 6 June 1940, the Prime Minister discusses large scale raids on French Channel ports adding (and the italics are his own) *Tanks and A.F.V.s must be made in flat-bottomed boats out of which they can crawl ashore.*

In fact this was nothing new. A system had been devised for just this purpose in 1917 and between the wars some experimental work had been done by the Royal Navy, but it was never properly funded. Now, with Prime Ministerial backing it assumed a new importance. Various training establishments were created under a Director of Combined Operations, among which Inverary on Loch Fyne in western Scotland, is significant. It was established in July 1941 as the first Combined Training Centre and included a naval base, aptly christened HMS *Quebec,* and a military establishment to work with it. Here, in time, the sailors in their landing craft and the soldiers with their tanks perfected beach landing techniques and we have already noted (Chapter 4) that 10th Royal Tank Regiment with their Churchills spent some time at Inverary in December 1941.

For a flavour of these operations it is worth taking a closer look at Exercise Schuyt I, which took place during the first week of May 1942. The exercise involved a landing by 36 Independent Infantry Brigade, supported by the Churchills of 12th Royal Tank Regiment (21st Army Tank Brigade) and it was based on the assumption that German troops occupied Scotland. The attacking force represented Westland, a mythical country in the Atlantic Ocean. The landing was to take place in Irvine Bay, north of the holiday resort of Ayr on the Firth of Clyde. It involved some twenty landing craft of various types which, presumably, would have sailed down from Inverary. To make things simpler the tanks, along with the infantry and transport, loaded at Ayr and Irvine, put out to sea and then came back to beach in Irvine Bay.

The interesting aspect, for students of amphibious warfare, is that the tanks were seen very much as a support element of the infantry and not the spearhead of the assault. Thus one finds that the 49 Churchills, along with three light anti-aircraft tanks and an assortment of Universal Carriers and Scout Cars were spread throughout the fleet. None of the Tank Landing Craft (LCT) carried more than three tanks and some of the loadings are decidedly odd. For example the leading LCT carried three Churchill Tanks, two scout cars and two 25 pounder field guns along with their Quad artillery tractors and ammunition limbers. The instructions issued to 12th RTR had the tanks forming up by squadrons upon landing while the anti-aircraft tanks raced ahead to protect specified bridges inland from enemy air attack. Once they had formed up the tanks would proceed, by squadron, over these bridges to an agreed rallying point where they were supposed to sit and wait, at five minutes notice, for orders from the infantry.

Quite how it all worked out is not recorded, but the planning seems to have been done by the staff of an infantry brigade with little idea of how tanks should be employed, especially infantry tanks. Schuyt I was followed, in July, by Exercise Dryshod which, as its name implies, was a theoretical landing conducted without real landing craft. Since this involved the entire 21st Army Tank Brigade it seems likely that it was a more realistic affair, at least in terms of how tanks should be used in such operations.

A smart line up Churchill tanks and their crews. Quite why all the turrets have been turned to 8 o'clock is not clear.

A Churchill Mark II of 1st Canadian Army Tank Brigade makes a run at the cameraman. The vast number of excellent quality photographs showing Canadians with their tanks implies a commendable effort to keep the folks back home well informed.

Meanwhile small scale raids were being staged up and down the Channel coast. Many were quite successful but they were little more than nuisance attacks as far as the Germans were concerned. 'Butcher and Bolt' was the uncompromising term Churchill had coined for these stunts but he had long kept something more impressive in mind. The target selected was the French port of Dieppe. Much has been written about the Dieppe raid which space prevents us from repeating here, but there are certain factors that should not be overlooked. The town and port of Dieppe occupies a narrow gap in the tall cliffs that fringe this part of the French coast and British military intelligence believed that its garrison was small and that it was poorly defended. Further they calculated that a strong raid, incorporating tanks, could be mounted and carried through before the Germans could mobilise sufficient reinforcements to prevent it, and that the troops and tanks would be on their landing craft and gone before any serious counter attack could be mounted.

The use of tanks was a considerable novelty and even now one is inclined to question the wisdom of committing Churchills to the operation. Even if the Germans were aware of the new tank they would not know any of its technical details so it could still be regarded as a Top Secret weapon. One must also consider that it was still proving unreliable in service and there was no reason to suppose that it would behave any better on a combat operation, especially where that involved the added rigours of an amphibious landing. On the other hand it was undoubtedly the toughest tank in service with the British Army and that would be an invaluable advantage in the kind of operation envisaged here.

Another posed shot from Exercise Spartan, but interesting for all that. *Bunty* appears to be from B Squadron, 12th Canadian Army Tank Regiment (Three Rivers). The picture symbolises the role of the infantry tank. The infantry, in this case, being represented by the Canadian Scottish.

The main matter for discussion, where tanks were concerned, was whether they should make a direct assault on the town or land on the flanks and, by advancing inland, pinch the town out until other troops had wrecked the port facility, stolen as many German invasion barges as they could and destroyed the adjacent airfield. Despite the fact that the latter proposal was inhibited by the local geography it was promoted at first. There was nowhere east of the town that favoured the landing of tanks but the small resort at Quiberville, eight

Inverary, on Loch Fyne, with Churchill tanks coming ashore from landing craft. The vessels are all Landing Craft, Tank 1, or LCT (1) of 372 tons displacement and each capable of carrying just three Churchills. The conditions look awful.

This photograph might not actually show the landings around Ayr but it may be taken as typical. The landing craft have grounded a long way out and the tanks have quite a run just to get ashore. The nearest tank, a Mark III, has extended exhaust pipes but those further out have the full deep wading gear fitted.

miles to the west, had a suitable beach and this was selected. Assuming the tanks got ashore they would have to cross two rivers and then climb to an airfield, on a plateau that also carried the main highway running south from the port. There was no guarantee that bridges over these rivers would not be blown by the Germans, and even if they were still standing, could anyone be sure that they were strong enough to support 40 ton tanks? Another argument against the pinching-out plan was that it gave the Germans time to demolish the port facilities themselves. And, although this was an object of the raid, it would not suit the major operation for which Dieppe was a rehearsal. On a major landing operation, where the attackers were there to stay, a demolished port was a serious handicap. So it was finally agreed to deliver the tanks right onto the beach in front of the town.

When it was originally conceived, as Operation Rutter in April 1942, the Dieppe raid was visualised as a British affair, with the tank element provided by 48th Royal Tank Regiment. In February 1942 the regiment had sent each of its three squadrons up to Inverary for combined operations training and in May B Squadron, with one troop from A Squadron, went across to the Isle of Wight to practise landing techniques with the Royal Navy. The location chosen for the trial was Thorness Bay, on the island's north coast. The craft they employed were Landing Craft, Tank Mark III or LCT (3); 640 ton vessels powered by two Davey Paxman diesel engines. Each LCT was capable of carrying five 40 ton tanks.

Another problem which had to be considered was the movement of tanks from the deck of the landing craft to the beach. Most tanks could pass through still water between eighteen inches and two feet deep without suffering any harm but when it got any deeper special arrangements had to be considered. At sea the problem was aggravated by tide and waves. Early work on waterproofing was carried out by the Department of Tank Design, Special Devices Branch at Westward Ho! in North Devon. When the Royal Electrical and Mechanical Engineers (REME) was formed in May 1942 it took over responsibility for applying the techniques.

Tanks and infantry ashore after landing during the same or a similar exercise. These tanks retain full deep wading gear which would normally be dumped at this stage.

A Churchill Mark II of A Squadron, 48th RTR, about to drive ashore from a landing craft. The absence of an air inlet, at least on this side, suggests that there was a risk of fouling the narrow entrance.

Above: Photographed from the shore at Thorness Bay on the Isle of Wight the same tank is seen climbing onto dry land. TLC 316 was an LCT (3). Further off shore TLC 151 was an LCT (2).

Right: The true meaning of Deep Wading. A Churchill at pretty well maximum depth but still going strong. The pennants on the wireless aerials may well have been intended as markers in the event of an accident, which is no doubt why the safety boat is staying close by.

The waterproofing of all military vehicles was devised to cover two requirements; shallow wading, which protected the vehicles in water up to three feet deep and deep wading, which allowed it to operate in six feet of water. Landing from the ocean, where waves would be encountered and the level of a beach could vary, demanded the latter, so that was the method adopted for the Churchills at Dieppe. Preparation of the tank demanded a tremendous amount of work. It involved filling virtually every seam between hatches, doors and even panels of armour with special plastic compounds and then further covering vulnerable apertures, such as gun mountings and ventilators, with waterproof sheeting made from balloon fabric. Electrical components like the starter motors, dynamos, distributors and even the instrument panel also had to be protected and, most important of all, something had to be done to permit the tank to breathe without being swamped. This was achieved on the Churchill by fitting three hollow, rectangular ducts over the two side air inlet and single rear outlet louvres. These ducts were about turret top level when fitted and open at the top, and in addition the exhaust pipes were extended and raised to about the same height.

Of course all this treatment severely limited the tank's ability to perform. Once it had all been applied it was unwise to drive the tank very far since waterproofing integrity rapidly degraded. And once the tank landed it was vital to get rid of it as quickly as possible. As soon as the tank was clear of the water, levers inside the crew compartment released the ducts, which fell aside, while small charges of Cordtex explosive were detonated to clear away sealant from weapon apertures, ventilators and the turret ring. Only then was the tank considered fit to fight.

The decision to make the Dieppe Raid largely a Canadian operation was taken on 30 April 1942 and the Calgary Regiment slipped into the role originally allotted to 48th RTR. They even took over some of that regiment's tanks; special versions of the Churchill that had been adapted as flamethrowers and carpet layers but these, coming under the heading of specialised armour, fall outside the scope of this study. Amphibious training was continued at Thorness Bay and at Seaford in Sussex where a section of sea wall similar to that at Dieppe could be found. Another site was the tiny fishing harbour of West Bay, on the coast near Bridport in Dorset. This last location was selected because, in a miniature way, it was a replica of Dieppe. A small harbour behind a typical shingle beach, overlooked by high cliffs on each side. Even so the choice of a shingle beach indicates that those planning the raid were by no means thorough enough, and this may be due to the excessive degree of secrecy applied.

Tests had already shown that Churchill tanks did not perform well on shingle. It got into the tracks and caused them to break, while on steeper slopes, even quite short ones, the tanks would dig in until they could not move either way. A member of the Royal Canadian Engineers, attached to the Calgary Regiment, designed the mat-laying apparatus to overcome this. However the beach at Dieppe was chert, not shingle. Chert is a form of rock found in chalk which appears as large boulders rather than small stones like shingle. If anything it is even worse than shingle, being just as liable to break tracks but even more difficult to drive over.

One curious change adopted for the actual operation was the decision to use Landing Craft, Tank, Mark II or LCT (2) instead of LCT (3) for the actual landing. The LCT (2) was a shorter version of LCT (3) displacing 453 tons (laden) and capable of carrying just three Churchill tanks and one other small vehicle as required. Ten of these craft were allocated to land thirty Churchill tanks. Eighteen of them were Churchill Mark III, five Churchill II and seven Churchill I, three of which were flamethrowers. Three Mark IIIs and two Mark Is carried carpet laying equipment and seven tanks were organised to tow Daimler scout cars ashore.

Operation Rutter, the original name for the Dieppe raid was scheduled for June, postponed until July and then cancelled. This caused dismay at Combined Operations headquarters and disappointment among the troops and there was continual pressure from British public opinion which was demanding a 'Second Front Now!' to support the Soviet Union. Taking all these factors into consideration the scheme was revived as Operation Jubilee which, weather and tides permitting, would take place on 19 August 1942.

Shortly before the Dieppe Raid 14th Canadian Army Tank Regiment (the Calgarys) parade during exercise Yukon II. Notice the regiment's Daimler Dingo scout cars lined up away over on the left.

Although the full operation involved landings at various locations on a ten mile front the main force of tanks and infantry would hit the town, square on, in successive waves. Once across the beach and over the esplanade wall the attackers would cross the broad promenade and gardens, enter the town and make for the heights beyond where they would support a flank attack on the airfield. Once a perimeter had been formed around the town various demolition and cutting out operations would begin, followed by a withdrawal to the beach and embarkation for England. In the event it did not work out anything like that at all.

The entire regiment, that is 58 tanks, mobilised for the operation but 28 of them, that is all of A Squadron and elements of C Squadron remained aboard their landing craft as a floating, offshore reserve. Twenty nine of the other thirty tanks rolled down the ramps of their landing craft. Two drowned in deep water and of the 27 that reached the beach 15 managed to climb the sea wall onto the promenade. And this was no mean feat. The carpet laying devices had been designed to assist this stage of the advance but the plan included a requirement that engineers should follow the first tanks ashore, carrying large baulks of timber which they would then use to build ramps for other tanks to climb over. Needless to say this idea failed completely. Engineers were also detailed to bring explosive charges ashore and destroy the huge concrete obstacles that had been constructed by the Germans to prevent tanks moving further into the town and, since none of these men were able to complete their tasks none of the tanks got further than the promenade. After fighting hard for most of the morning, ten tanks made their way back to the beach to cover the infantry in their withdrawal but enemy fire was so heavy that relatively few men actually got away.

If the scheme had gone to plan the tanks would now have re-embarked, along with the troops, leaving the facilities devastated and the Germans, presumably, bewildered. But with hindsight it is obvious that the plan was over optimistic and British intelligence seriously faulty. Although they were not forewarned of the raid, except by a brief naval action a few minutes before the landing, German troops on the Channel coast were generally alert to the threat of invasion and well prepared to meet it. The first wave of troops met a withering fire which they were never able to suppress and as more men and tanks arrived, casualties mounted.

On the other hand the tanks stood up well to enemy fire and the main problems were caused by tracks breaking and turrets becoming jammed. In all fourteen of the Calgarys were killed, but only two of those that got ashore made it back to England and, needless to say the only tanks that returned were those in the floating reserve and one from regimental headquarters which never managed to get ashore.

The Allied lessons of Dieppe are beyond the scope of this work but it gave the Germans a chance to evaluate the Churchill at their leisure. As the prisoners were rounded up, wounded and dead removed and the fires died down, the tide crept in to swamp some of the tanks. Those that were

An unusual photograph of an early Churchill on a shingle beach. The circumstances are not recorded but the tank is in a sorry state. Shingle and Churchill tanks never did mix well.

still running were used to drag others from the beach and they were examined briefly in the Dieppe area.

A copy of the German report came into British hands in the Middle East and it was sent to London for translation. As a Colonel from the Military Intelligence branch M.I.10 remarked "the comparisons made here are not flattering". The Germans compared the Churchill not only to their own tanks but to the Russian machines they were now encountering in the east. And it did not compare favourably. 'The vehicle' says the report 'offers nothing worthy of consideration by technical personnel, nor has it any new constructive features either in the metallurgical field, or in the field of weapon technology'. The 3 inch howitzer was 'bad and old fashioned'; the 2-pounder 'left behind both in construction and effectiveness' while the 6-pounder's performance 'does not approach that of Russian guns of the same calibre'. Ammunition revealed no new or noteworthy features and, as for armour it was seen as very thick but of poor quality and did not compare well with German or Russian plate. They also claimed that the tracks were brittle and of clumsy design which fractured ev-

Abandoned tanks and a burning landing craft. Typical scenes in the aftermath of the Dieppe Raid. Broken tracks became something of an epidemic on this beach.

ery time it received a direct hit. Photographs seem to indicate that the Churchills abandoned at Dieppe had the heavy, spudded cast track made from a material described as B.T.S. 3.

Discussing performance the Germans noted that the tracks made so much noise that they believed it would be impossible for anyone to use the radio while the tank was moving. No similar complaint is heard from British sources but the Germans made the point that when the tank halted to use its radio it provided a good opportunity to knock it out. Not that this was regarded as a problem. As the report sums up ' The shape is also not modern. In conclusion it may be said that the English Churchill tank, in its present form, is easy to combat'.

The survey then went on to try and discover how the various tanks were knocked out during the action. Only two out of the 29 tanks showed signs of having their armour penetrated by anti-tank rounds, which were estimated at between 3.7 and 5cm. Many tanks had been struck, but the limited number of penetrations might have given the Germans pause for thought. The vulnerability of the tracks has already been

Right: German soldiers examine *Betty*, abandoned in the collapsed bunker on the esplanade at Dieppe. Clearly it was the nearside track which broke. This angle gives a good view of the extended exhaust pipes.

Below: *Blondie* captured at Dieppe, restored to running order and seen here with a German crew. Certainly at this stage Britain's latest heavy tank did not strike any fear into German hearts.

noted but this damage was not restricted to enemy fire. The inspection teams reported that tracks had been broken by stones which became jammed in the links and carried around the sprockets. They also noted that many of the tanks had become bogged down in the beach and decided to carry out a comparative test with one of their own tanks, although the type is not specified. This showed that on beaches with a slope of between 15 and 20 degrees the German tank could manage quite well but where the slope increased to between 30 and 40 degrees the tank started to slip and then dug itself in until the tracks ceased to function. However they also noted that after repeated attempts to tackle the same slope their tank gradually reduced the gradient until it could get up, and that caused them to recommend that one should not rely on steep shingle beaches to keep tanks at bay.

Another interesting comment in this area concerns a tank which apparently fell into a hole on the promenade. The report refers to a tank 'falling into a large ditch, covered with a camouflage net at the rear entrance to the pill box' which sounds like a purpose made trap. The closest one can find to this in Canadian accounts concerns the Mark III tank *Betty* which, after roaming the promenade for some time plunged into what appears to have been a German dug-out. Attempts were then made, using other Churchills, to tow *Betty* out but no sooner was this done than it broke a track and was driven back into the hole to give the crew more cover while they evacuated.

The ultimate fate of the captured Churchills is not entirely clear. Some must have been damaged beyond repair following immersion in the sea and the number of track breakages probably meant that it was impossible to refit all of them. Some were definitely sacrificed as hard targets and the German report gives basic figures from the results of firing a whole range of weapons at the tanks, in the Dieppe area, shortly after the raid. Smallest of all, and least effective, was the armour piercing rifle grenade fired from a 98K carbine. The largest a 15cm medium howitzer which was described as having a destructive effect on both hull and turret. Between these extremes anything below 5cm calibre was relatively ineffective but the three most outstanding, all firing at 1,000 metres, were:

> The75mm PAK 40 tank gun firing APCBC which could penetrate the turret and hull sides.
>
> The 88mm FLAK 18 anti-aircraft gun firing APCBC which penetrated everywhere.
>
> The 100mm medium gun K18 firing APCBC which effected penetration of the turret and hull at front and sides.
>
> Unfortunately this report does not say whether the target tank(s) had the cast or welded turret. Further trials, carried out in Germany, added the Russian 76.2mm gun to the list of those that could penetrate the Churchill at 1,000 metres.

The normal German practice with captured enemy tanks was to give them an official designation and incorporate them into the Panzer force as *Beute* (or Booty) Panzers but there is no firm evidence for this in surviving records. In all probability there were simply too few Churchills, and they of doubtful reliability, to justify it. German engineers, according to some sources, modified one tank by using extra bogie wheels to create return rollers, instead of having the tracks run back along skid rails. This, by all accounts, resulted in a useful improvement in performance, but it was one the British did not learn about until it was too late. There is no surviving evidence to suggest that it was ever considered by engineers in Britain independently of the German work.

7

The Great Rework Scheme

The idea of organising a major rework scheme to remedy some of the more glaring faults in Churchill tanks had been discussed almost as soon as production began. Nothing was agreed for some time because the various authorities could not decide whether it would be wiser to cancel production altogether and create a new infantry tank, or try to salvage something from the existing design. The final decision seems to have been taken at a meeting, held in the office of the Controller-General of Research & Development (CGRD) at the Ministry of Supply. The date of this meeting was 3 November 1941, at 3 o'clock in the afternoon, if that makes any difference.

Vauxhall Motors had prepared a statement for the meeting which listed three major changes that they wished to introduce and posed five questions to which they wanted answers. The three changes were:

To fit mudguards over the top run of the tracks.
To alter the shape and flow of the air inlets.
To alter the air outlet at the rear of the hull.

The original air inlets had certainly created problems and the lack of mudguards was a nuisance but no complaints about the air outlet can be traced in surviving documents, beyond the fact that it generated a dust cloud. Compared with the continual problems being posed by engines, gearboxes, suspension units and tracks they hardly appear to be significant; yet none of these other matters were discussed at this meeting. The mudguards, which would be made from thin steel, were probably time consuming to manufacture but one imagines that they could have been made anywhere, delivered to units and fitted by them without too much difficulty. The new air inlets, which drew air in from the top, would also have to be manufactured but it is not clear if they could be bolted to the hull directly in place of the older type, or whether the hull might need to be altered in any way. Certainly the hull would have to be modified if the air outlet was to be altered but there seems to be no good reason why the first two could not be introduced easily.

The nature of the proposed change to the air outlet is not clear but in any case it was agreed to drop this from the rework programme and only introduce it with the next model of Churchill which would be produced, according to this report, after the 1001st Churchill had left the factory. So now the rework programme seems to have been reduced to the point where it only concerned two processes, neither of which appears to involve any major construction problems And it certainly did not include the kind of remedial work one might expect from the scale of complaints. Yet the meeting agreed that two firms, Vauxhall Motors themselves, and Broom & Wade, would turn their production lines over to the rework programme from 1 March and 1 April 1942 respectively. This raised another question. If two firms were taken out of production would the supply of new tanks suffer? In other words would it be more sensible to set all production lines to work on new tanks, incorporating these modifications, and leave those already built as they were? The general view was that production would not suffer. Those manufacturing bottlenecks already identified did not occur on the production line itself and the loss of Vauxhall and Broom & Wade would only reduce output by 30 tanks a month. Yet with both firms engaged on rebuilding they could guarantee to turn out 60 reworked tanks a month as good as new. Thus the gain in production of battleworthy tanks would be 30 a month over existing deliveries. On the other hand it was agreed that there was little point in trying to rework the first 300 Churchills which, presumably, were regarded as beyond redemption. Units that had these tanks would be first in line for replacement since they were the most experienced Churchill equipped regiments.

The five questions posed by the team from Vauxhall Motors were mainly formed to confirm the above decisions, but the first one is interesting. Vauxhall wanted to know if, once the rework programme began, production of new tanks should be slowed down in order to incorporate these changes.

The prototype Churchill T30971, modified but not reworked. The crude front mudguards and raised panels along the sides of the tracks appear to be an early attempt to defeat the mud menace.

The meeting said no; indeed they added, in answer to another question, that they planned to drive the other firms harder still in order to increase production.

When this programme was explained to the Tank Board at their next meeting on 25 November General Richardson became rather optimistic. He stated that the War Office would require 261 of the reworked tanks, fit to proceed overseas, by 1 January 1942, and 1,304 by 1 April. He was quickly told that newly built Churchills, finished to the latest standard, would not start coming off the production line before March and that the 700 scheduled for rework, tanks 301 to 1,000, would not appear for at least twelve months after that. The chairman, Geoffrey Burton, added helpfully that when it was designed the Churchill had not been regarded as the ultimate infantry tank and that it had only been designed for short approach marches. This peculiar statement, something of a *non sequitur* as it stands, appears to have been intended as a means of damping enthusiasm, but what did it mean?

The approach march is, of course, the distance a tank has to travel by road from its immediate base to the start line for a battle. Obviously there is no point in subjecting any tank to a longer approach march than necessary because it uses up a lot of fuel and wears out the tank and its crew. Furthermore if the tank in question is slow moving, like an infantry tank, then there is a time factor to consider. Long columns of slow moving tanks, crawling up to a battle, not only stretch preparation time but block traffic routes and increase the chances of detection from the air, warning the enemy of the impending attack. In the First World War, when tanks were incredibly slow and vulnerable to damage, long approach marches were hardly ever considered. At the Battle of Cambrai, in November 1917, miles of extra railway track were laid, as close to the start point as they could get, in order to deliver the tanks. But this could only be done for large, set-piece battles. In World War II flexibility was a key factor for success and the task of delivering tanks to the front line devolved upon tank transporters when they were available. Burton's strange statement more or less relegates the Churchill to First World War status, which is clearly ridiculous.

A reworked Mark II with new style air intakes and full length mudguards trying its luck in some particularly glutinous mud. The man at the front does not look particularly happy so maybe it is stuck.

The full rework scheme, when it was ultimately established, involved more, much more, than the three items mentioned in the original discussion. This is made clear from a sixteen page booklet which, unfortunately, is undated, but since it does not mention Churchills beyond the Mark IV it can reasonably be associated with a period around late 1942, early 1943. It is sub-titled 'Built to Major Re-work Specification' and goes on to explain:

On Vehicles built to the Manufacturers' "Re-work" Specification the W.D. Number is suffixed by the letter "R".

It would require too much space to cover every item in the booklet but the following should be mentioned briefly. Air flow through the engine compartment was improved, including the new inlets. Full length mudguards were incorporated as were continuous track return skid rails. A baffle plate was added below the rear outlet to prevent dust from being blown forwards. Blast shields were fitted to the ends of the new mudguards and the side escape doors were provided with catches that enabled them to be secured in the open position.

Some of the suspension bogie frames were strengthened and the rebound pads redesigned to give a smoother ride. Rubber tyres were removed from drive sprockets and the front idler axles made stronger. Stowage was improved, especially for ammunition, and the driver was provided with a detachable windscreen that fitted into his visor aperture when the armoured door was open. In the engine compartment improvements were made to the petrol pumps, radiator caps, exhaust system and the actual engine mountings. New carburettors were supplied and the starter motor improved. Gearbox improvements were introduced in two stages and the boxes marked accordingly and plans were announced to improve the steering brakes. In all 71 changes are listed - if all the stowage modifications are taken as one task - but to give some idea of the scale of work involved these include a waterproofing process carried out on all hull seams which are sealed with a mastic compound called Bostik. This suggests that every tank which went through the full rework programme had to be stripped right down to the basic hull assembly. And that would certainly demand that they went right back to the production line.

At a meeting of the Tank Design, Supply and Liaison Committee, which was held on 24 April 1942, the Director of Tank Design put forward a suggestion. He said that, in addition to a pair of selected, reworked Churchills which were being tested at Chobham, others should be tested by units in the field. The matter was approved and the object of the trial, codenamed Trent, was defined as follows:

It is required to carry out intensive mileage trials on 25 Churchill Tanks which have been completed to rework specification in order to satisfy ourselves that this rework makes the tank mechanically reliable and in every way battleworthy.

A fine example of a fully reworked Mark III. The enormous notice on the side, although difficult to read, appears to be an exhortation to factory staff to try harder for the war effort. Presumably the firm that built the tank also manufactured ammunition for they have placed various samples upon the tank.

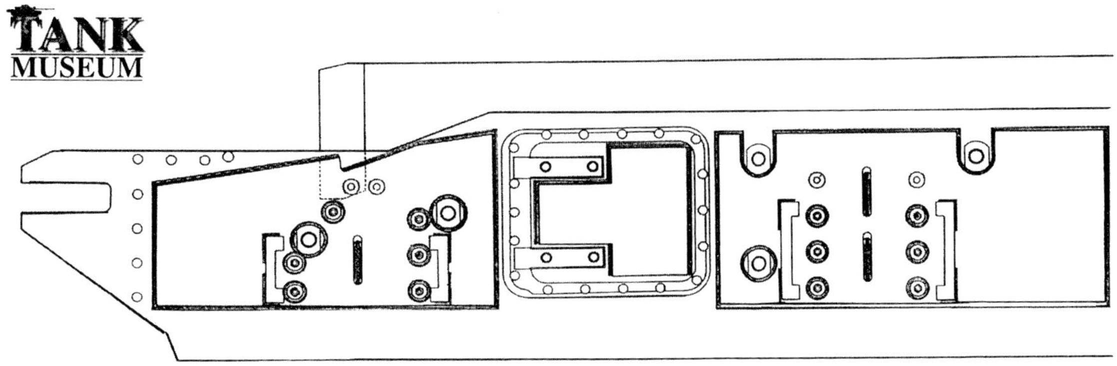

CHURCHILL PANNIER APPLIQUE ARMOUR

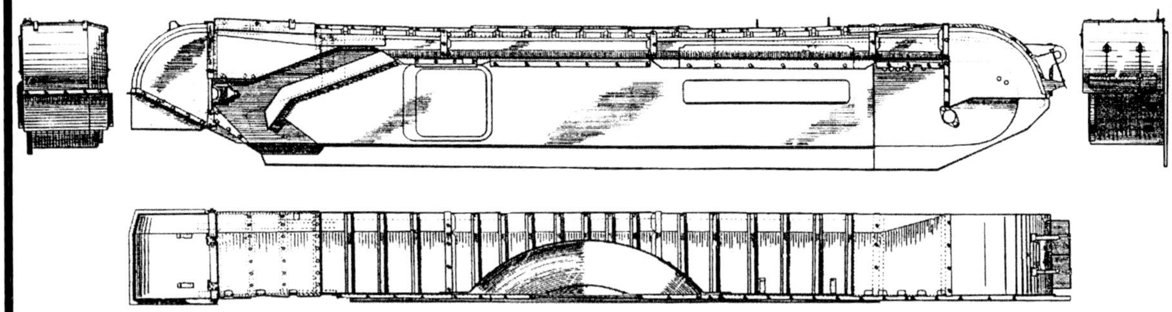

CHURCHILL TRACK MUDGUARDS

This page and top of page 86: Drawings showing revised air intakes, track guards and track skid rails.

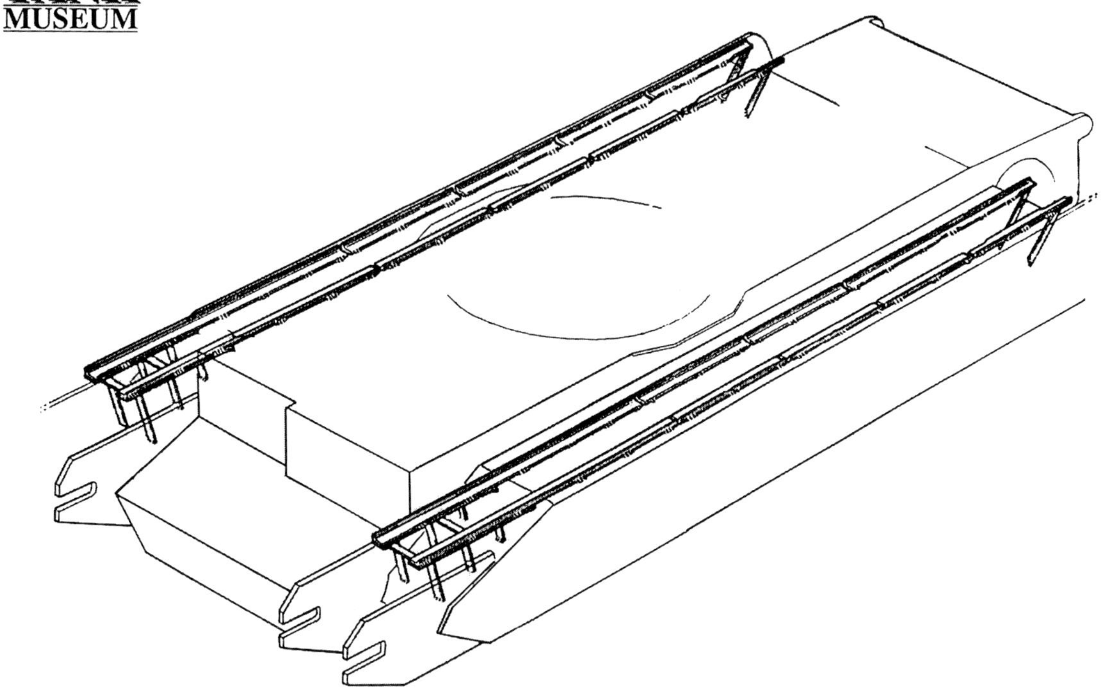

CHURCHILL TRACK SKID RAILS

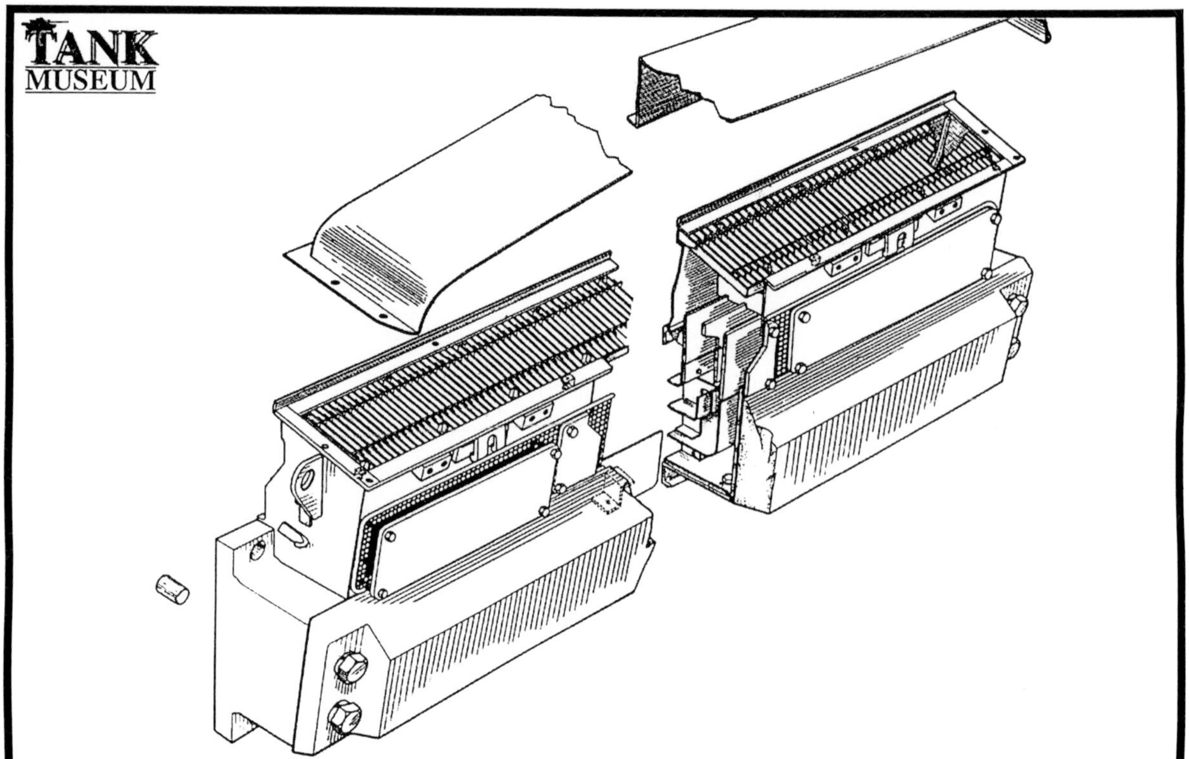

CHURCHILL INLET LOUVRE (FOR UPGRADED MK I & II)

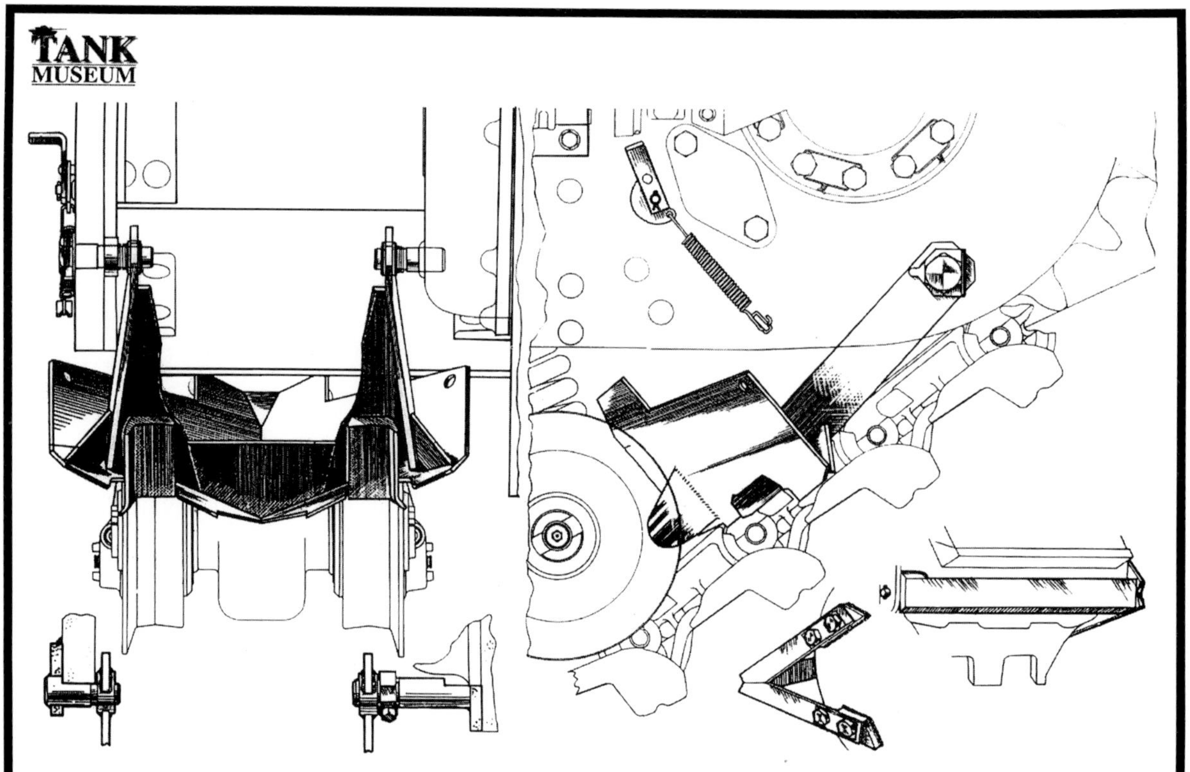

CHURCHILL TRACK MUD PLOUGHS & STRIPPER PLATES

Drawing of the mud plough fitted to Churchill tanks.

Which was a laudable aim, if nothing else. The first units to receive reworked tanks, as it turned out, were the three regiments of 21st Army Tank Brigade; that is 12th RTR, 48th RTR and 145 RAC. They were already in Scotland, having taken part in the landing exercise recounted in the previous chapter, so it was decided to carry out the exercise around the Scottish town of Hawick, in Teviotdale near to the English border. In all 25 tanks were supplied for the trial. eight each from 12th RTR and 145 RAC and nine from 48th RTR. It was supposed to extend over six weeks from 12 July until 23 August 1942 and each regiment appears to have operated independently over different routes which mixed road and cross-country driving. Although it was high summer the British weather lived up to its reputation. The first two weeks were wet and the ground became soaked. Then it dried up for two weeks and the cross-country courses became dusty. The following week it was showery and the final week was wet again. However as if to ensure that the trials could not be conducted in a proper manner, the War Office announced half way through the trial that the tanks would be needed for an exercise.

Thus, after ten days the brigade withdrew from the trial in order to take part in Exercise Dryshod and when it came back, ten days later, the trial resumed. Two weeks after that the War Office stepped in again to announce that the trial must be finished by a certain date, so the organisers decided to stop using the cross-country courses and complete the trial on the roads in order to build up as much mileage as possible. Despite these unfortunate attempts to reduce the whole thing to the level of a farce the Department of Tank Design appears to have taken the trial seriously and their report, which was issued in October 1942, was very thorough. Each regiment filed its own report, every tank was identified by number and its performance well documented. Twelfth RTR appear to have scored best, although it was not a contest. All of their tanks exceeded 1,000 miles during the trial, of which 500 miles (exactly, for every tank) was across country. However they ran over good, wide roads with easy gradients, except around Hawick itself, and their cross-country course, which was on top of a hill, remained fairly dry and firm.

Things were not so good for 48th RTR. The roads they had to use were narrow lanes, which restricted the speed to about 10 mph, while their cross-country course included some heavy going and long, boggy hills. This meant that more than half of the driving was done in second gear and in fact they gave up after about 150 miles had been covered by each tank and stuck to the roads for the remainder of their trial. Only four of their tanks exceeded 1,000 miles. By comparison 145 RAC came off lightly. They had both lanes and wide roads to negotiate and the choice of two cross-country areas, both of which stayed quite firm in the wet. However one of their tanks was involved in an accident and had to be withdrawn while only one exceeded 1,000 miles (it did 1,001) three others did 1,000 exactly and the rest about 800 miles. As an aside to these results it is interesting to see a comparison between the mileage claimed for the trial and the total odometer readings after the trial was over (and one must assume that the mileage on Dryshod was included in the trial). Most tanks appear to have done about 80 or 90 miles before Trent began and one assumes that, as part of the rework programme, the clocks were reset to zero, thus providing inspiration for generations of used car salesmen.

Picking through the report on Trent, various factors stick out. After 250 miles there were eight tanks which had not suffered any stoppages and there were no reports of suspension or gearbox failures. Most of the troubles seem to have been with the engine and tracks. By 500 miles only one tank recorded no stoppages. Of the rest engine and track failures had continued at the same rate while brake failures had caught up. Instances of gearbox or suspension failure were still few. At 750 miles every tank had suffered at least one stoppage and the failure rate remained constant except that there was a notable increase in final drive and gearbox failures. When the 1,000 mile mark was reached eight tanks had dropped out, plus the one involved in the accident, but gearbox failures did not increase at the same pace and suspension failures were never serious. Of course the fact that the second part of the trial was carried out entirely on roads was seen as having something to do with the low failure rate. Since so many problems were associated with the engine it is worth pointing out that the majority, by far, were due to the fuel pump drive which was only an accessory failure.

All but three of the tanks had the new manganese tracks and where these failed it was due almost entirely to a massive loss of retaining plates, which held the track pins in place, so again it could not be regarded as a failure in the track itself. The biggest problem of all, and the one which seems to have surprised the DTD most, concerned the Lockheed hydraulic system. The log for every tank involved records endless halts to drain or bleed some part of the system which, according to the DTD, was due to the poor fit of plugs and unions. Comments on the suspension appear to be far more harsh than the initial evidence seems to merit. However the report points out that the loss of one or more track bogies did not necessarily stop the tank or disrupt the trial so it was not noted as serious. Even so the report refers to suspension troubles as intense and claims that a lot of maintenance and replacement was required to keep the tanks on the road. The statistics, for what they are worth, reveal that axles seizing or breaking, and bearings running hot were the biggest problems, and it is worth noting that the most troublesome bogies were stations 6, 7 and 8 which, if nothing else, suggest where the centre of gravity was. Finally it is worth noting that bogie bracket failures became endemic in 48th RTR, which did most of the road running, and this took the form of broken welds along the top of the bracket, which the DTD put down to poor workmanship.

The report on Trent, well presented as it is, never does attempt to answer the main question; whether or not the rework scheme had succeeded. Beyond cataloguing the various faults it expresses no views at all. The closest one comes to any conclusion is in an appended letter, dated 12 October

1942, by a senior instructor at the Driving & Maintenance Wing of the AFV School at Bovington. Having studied the Trent report he compared it with similar trials held at Bovington. He is obviously not happy with the way Trent was conducted. He criticises the high road speeds, claims that the cross-country courses were too easy and is scathing about the standard of parade maintenance. He agrees that in the main, stoppages on the Hawick trial mirror those at Bovington but suggests that the hydraulic failures were due entirely to poor maintenance standards. Once again he is not prepared to say whether the reworked tanks are an improvement or not, but he does claim that slower running and good preventative maintenance will reduce the instances of failure, which is probably true of every machine.

In an earlier chapter it may be recalled that Vauxhall Motors suggested the use of rubber tyred road rollers but that this was treated as low priority due to the general shortage of rubber. Even so it is clear that something was done because by December 1941 Vauxhall is reporting to the Ministry of Supply that a Churchill with rubber tyred suspension is putting up a remarkable show. They claim that tyres improve the behaviour and life of the tank along with the fighting efficiency of the tank and its crew. Even so they admit that it would take some time to institute the new design. Pressed by the Tank Board in January they confirm that it would involve a major redesign of tracks and suspension, which they expected to take nine months once the design had been released. Even so two members of the Board, Lucas and Colonel Green, go so far as to state that A22 can never be really reliable unless it does get rubber tyres.

One imagines that, faced with this sort of delay, the project would then be quietly dropped but there is evidence to suggest otherwise. In September 1942 the Department of Tank Design issued a report compiled by the Fighting Vehicle Proving Establishment on a project known as Pussyfoot (but spelled Pusseyfoot in the report) which is described as an Infantry Tank Mark IV fitted with rubber tyred road rollers. The tank in question is T32144, a Mark II which comes from Vauxhall Motors. Whether it is the same tank as reported on in the previous December is not clear, but photographs from the September 1942 report show it with revised air inlets and track guards, indicating a reworked machine.

A reworked Mark III at speed, churning up the dust. It is a problem which plagued the Churchill for its entire service life.

Tyres were supplied by two firms, T B Andre Ltd and Firestone and the tanks is also fitted with special short pitch, cast manganese steel tracks by Firth Brown. The final drive and sprockets were also modified to suit the new tracks. For the initial trials the tank's original rollers were fitted with twin rubber tyres, 7 inches wide and about 2 inches thick, but these gave a lot of trouble during road runs. Subsequently new roll-

The reworked Mark II tank fitted with rubber tyred rollers and known by the codename Pussyfoot. It is interesting to observe how the addition of mudguards and larger air intakes suddenly makes the turret look under sized on these tanks.

A cross section drawing of a Mark IV. The horseshoe shaped fitting inside the turret is the gunner's shoulderpiece, indicating a tank with free elevation.

ers of larger diameter were fitted with thinner rubber tyres to match, and these appear to have worked a lot better.

The report is quite comprehensive but we only need concentrate on two factors here. First heat. This appears to have been the biggest problem. At speeds above 10mph the tyres got quite warm and at 16mph became so hot that some of them broke up. Further it was noticed that if the tank slowed down, after a spell at 16mph, the tyres did not cool down too quickly. The other matter was noise. As expected this was reduced, but not dramatically. Tests were carried out by specialists, in this case the Post Office Research Station at Dollis Hill in north London. Using what is described as a noise meter microphone they tested the special Pussyfoot tank against a conventional one with steel rollers on a concrete road. Each tank went by the microphone at a distance of six feet, doing 10mph in top gear. The conventional tank registered 116 decibels yet the rubber tyred machine gave a reading of 98db.

The same appears to have been true of most other factors tested. Climbing ability and top speed were almost identical, while oil consumption showed a modest improvement. It is also interesting to note that as the noise from the suspension diminished so the sound from the engine and ex-

A new Churchill Mark IV equipped with the original, shorter (Mark 3) version of the 6-pounder.

Another Mark IV, photographed at MEE Chertsey. This one is fitted with the longer Mark 5 gun.

haust increased. At speeds over 14mph an unpleasant high frequency noise was emitted which, according to the report 'rendered intelligible conversation difficult' and this was put down to the combined effect of shorter pitch track and the spacing of suspension units which meant that all bogies were passing over joints in the track at the same time. In their final summary the testing authorities praised the reduction in noise but explained that the tank was unsuitable for production in this form because the tyres could not withstand road speeds above 14mph or cross-country running above 10mph. They finished by recommending that the project be closed and the tank disposed of.

In the meantime a new version of the Churchill was being considered. Back in October 1941 firing trials were undertaken against examples of the welded Mark III turret, and the results described as unsatisfactory. In the first place it was reported that supplies of plate suitable for welding were in short supply and a War Office summary announced that up to that date 389 turret had been delivered. Even so firing trials continued into November and in far too many instances the armour plate was described as flaking. This meant that if the external face of the armour was struck by a projectile, even if it did not penetrate, a sizeable chunk might come away on the inside and do a considerable amount of damage. The Bullet-Proof Plate Technical Committee stated that they were unable to guarantee supplies of plate that were entirely flake free, and the only solution they could suggest was to employ double skin construction. This would involve creating an inner shell of mild steel, which should cure the effect of flaking, but it would also require substantial alterations to the design. Given the reported shortage of plate suitable for welding the only alternative was to initiate an entirely new design. A cast 6-pounder turret for A22 is first mentioned in a Tank Board report dated 23 January 1942 which reveals that trials with 6-pounder shot had proved successful, showing that British techniques in producing cast turrets had improved. The report went on to note that it was now arguable that our castings were in any way inferior to American practice. It was therefore agreed that the obvious move was to return to a cast turret, but bearing in mind the greater size of a 6-pounder turret compared with the original pattern, it had to be accepted that in order to produce a turret of the same weight some protection would have to be sacrificed.

Yet, as the relevant Plate Thickness diagrams reveal, the cast turret was actually thicker on its upper surfaces than the welded version and no thinner on other surfaces. Notice also that the Izod Test (I.T.) figure for the cast turret is higher in some places, suggesting a more resilient structure. This was one of the principal problems with castings, that of controlling the thickness of different surfaces. It required a good deal of shop floor skill and production time to grind excess material from large castings so it was probably no quicker to produce these cast turrets than the earlier welded type. It seems to have added something in the region of half a ton to the over-

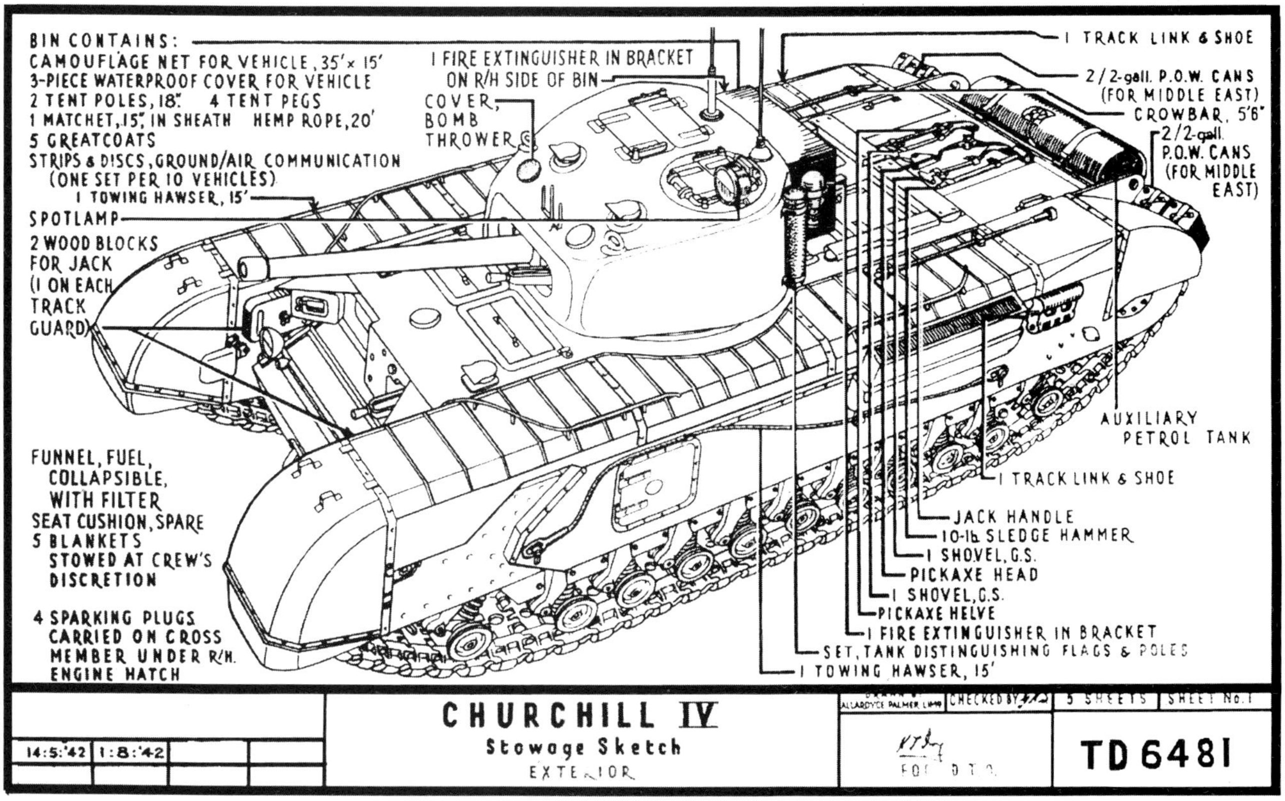

Pages 91-92 and top of page 93: Stowage, plate thickness and other diagrams related to the Churchill Mark IV.

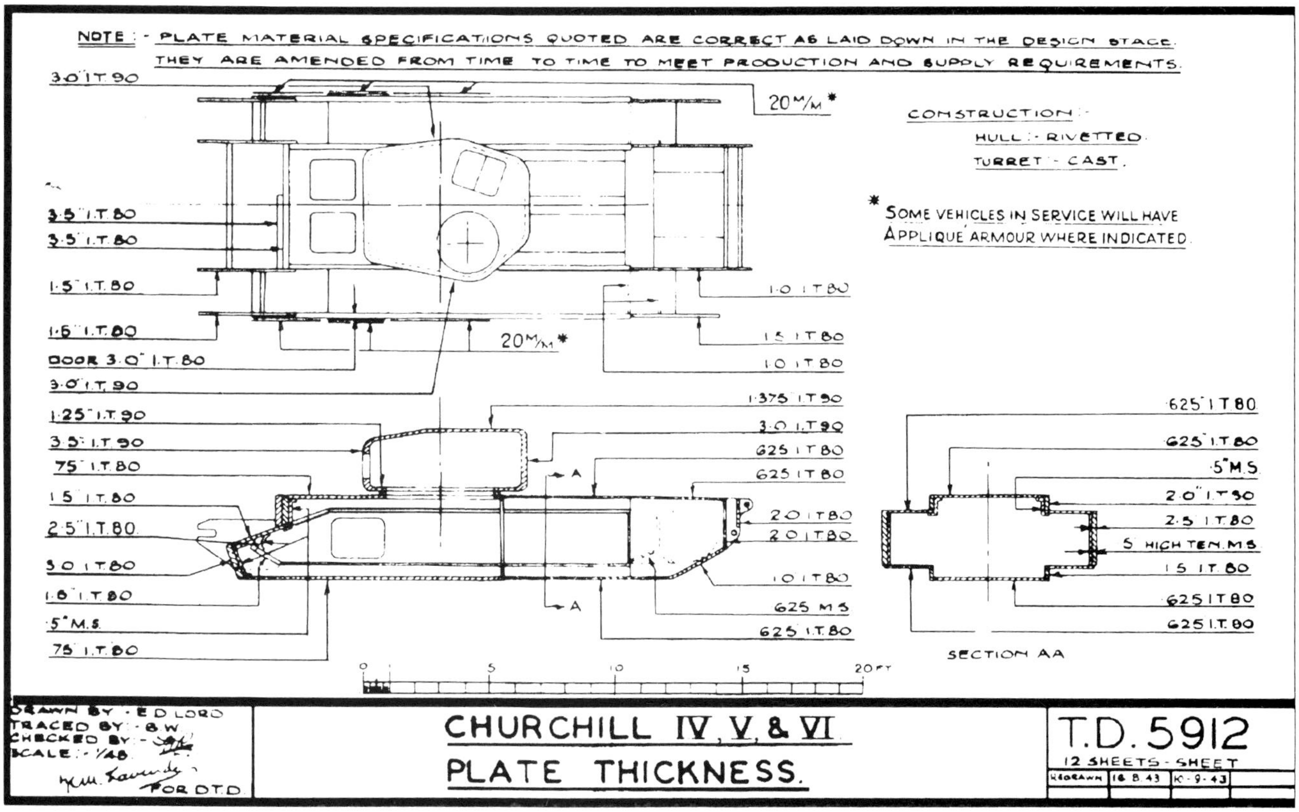

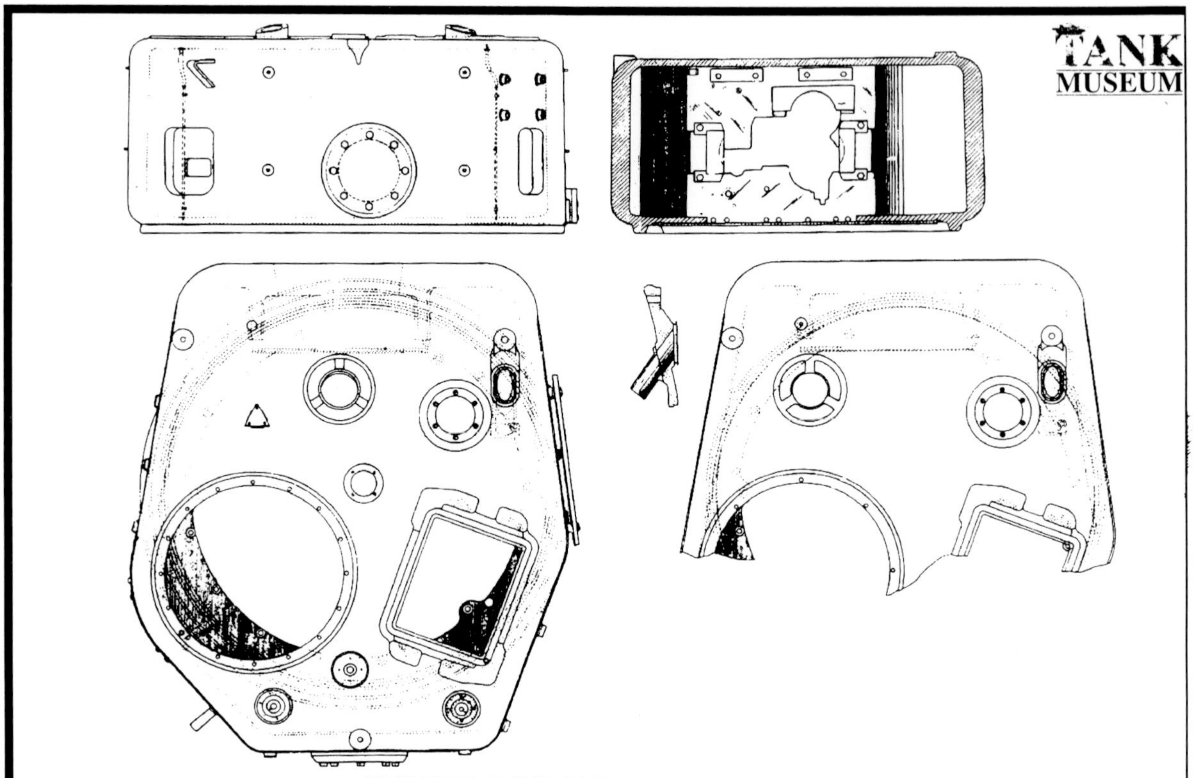

TURRET CHURCHILL IV & IV (75MM)

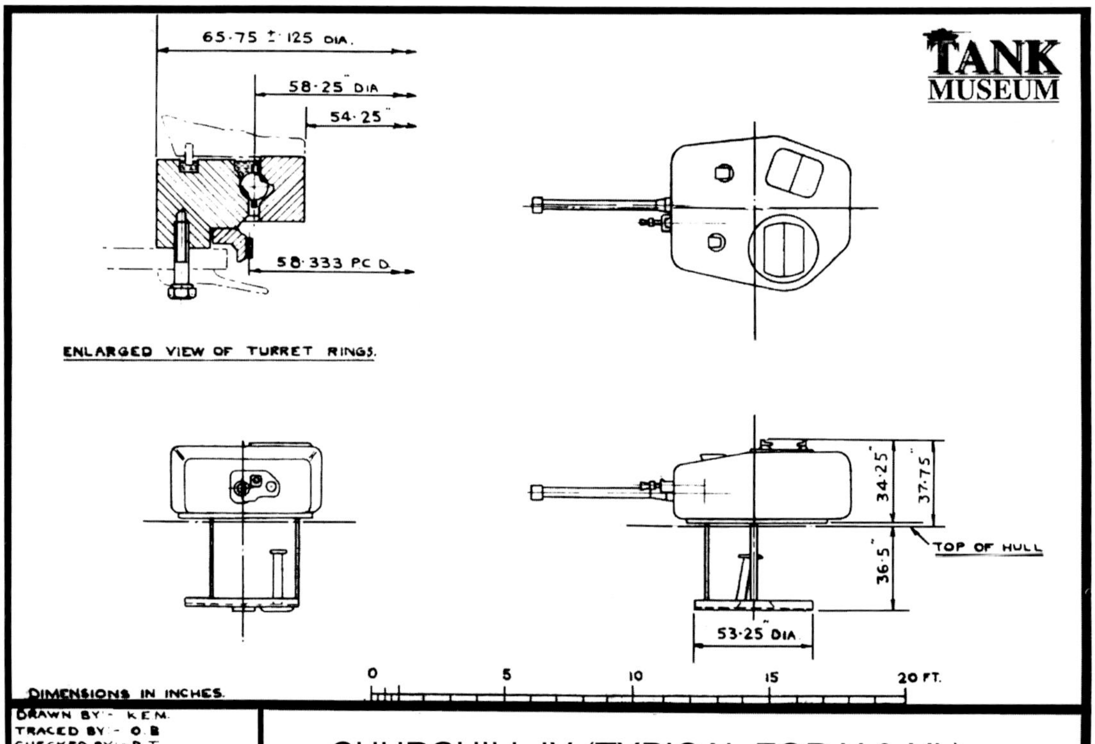

CHURCHILL IV (TYPICAL FOR V & VI)

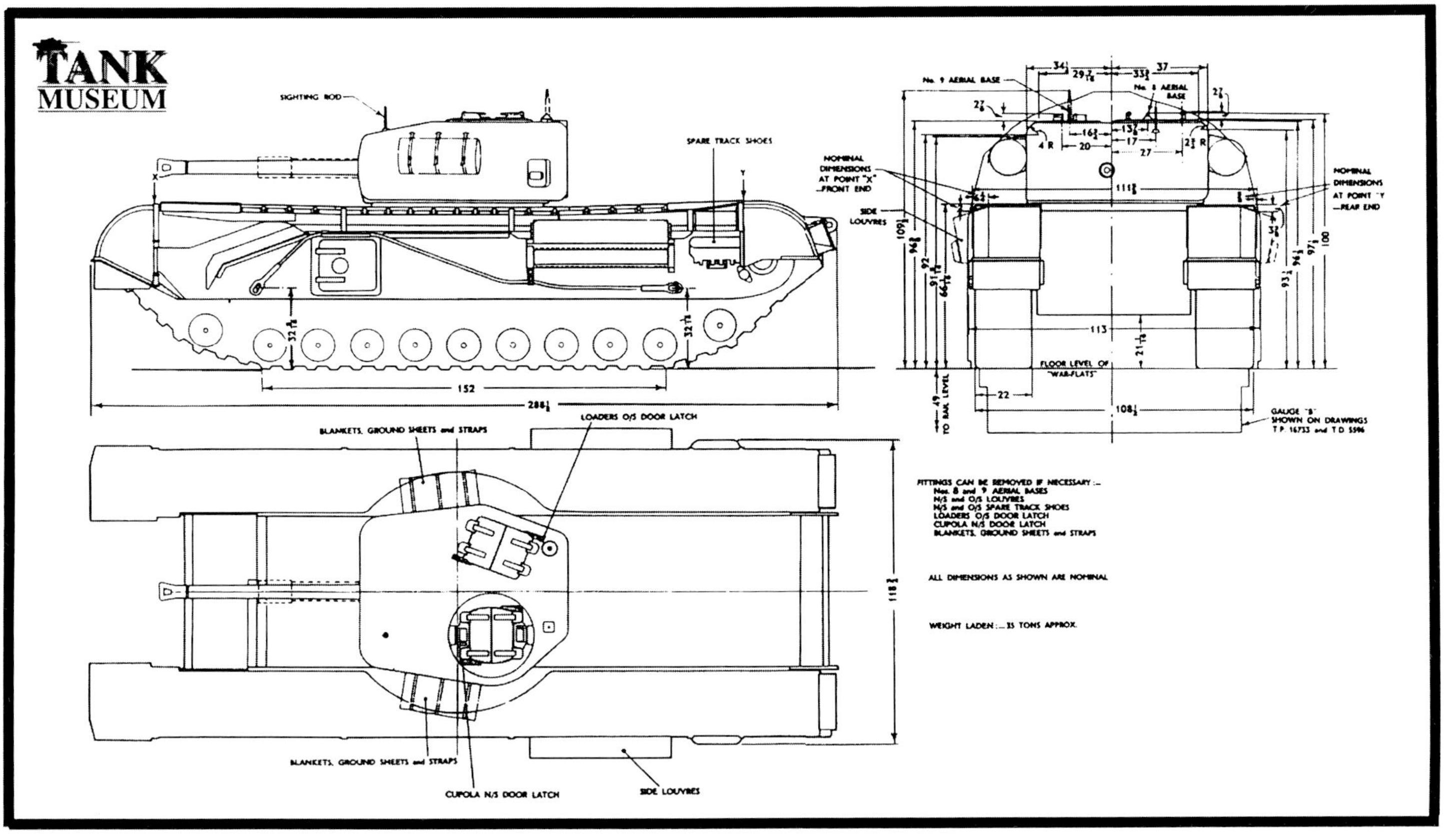

CHURCHILL V. & VI. TRANSPORT DIAGRAM

all weight of the new Mark IV Churchill but this was rapidly swallowed up in a more extensive up-armouring process which will be dealt with later.

Meanwhile the matter of Churchill production was still causing a lot of problems. In May 1942 it was reported that 3,500 tanks were on order and that they should be finished by March 1943, by which time a replacement design must be ready. Within a few days the completion date had been extended by a month and two other matters were dealt with. It was reported that 30 per cent of British tanks then in production were Churchills and that the last 500 were supposed to be fitted with rubber tyres. It was also explained that the replacement assault, or infantry tank should be based on the A27 design, that was the Cromwell cruiser tank.

Vauxhall Motors were informed that they would be expected to turn over to Cromwell production when the last Churchill was finished, and that brought a swift reaction. Under pressure from the Tank Board the company circulated a report in July 1942 which dwelt upon the recent success of trials on two Churchills in an effort to stave off what appeared to be inevitable. Yet it seems clear that the authorities were undecided. The General Staff asked the War Office for a statement onA22, based on user experience but the Tank Board continued to make it clear that 3,500 would be the final figure. In the meantime it was announced that six Churchills would be sent to the Middle East for what was described as 'operational experience'.

In October 1942 the Tank Board again stated that the Churchill was obsolescent and going out of production. But now they were faced with another problem. The only British weapon suitable for the close support role in tanks up to 1941 was the 3 inch howitzer which, as the Germans had reported on examples captured at Dieppe, was an inferior piece which was already completely out of date. Design work on a replacement began in January 1942 and it appears to have been a very odd design. The plan seems to have been to use part of the barrel from the famous 3.7 inch anti-aircraft gun, mated to the breech assembly of the equally successful 25-pounder gun, to produce a 95mm howitzer which would fire ammunition from the old 3.7 inch mountain howitzer. The gun would be produced in two forms, as an infantry weapon on a towed carriage and as a close support tank gun which could replace the 6-pounder.

The towed version seems to have been a total disaster, although it took the authorities a long time to wake up to this, while the tank weapon caused its own problems. In November 1942 a meeting of the AFV Liaison Committee was told that production of 6-pounder turrets for A22 was almost complete (and we must assume that they were referring to welded turrets) but that it was now revealed that the 95mm would not, after all, be interchangeable with the 6-pounder in the Mark III turret. This was presented in such a way as to assume that the 95mm would not be used, since the scrapping of new turrets appeared to be unthinkable. A week later the General Staff told that Committee that they considered the 95mm essential, even if it meant developing a special turret, and that the scrapping of some 6-pounder turrets might be inevitable. On the 24 November it was announced that the requirement for 95mm tank guns would be 250, but whether this applied to the Churchill only, or included A27 versions is not clear.

Surviving evidence is by no means complete but it does appear to reveal a state of confusion which might not be unfamiliar to students of British tank design. Since we know that in due course the 95mm howitzer fitted comfortably into the cast 6-pounder turret it must be assumed that the problems, whatever they were, applied to the welded version only. There is certainly no evidence that a 95mm version of the Churchill with welded turret ever existed but, if that is where the trouble lay, what was meant by the statement in November 1942 that turret production for A22 was almost complete? Are we supposed to believe that all turrets then being built for the balance of the 3,500 Churchills were of the welded type? If so what were the cast ones for? The fair answer may be that those who made the statement were not aware of the cast turret, but a cynic might suggest that this was just one

Above: A Mark IV without track guards is an unusual sight. This one has the Mark 5 gun fitted with a counterweight at the muzzle end of the barrel, which suggests that the weapon is on a free elevation mounting.

Below: An example of the Churchill Mark V, fitted with the 95mm howitzer.

problem too many. That the committee members thought they could dispose of another complication by being a bit awkward.

Late in December 1942 it was announced that 95mm gun mountings would be ready by January but that there would be delays in the supply of guns and ammunition. Since it was intended for use in a tank, rather than a field gun, a new fixed round was developed. The gun itself, known as the 95mm Mark I tank howitzer, when mounted in the cast A22 turret had a full elevation of just 37 degrees, which limited the effective range to 6,800 yards, which was adequate for a tank. High explosive and smoke rounds were provided and the version of the Churchill mounting a 95mm howitzer in the cast turret was classified Mark V.

The rework programme was continuing to make waves. On 21 November 1942 the Director of Armoured Fighting Vehicles at the Ministry of Supply told the AFV Liaison Committee that modifications were continually being embodied in new and reworked Churchill tanks, making a lot of extra work for REME personnel. He suggested that a rolling programme should be initiated whereby tanks now issued to units should return to manufacturers on a continual basis so that they could carry out changes. He went on to say that it seemed unlikely that Churchills could ever be made reliable when travelling long distances but finished up by asking the Committee to reconsider their intention to terminate production at 3,500 machines.

A paper issued a few days later reported on the results of Trent I, which seemed to show that problems had, to use the report's word, been 'Localised'. This, presumably, meant that they were not rampant any more and might be regarded as under control. It was further announced that by this date - 24 November 1942 -1,205 Churchills had been reworked while the first 247 tanks would only be used for training, and therefore might not have been reworked. The scheme was running smoothly under the control of the Director of Mechanical Engineering at the War Office who arranged, through DAFV, for tanks to be removed from units and delivered to Vauxhalls, or Broom & Wade for reworking.

On 3 December the Tank Board announced that all reworked tanks should be fitted with the 6-pounder, which seems to have been wishful thinking although it might have been one way of using up all those extra turrets. Five days after that the General Staff confirmed that they would stick at 3,500 as the final figure for Churchill production but, while they were talking statistics, wished it to be known that there were now 169 official modifications incorporated into the rework programme; which probably explains how it was that, in addition to the original rework scheme for new tanks there was now a rework scheme for reworked tanks. The General Staff was agitating for progress.

(text continues on page 99)

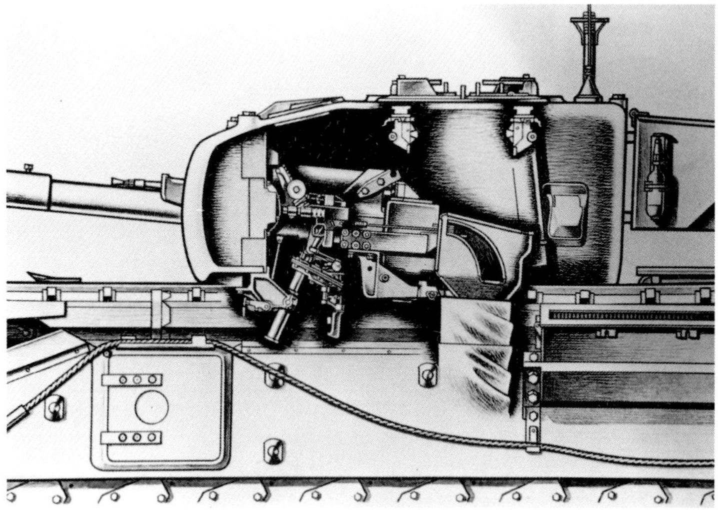

Detailed section drawing of the 95mm howitzer turret of a Churchill V. In this case the geared elevation mechanism can clearly be seen beneath the gun.

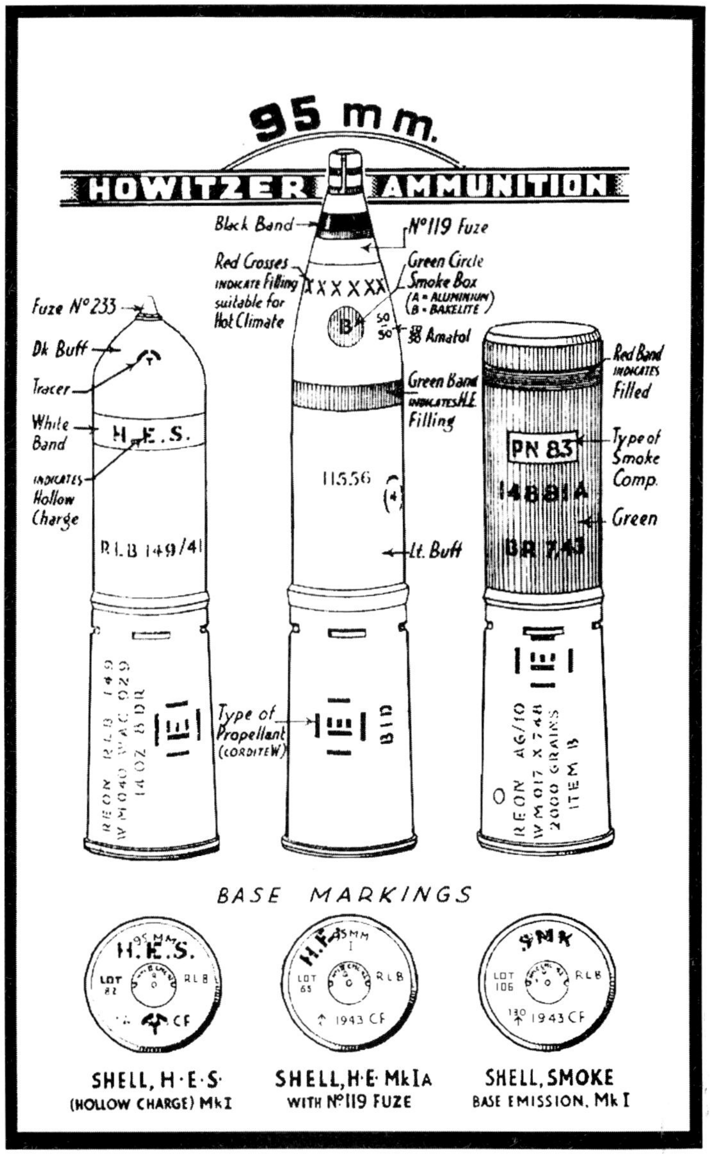

Right and pages 96-98: Stowage diagrams and ammunition details for the Churchill Mark V.

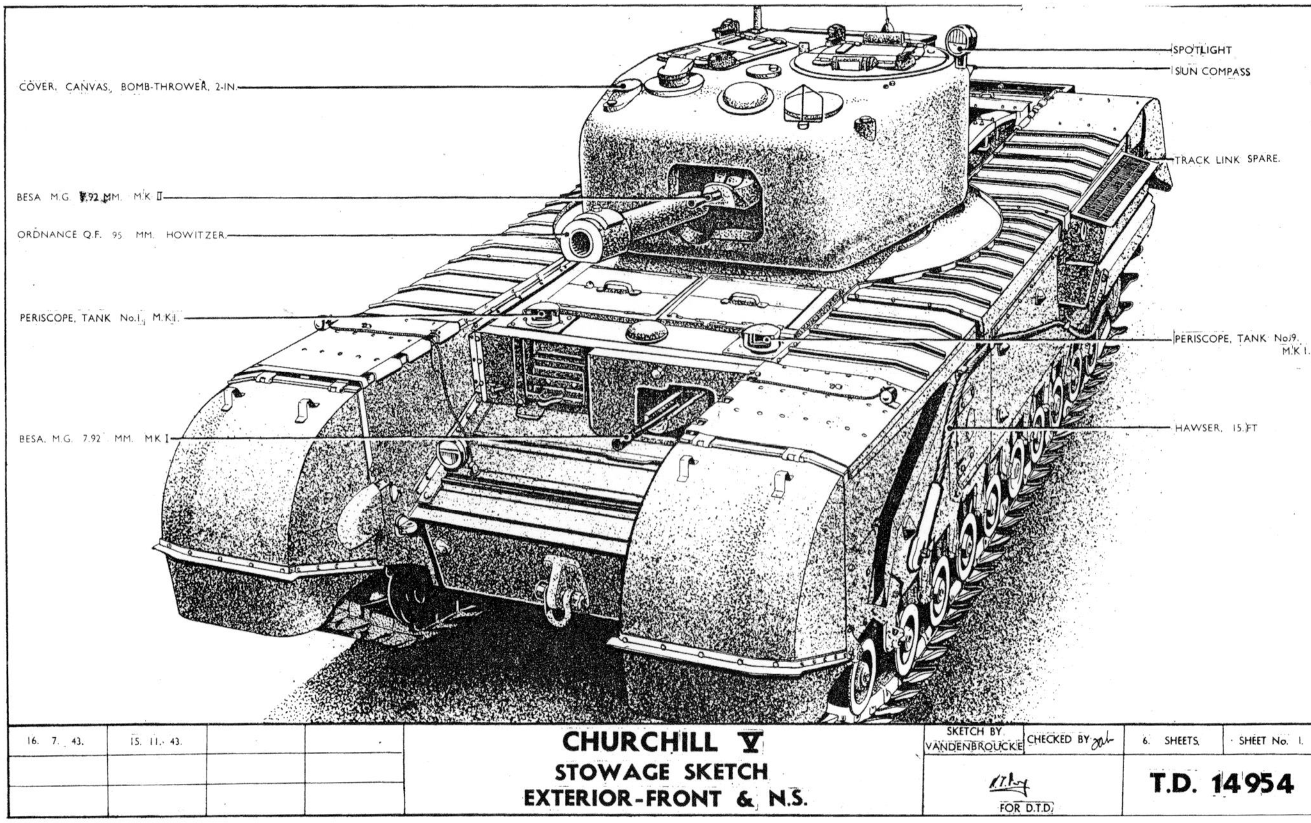

SPOTLIGHT
SUN COMPASS
COVER, CANVAS, BOMB-THROWER, 2-IN.
TRACK LINK SPARE.
BESA M.G. 7.92 MM. MK II
ORDNANCE Q.F. 95 MM. HOWITZER.
PERISCOPE, TANK No.1, M.K.I.
PERISCOPE, TANK No.9. M.K.I.
HAWSER, 15 FT
BESA, M.G. 7.92 MM. MK I
16. 7. 43.
15. 11. 43.
CHURCHILL V
STOWAGE SKETCH
EXTERIOR-FRONT & N.S.
SKETCH BY VANDENBROUCKE
CHECKED BY
FOR D.T.D.
6. SHEETS.
SHEET No. 1.
T.D. 14954

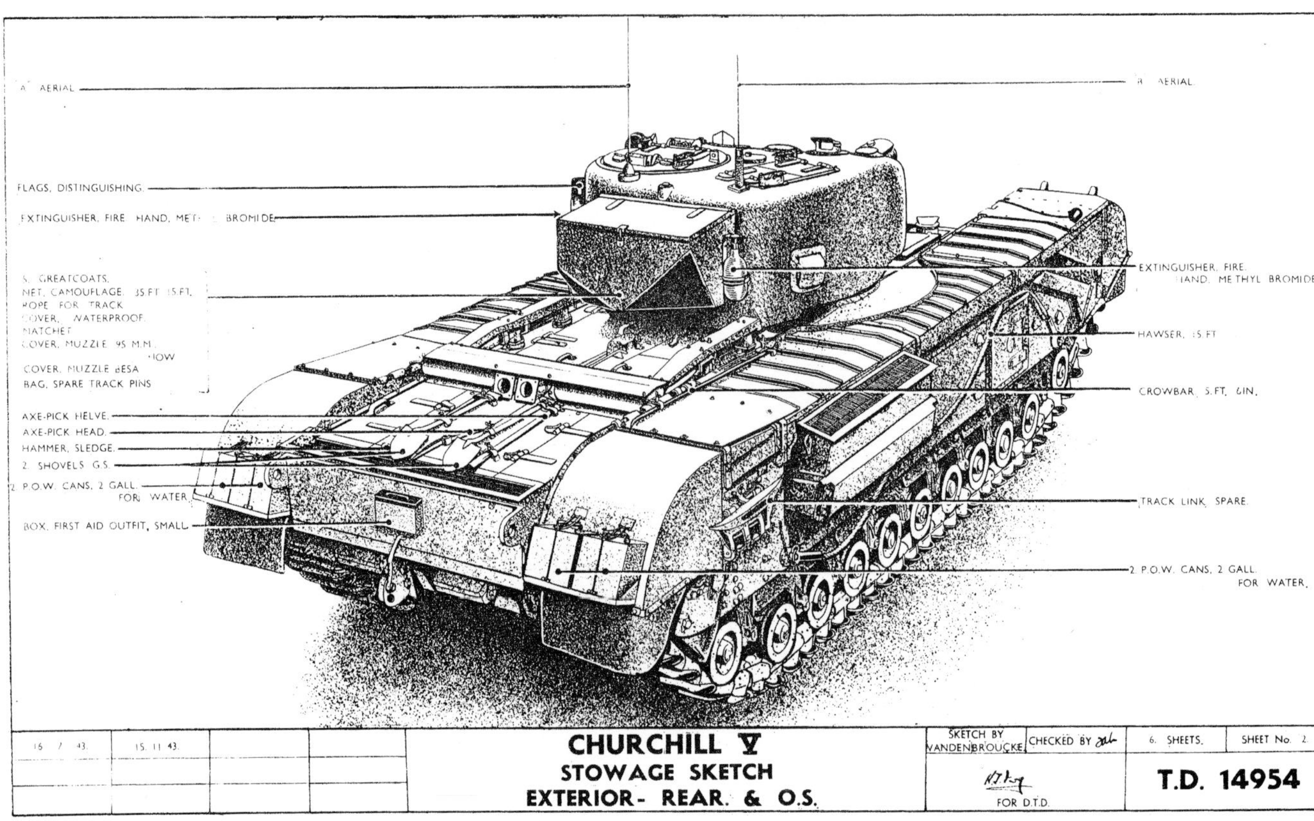

"A" AERIAL
"B" AERIAL
FLAGS, DISTINGUISHING.
EXTINGUISHER, FIRE HAND, METHYL BROMIDE
5. GREATCOATS.
NET, CAMOUFLAGE. 35 FT 15 FT.
ROPE FOR TRACK
COVER, WATERPROOF.
HATCHET
COVER, MUZZLE 95 M.M. HOW
COVER, MUZZLE BESA
BAG, SPARE TRACK PINS
EXTINGUISHER, FIRE HAND, METHYL BROMIDE
HAWSER, 15 FT
AXE-PICK HELVE.
AXE-PICK HEAD.
HAMMER, SLEDGE.
2. SHOVELS G.S.
CROWBAR, 5 FT, 6 IN.
2. P.O.W. CANS, 2 GALL. FOR WATER.
BOX, FIRST AID OUTFIT, SMALL.
TRACK LINK, SPARE.
2. P.O.W. CANS, 2 GALL. FOR WATER.
16. 7. 43.
15. 11. 43.
CHURCHILL V
STOWAGE SKETCH
EXTERIOR- REAR. & O.S.
SKETCH BY VANDENBROUCKE
CHECKED BY
FOR D.T.D.
6. SHEETS.
SHEET No. 2
T.D. 14954

2 PERISCOPES TANK No. 8 MK
6 GRENADES HAND
BINOCULARS
COMMANDER'S WALLET
CAN OIL M.G. MK II
2 PLUGS CLEARING BESA M.G.
LAMP HAND HELLESEN
BRUSH CLEANING PERISCOPE
2 PRISMS TANK PERISCOPE
COMPOUND ANTI-DIM AND FLANNEL
OINTMENT ANTI GAS
24 MAGAZINES THOMPSON (20 RDS) OR
8 MAGAZINES STEN (32 RDS)
BOTTLE WATER
12 BOXES BESA 7.92 M.M. AMMN
CAPE AND GLOVES ANTI GAS
6 PROTECTORS SPLINTER GLASS
CHURCHILL V
STOWAGE SKETCH
INTERIOR-TURRET-REAR & N.S.
6 SHEETS
T.D. 14954

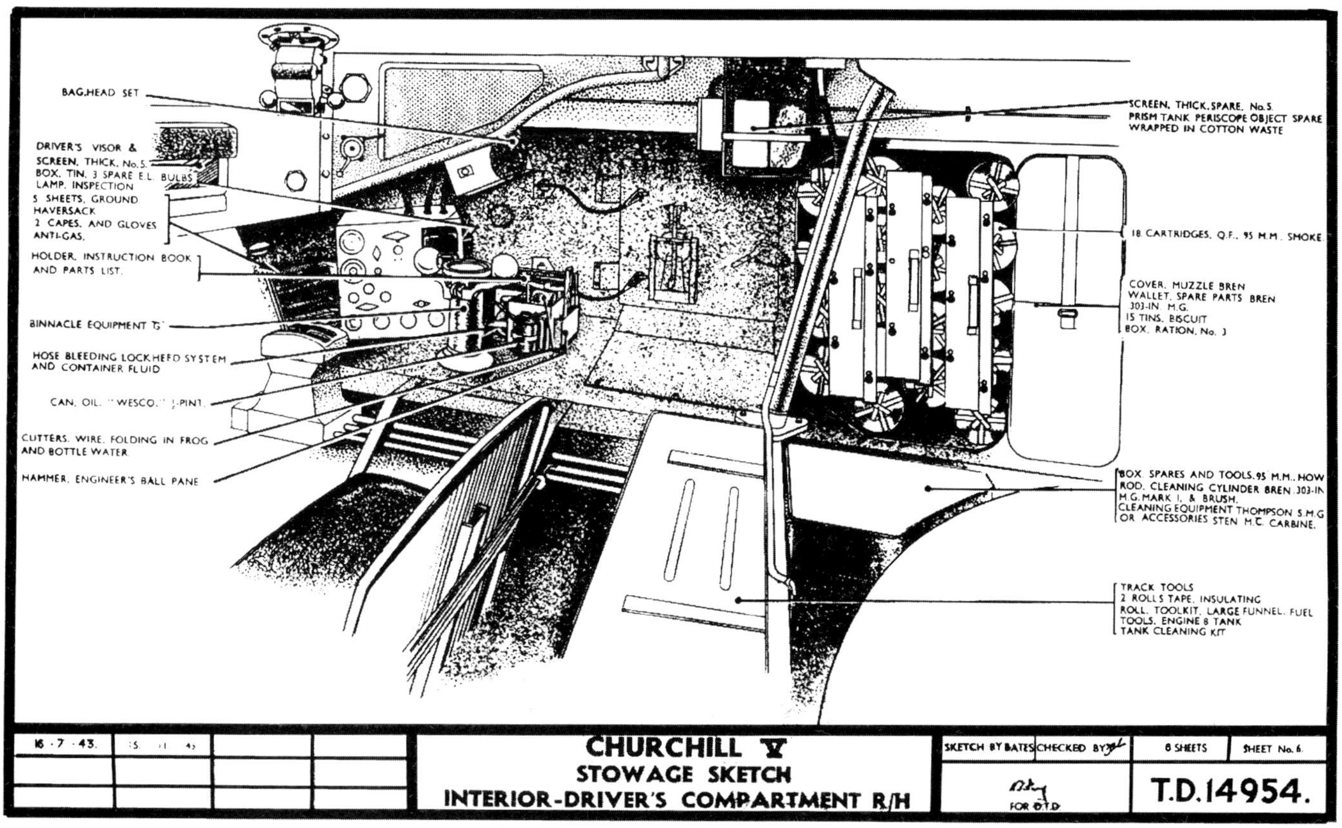

BAG, HEAD SET
DRIVER'S VISOR & SCREEN, THICK, No. 5.
BOX, TIN, 3 SPARE E.L. BULBS
LAMP, INSPECTION
5 SHEETS, GROUND
HAVERSACK
2 CAPES, AND GLOVES ANTI-GAS.
HOLDER, INSTRUCTION BOOK AND PARTS LIST.
BINNACLE EQUIPMENT 'D'
HOSE BLEEDING LOCKHEED SYSTEM AND CONTAINER FLUID
CAN, OIL, "WESCO," ½-PINT.
CUTTERS, WIRE, FOLDING IN FROG AND BOTTLE WATER
HAMMER, ENGINEER'S BALL PANE
SCREEN, THICK, SPARE, No. 5
PRISM TANK PERISCOPE OBJECT SPARE WRAPPED IN COTTON WASTE
18 CARTRIDGES, Q.F., 95 M.M., SMOKE
COVER, MUZZLE BREN
WALLET, SPARE PARTS BREN .303-IN. M.G.
15 TINS, BISCUIT
BOX, RATION, No. 3
BOX SPARES AND TOOLS, 95 M.M. HOW.
ROD, CLEANING CYLINDER BREN .303-IN M.G. MARK I, & BRUSH.
CLEANING EQUIPMENT THOMPSON S.M.G. OR ACCESSORIES STEN M.C. CARBINE.
TRACK TOOLS
2 ROLLS TAPE, INSULATING
ROLL, TOOLKIT, LARGE FUNNEL, FUEL
TOOLS, ENGINE & TANK
TANK CLEANING KIT
16-7-43.
CHURCHILL V
STOWAGE SKETCH
INTERIOR-DRIVER'S COMPARTMENT R/H
SKETCH BY BATES
CHECKED BY
FOR D.T.D.
6 SHEETS
SHEET No. 6.
T.D. 14954.

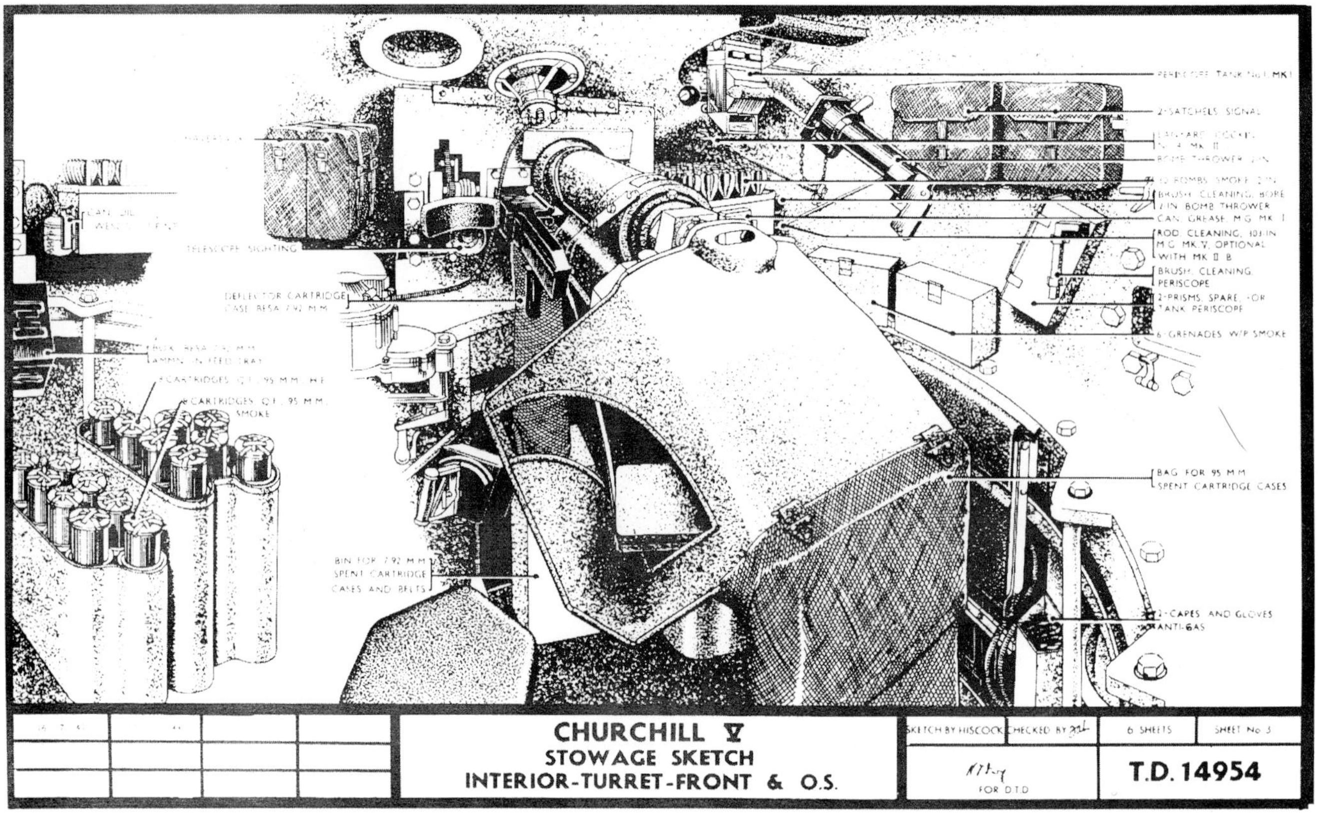

PERISCOPE TANK No. 1 MK I
2-SATCHELS SIGNAL
BOMB THROWER 2-IN
BRUSH CLEANING BORE 2-IN BOMB THROWER
CAN GREASE M.G. MK I
ROD CLEANING 30½-IN M.G. MK.V, OPTIONAL WITH MK II B
BRUSH CLEANING PERISCOPE
2-PRISMS SPARE FOR TANK PERISCOPE
6-GRENADES W/P SMOKE
BAG FOR 95 M.M. SPENT CARTRIDGE CASES
2-CAPES AND GLOVES ANTI-GAS
TELESCOPE SIGHTING
DEFLECTOR CARTRIDGE CASE BESA 7.92 M.M.
BOX BESA 7.92 M.M. AMMN. IN FEED TRAY
8 CARTRIDGES Q.F. 95 M.M. H.E.
8 CARTRIDGES Q.F. 95 M.M. SMOKE
BIN FOR 7.92 M.M. SPENT CARTRIDGE CASES AND BELTS
CHURCHILL V
STOWAGE SKETCH
INTERIOR-TURRET-FRONT & O.S.
SKETCH BY HISCOCK
CHECKED BY
6 SHEETS
SHEET No. 3
FOR D.T.D.
T.D. 14954

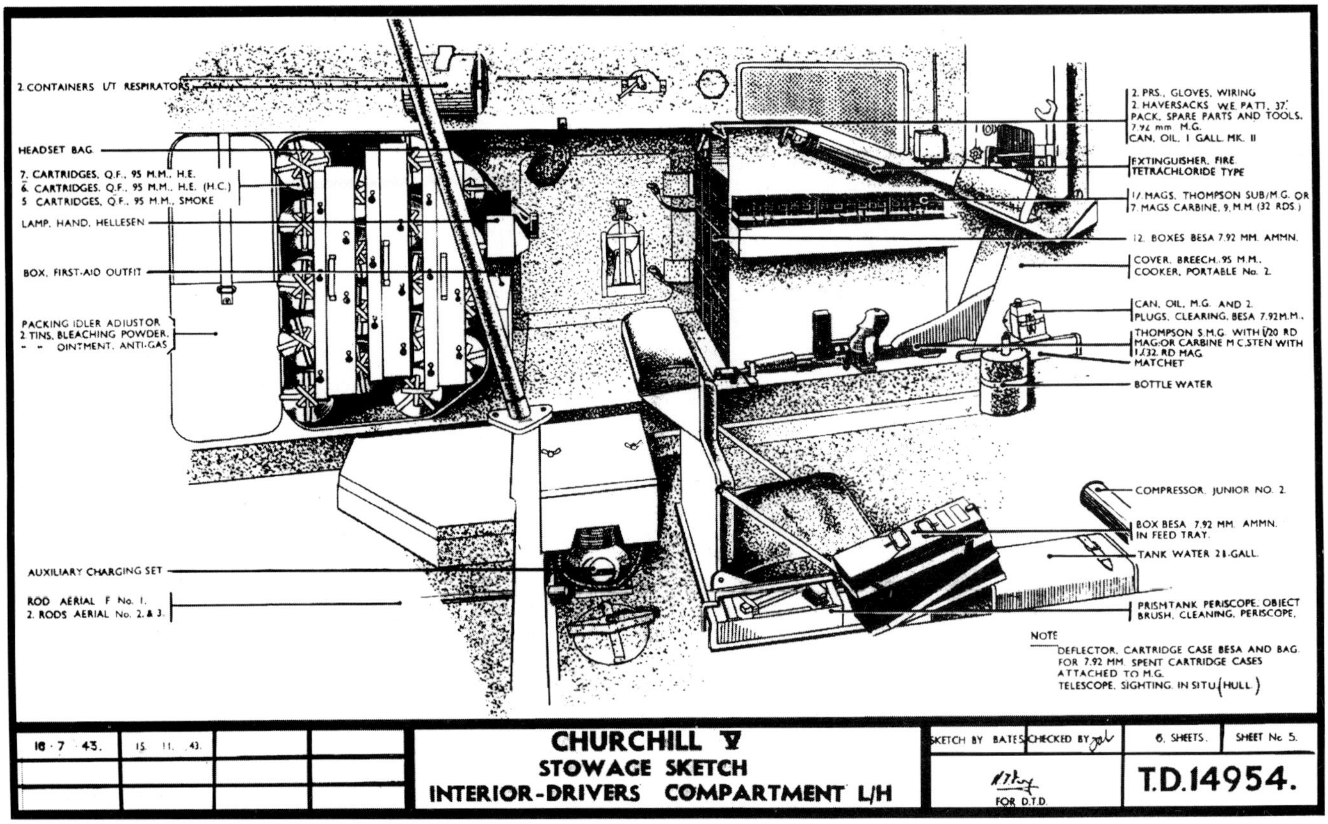

2 CONTAINERS U/T RESPIRATORS
HEADSET BAG
7. CARTRIDGES, Q.F., 95 M.M., H.E.
6. CARTRIDGES, Q.F., 95 M.M., H.E. (H.C.)
5 CARTRIDGES, Q.F., 95 M.M., SMOKE
LAMP, HAND, HELLESEN
BOX, FIRST-AID OUTFIT
PACKING IDLER ADJUSTOR
2 TINS, BLEACHING POWDER
- - OINTMENT, ANTI-GAS
AUXILIARY CHARGING SET
ROD AERIAL F No. 1.
2. RODS AERIAL No. 2. & 3.
2. PRS., GLOVES, WIRING
2. HAVERSACKS W.E. PATT. 37.
PACK, SPARE PARTS AND TOOLS, 7.92 mm M.G.
CAN, OIL, 1 GALL. MK. II
EXTINGUISHER, FIRE TETRACHLORIDE TYPE
11 MAGS. THOMPSON SUB/M.G. OR 7. MAGS CARBINE, 9. M.M. (32 RDS.)
12. BOXES BESA 7.92 MM. AMMN.
COVER, BREECH, 95 M.M.
COOKER, PORTABLE No. 2.
CAN, OIL, M.G. AND 2.
PLUGS, CLEARING, BESA 7.92M.M.
THOMPSON S.M.G. WITH 1/20 RD MAG OR CARBINE M.C. STEN WITH 1/32 RD MAG
MATCHET
BOTTLE WATER
COMPRESSOR, JUNIOR NO. 2.
BOX BESA 7.92 MM. AMMN. IN FEED TRAY.
TANK WATER 2½-GALL.
PRISM TANK PERISCOPE, OBJECT
BRUSH, CLEANING, PERISCOPE.
NOTE
DEFLECTOR, CARTRIDGE CASE BESA AND BAG FOR 7.92 MM. SPENT CARTRIDGE CASES ATTACHED TO M.G.
TELESCOPE, SIGHTING, IN SITU (HULL)
16 - 7 - 43.
15. 11. 43.
CHURCHILL V
STOWAGE SKETCH
INTERIOR-DRIVERS COMPARTMENT L/H
SKETCH BY BATES
CHECKED BY
6. SHEETS.
SHEET No. 5.
FOR D.T.D.
T.D. 14954.

Much of what happens in this period is tied up with yet another change in armament which it will be more logical for us to study later. Suffice it to say at this point that on 20 January 1943 the War Cabinet approved production of another 500 Churchill tanks. Such moves can only be attributed to expediency, for at this time the Churchill still had to prove itself, and perhaps some Prime Ministerial pressure. It certainly flies in the face of the declared intentions of the Tank Board, War Office and General Staff. Seeking to qualify the present position the Tank Board reported on 25 January that of the 1,205 Churchills reworked so far, some of the first 289 would still carry the 2-pounder but at least 100 would be 6-pounder. Out of the next 200 there would be 100 6-pounder and 100 equipped as Canal Defence Light (CDL) tanks and all the remaining tanks would be 6-pounder.

In an effort to re-establish some semblance of authority the Tank Board produced a Statement of Policy which announced that Churchill production would be limited to 4,000 machines and when they were completed Vauxhall Motors and its subsidiary companies would be turned over to Cromwell production. Little did they know.

Since nobody could claim that the Dieppe operation was a representative example of the employment of Churchill tanks there was still plenty of scope for tactical evolution. Certainly events in the Western Desert were not providing much evidence; apart from the use of Matildas against Italian positions on the Libyan frontier most subsequent operations by their successor, the Valentine, were almost indistinguishable from cruiser tank actions. Indeed the pure infantry tank battle always seems to have been something of a myth. However this is not to say that it might not happen and contingency plans had to be prepared, and if this was the case it clearly had to be done in close co-operation with the infantry themselves. Thus by 1942 the School of Infantry, at Warminster in Wiltshire, was engaged in trials to examine various options.

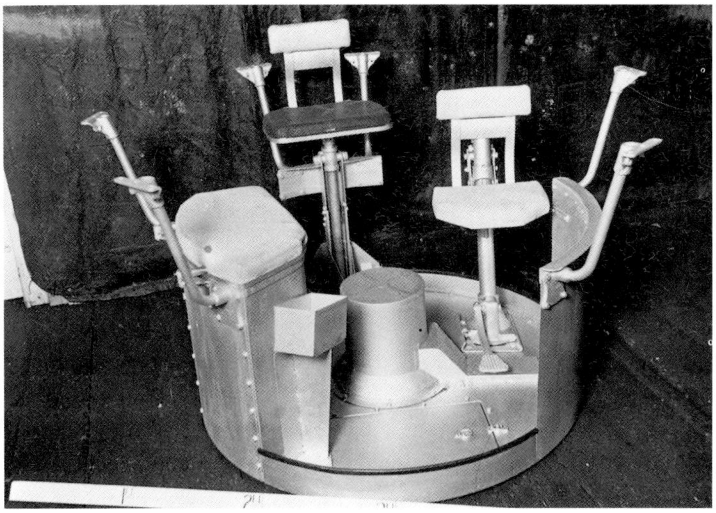

A factory photograph of the turret turntable of a Mark V Churchill showing, from the left, the loader's, commander's and gunner's seats. Notice that two of them are adjustable for height. The loader is actually seated on an ammunition bin.

Many hands can get in the way. Women soldiers of the Auxiliary Territorial Service apparently helping to maintain a Mark IV Churchill. One gets the impression that the photographer has rather overdone it in an effort to get his message across. The various combinations of tools and poses seem highly improbable.

Left: A more plausible scenario in a Churchill factory, moving a turret, complete with basket and turntable, onto a waiting tank.

One result of this was the discovery that in a combined infantry/tank attack the crucial time came when the preliminary artillery barrage lifted from the enemy position to allow the infantry to move in. This gap, no matter how brief it was, gave the enemy time to man his anti-tank guns and pick off the tanks as they moved in at the spear head of the assault. It was virtually impossible to locate every hostile gun before the attack began and indeed the tanks themselves would not know where individual guns were until they unmasked, which in effect means when the gun in question fired, by which time of course the tank was already vulnerable, if not destroyed. It was therefore suggested that infantry tanks could advance within the barrage to overcome this problem. The question was, of course, what might happen to the tanks if they attempted to advance within a barrage, rather than behind it.

An experiment was organised and it took place at Larkhill on Salisbury Plain, on 5 June 1942, involving a squadron of sixteen Churchills from an unidentified regiment. Two trials were arranged. In the first, three field regiments of 25-pounders, that is 72 guns, fired at the rate of 72 rounds per minute into an area 600 yards by 400 yards, firing air-burst high explosive shells. The sixteen tanks concentrated within this area, in a box 300 yards by 400 yards, moving at just 5 mph. It must have taken a fair amount of courage, on the part of the tank crews, to drive in formation into this zone but in fact it proved perfectly safe and there were no personnel casualties whatever. The guns continued firing for eight minutes while the tanks were in the impact area. The tanks discovered that they could not only move about within the fire swept zone but they could still see clearly and deal with any target easily. A number of tanks were hit, and one had its track broken, but the remaining fifteen kept moving. Another tank lost one suspension bogie and broke down shortly afterwards but even then, fourteen tanks survived and the other two were swiftly repaired, so the trial was regarded as a success.

In the second trial the barrage was stepped up to a more realistic intensity. This time as the tanks moved up to the barrage zone the guns fired a concentrated burst for two minutes at the rate of 180 rounds per minute. As the tanks moved into the danger zone this rate reduced to 108 rpm and continued for five minutes. On this occasion there were no tank casualties at all. Crews found that they were not affected by the detonations, nor did they report any difficulty in observation so in a real action they would be able to hunt down the enemy's anti-tank guns with ease. This would mean that many of the guns would be taken out while their crews were sheltering from the barrage in slit trenches. Even if they did emerge to man their guns under fire the chances of picking off approaching tanks with shells exploding all over the place were reduced considerably.

Ultimately the results of the trial were inconclusive. Some pointed out that the conditions on Salisbury Plain were hardly typical since the soil was rich and damp. In drier climates, they argued, huge clouds of dust would be thrown up, obscuring the targets and making navigation difficult. Subsequent examination of the tanks revealed more damage. In particular shrapnel from the shells had found its way into the air inlets and it was suggested that the Churchill was unsuit-

This may not show the event described in the text but at least it gives a flavour of Churchill tanks operating in an artillery barrage.

A Churchill crew from 43rd RTR demonstrating the Lakeman anti-aircraft mount with a Bren gun fitted with the drum magazine. Notice on the left the three-pronged vane sight used by the commander to align his tank on a target.

able on this account. It is surprising to find no reports of damage to vision equipment. Certainly it was possible to change damaged prisms in periscopes but if this happened in the heat of action it might reduce combat effectiveness and one wonders whether communications might suffer if aerials were damaged. There is no evidence that the practice was ever employed in action, at least not deliberately, and it would be interesting to know if any consideration was given to the system used in the First World War where the barrage was planned around shell-free lanes along which the tanks could move.

As a footnote to this period there is a complaint by the AFV Liaison Committee, dated 16 March 1943, that up to that period, no 95mm gun tanks, either A22 or A27, had arrived at the Lulworth Gunnery School for testing; and Lulworth was also interested in discussing the PLM mounting. The PLM mount was a truly awful thing which must have been cursed by every tank crewman who had anything to do with it. The purpose of the PLM mount was to provide the tank with what was known as local anti-aircraft defence. Before it appeared tank crews relied on something called the Lakeman mount which, if anything, was even worse.

Experience in France had convinced British tank crews that they needed some sort of weapon to deal with marauding enemy aircraft and Major Tom Lakeman had provided a solution - in haste as usual. It was simple enough. Just a portable arrangement of steel tubes and springs which, when fitted to the turret, enabled the tank commander to aim and fire a Bren light machine-gun at attacking enemy aircraft. In practice the chances of hitting an aircraft were slim; the chances of being hit by the aircraft somewhat greater since the gunner had to expose a good deal of his upper body. The PLM mount probably did nothing to improve the former but it did at least enable the gunner to increase his chances since it was operated from inside the turret.

The mounting consisted of a central pillar, bolted to the top of the turret, which carried a pair of Vickers K light machine-guns fitted with drum magazines. A downward extension of the pillar, passing through a hole in the turret, terminated in a pair of handlebars, like those on a bicycle, with a trigger built into the handle. If an aircraft attacked, the gunner ducked down inside the turret so that he could just see out over the rim of the hatch. Inside, protected by armour, he gripped the handlebars which traversed, elevated and fired the guns. Even if it did prove to be of some use in action the mounting was a dreadful nuisance at all other times. Even with the handlebars folded away they occupied space inside the turret and gave the gunner and loader something else to snag their clothing and bang their heads on. Outside it obstructed the commander's view. Got in his way as he climbed in and out, and reportedly interfered with radio reception. Unfortunately it has proved impossible to discover what the letters PLM stand for.

Vauxhall Motors clearly did not like the thing but chose to complain about its weight, which cannot have been significant. However the Director, Royal Armoured Corps (DRAC) told them that if weight was a problem they might delete the auxiliary fuel tank, fitted on the back of the hull. It seemed that there was no getting away from the PLM mount. It was certainly available for a range of British tanks and armoured cars but if photographs are any guide the only vehicle that ever carried it in action was the Humber Scout Car. If it was supplied for tanks going overseas the chances are that crews managed to lose it before they arrived in the theatre of war. British tank crews have always had an ambivalent attitude to externally mounted machine-guns, especially for anti-aircraft use. The general view seems to have been that if enemy aircraft were about the sensible thing to do was to stay inside the tank, close all the hatches, and wait until they flew away.

8

Churchills in North Africa

Tanks had seen action in Palestine in 1917 and they had been operating with British forces based in Egypt since 1926. It was a region fraught with problems which rapidly threatened to get worse as Mussolini's aspirations became clear. Thus, of all the tank producing nations Britain, with France and Italy, had every reason to consider the use of tanks in desert conditions. Yet there is very little evidence to suggest that this had any serious effect on the design of any armoured fighting vehicle built in the United Kingdom either before or during the Second World War.

Of course, when the design of the Churchill was being rushed through there were far more immediate concerns and, as we have already seen, it was produced at first with the defence of the homeland in mind. So it is probably fair to say that of all British tanks produced since the end of the First World War the Churchill was the least prepared for desert service. Yet when, after the Fall of France, the main focus of British attention shifted to the defence of Egypt its deployment out there was only a matter of time. Indeed we have already noticed that the Prime Minister was advocating that very thing.

We have also seen how, by late 1940, two Churchills had been prepared for desert trials. Quite what this preparation involved is not clear, beyond the fact that both tanks were fitted with track covers which, in this context were referred to as sand guards. We also know, from subsequent reports, that one of the tanks was fitted with full length skid rails supporting the upper run of the tracks while the other had the dome shaped supports spread out along the top of each sponson which only supported the tracks at intervals. The army in Egypt referred to these, appropriately enough, as camel backs. The two tanks were Vauxhall built machines T30999 and T 31000. Photographic evidence reveals that they were of the Mark II type. The task of testing them would fall to the Middle East branch of the Mechanisation Experimental Establishment which was based in Cairo.

With U-Boats waiting off the Straits of Gibraltar and the threat from Axis aircraft over the Mediterranean it was normal practice to route convoys bound for Egypt well out into the Atlantic, around the southern Cape of Africa and up to Alexandria via the Red Sea and Suez Canal. It was a long voyage, relatively safe from enemy action but at the mercy of every kind of weather that nature has to offer, especially in the Atlantic in winter. The two tanks travelled as deck cargo, as we know from the first report filed on them by the authorities in Egypt. They are described as arriving in deplorable condition, without even the benefit of tarpaulin covers to keep out the worst of the rain and sea spray, which all played havoc with the two tanks.

Surface rust must have been sprouting up everywhere and it is claimed that many alloy castings had simply dissolved during the voyage. Even before trials could begin cables had to be sent to Britain ordering replacement parts and they were not delivered for three or four weeks. It was a period of relative stability in the Western Desert. The British had fallen back to strong positions on the Gazala Line and even the daring Germans had to resist the urge to attack while they recruited their strength. Even so, as both sides knew only too well, it was just a breathing space and in workshops on both sides engineers were working flat out to return as many old tanks as possible to battleworthy condition, in addition to preparing new ones. It is quite surprising to find that anyone had time to conduct trials on experimental types, especially since the work involved a good deal of maintenance.

Reading between the terse lines of the report it would appear that instructions had been received from Britain that the two tanks should be limited to speeds of 10mph and below. Yet the party in charge of desert trials for the Churchills only honoured this in part. Only one was adapted, by blanking off fourth gear. The other, referred to as the unrestricted machine, began trials at once. Those concerned with the Churchill's trials at home might have been surprised to read that general performance in the desert was good, and that steering was described as excellent. Of course, when it comes to steering, sand is much kinder to tanks than mud, although

One of the pair of Churchill Mark II tanks shipped out to the Middle East early in 1942. Seen here being prepared for trials at a workshop in the Nile Delta. This is T31000.

T30999, the other Churchill II sent out to Egypt. Notice that on both tanks the air intakes appear to have been modified from the original design.

it has other ways of harming them. Even so this would almost certainly have been the first tank operated in desert conditions with the Merritt-Brown transmission, which obviously performed well on a good surface. They certainly had suspension problems and the tracks were reported to be breaking up.

By February the unrestricted machine had been withdrawn to be fitted with a stronger suspension, designed locally. Which says something about the workshop resources available to British forces in Egypt. Unfortunately, shortly after the work on this modification began, drawings arrived from Britain to show how Vauxhall Motors proposed to improve the suspension and it was feared that the local design would not prove strong enough. Meanwhile the second tank continued with its trials, still restricted to 10mph. It experienced identical failures with the suspension and had soon worn out its gearbox but in neither case, the report noted, had the tracks yet failed entirely. They did complain that the two tanks were incapable of running straight, and continually tried to pull over to the left.

And so the trials go on, until mid-June. There is an occasional engine failure and the suspension is a continual source of trouble but nothing terminal. Despite the sand guards great clouds of the stuff are blown up in the driver's face, just as had been the case with dust in Britain, but it is notable that the camel back track supports had all worn out within about 700 miles while the continuous skid rails showed no signs of failure. All things considered these trials must be regarded as moderately successful. It is not clear whether the trials crews were Churchill trained men who travelled out with the tanks or local crews who more or less learned on the job, but they seem to have kept both tanks running, after a fashion, for around six months on a modest stock of spares. There are no complaints of steering problems or, perhaps more surprisingly, any adverse comments on cooling. And all this, it should be noted, is before the Dieppe raid had taken place.

One way to keep dust from blowing up into the driver's face was to stretch a canvas screen or dodger, across the front horns. As this shot shows it seems to work.

Another matter recorded in the last chapter was the decision to send six more Churchills out to the Middle East for what was described as 'operational experience'. This appears to have been agreed in July and once again it would be reasonable to suspect some influence being exerted by the Prime

One of the Kingforce Churchills being prepared for action. It is parked with its turret reversed. The camouflage scheme has been described variously as black, brown or green painted in a disruptive pattern with sand. Comparing all six tanks one gets the impression of a strictly standardised design.

A Kingforce Churchill at the time of the Alamein battle. This tank is starting to sport the accumulation of unofficial stowage so beloved of the Eighth Army and it also has a canvas dodger slung at the front.

Minister. Whatever the impulse the six tanks, all reworked Mark IIIs, arrived in Egypt on 1 October 1942, destined for the MEE at Cairo. Records reveal that the six machines were:

T31665/R
T31950/R
T31954/R
T68186/R
T68189/R
T68714/R

... representing four of the firms building Churchills under the Vauxhall umbrella. Work began at once on preparing them for desert service, which included suspending a canvas screen, or dodger, between the front horns to keep down the cloud of sand which blinded the driver. While this was going on an operational unit, known as the Special Tank Squadron, was being formed under the command of Major Norris King MC, of the Royal Gloucestershire Hussars. His second-in-command was a Lieutenant A L Deans of the Royal Tank Regiment with four more officers and 52 other ranks. The men who operated the tanks had all come out from Britain and had Churchill experience, although none of them had ever seen the desert before. The officers and NCOs, by way of contrast, were all old desert hands although the latter were described by the second-in-command as being 'not very bright in mechanical matters'.

The Special Tank Squadron was formed on 14 October 1942 and just four days later the men and tanks were loaded onto a train and carried west into the desert. Unloaded at the Burg el Arab railhead on the 20th the tanks were then loaded onto tank transporters and driven forward to Tel el Eisa. All around them were signs of great activity. With as much secrecy as could be maintained the British Eighth Army and its Allies, under the direct command of General Bernard Montgomery, was about to launch its greatest desert offensive, Operation Lightfoot, the opening phase of the second battle of El Alamein.

On the night of 23 October, as the Churchill tanks prepared for their final move forward, the western sky was alight with the flashes from 1,000 guns. The kind of barrage which had not been seen since the dreadful days of the First World War. The six tanks were directed along a route known as the Moon track, at the end of which, they were told, they should report to headquarters of 7th Motor Brigade, itself part of the 1st Armoured Division. Beneath the barrage Montgomery's troops had opened two broad paths through the extensive German minefields and the task allotted to 1st Armoured Division was to advance through the northernmost lane. But it was hard going. Strong enemy forces were waiting and, within the narrow confines of the cleared lane the attacking armour had no room to deploy. By the evening of 24 October the leading British tank regiment, the Queen's Bays, was down to a composite squadron of twelve tanks, a mixture of British Crusaders and American Shermans.

British planners had named a feature on their maps as Kidney Ridge, due to the shape of its contour line. But there was no precise agreement as to where it was or, indeed, whether it was a ridge at all. North and south of Kidney Ridge were two enemy positions nicknamed (by the British) Woodcock and Snipe respectively. Each one was well defended by Axis tanks, some of which were dug in, that had a clear field of fire for some 2,000 yards. They had taken a battering, but some 88mm guns had been moved up in support. As the situation deteriorated the Churchills were called up to deal with the dug in tanks.

The Special Tank Squadron was by now being referred to as Kingforce after their commanding officer and it was Norris King who led the way in T68189/R, followed by 2nd Lieutenant Appleby in T31665/R and Corporal Kelly in T68186/R. Corporal Kelly did not last very long. When his 6-pounder gun recoiled after the first round had been fired it stuck, refusing to run out again. Unable to contribute more to the action Corporal Kelly ordered his driver to engage reverse and quietly withdrew from action. Nobody knows precisely what happened to 2/Lt Appleby. His tank was seen to advance into the thick of the action. For a while it disappeared over the edge of the ridge but a short time later it was seen reversing slowly out of action. Then it stopped, smoke appeared and it finally burst into flames and continued to burn for the rest of the night. Only one man, and he wounded, is believed to have survived from the crew of five. When it was examined later it was seen that 38 rounds of 50mm calibre had struck the front of the tank, one of which had penetrated. There was damage from one high explosive round and six strikes from 75mm guns, one of which had penetrated. This was probably the shot that did the damage. However on closer inspection eight strikes were found on the back of the tank, one of which had gone through the gearbox and three into the turret. They were of 57mm calibre and had obviously come from British six-pounder guns.

Putting the facts together after the event its seems as if Lt Appleby had gone further into the German position than was wise, and taken a lot of punishment. As he reversed out his tank, wreathed in smoke, was spotted by an Australian anti-tank battery. Being unfamiliar with the shape they took the sensible precaution and opened fire. It was an unfortunate turn of events but there is no doubt that Appleby's tank absorbed a tremendous amount of punishment before it was destroyed. Major King himself, on the other hand, had an excellent day. He fired 45 rounds of 6-pounder, claimed hits on four enemy tanks and took eight enemy rounds on his Churchill, none of which penetrated. He withdrew in good order, describing his tank as fighting fit.

Kingforce spent seven days in reserve, getting the five remaining Churchills back into working order. Meanwhile with Rommel back from Europe Axis forces started to draw back to a new defensive line, but they did not go easily. Their position at Tell el Aqqaqir proved very awkward and British 2nd Armoured Brigade, supported by 7th Motor Brigade, failed to take it. Kingforce received instructions to move forward to take the pressure off some Shermans which were in trouble. Moving forward along the Boomerang track they rolled into action, and into a storm of fire. Guns and turrets caused most

Another popular feature of desert tanks was this sunshade attachment, intended to make the tank look like a heavy lorry from a distance. It is very rare to see it on a Churchill.

of the problems. One tank suffered traverse failure without firing a shot while another received nine 50mm rounds which jammed the turret. A similar fate awaited Corporal McCaffery's tank, T68714/R while Lt Howard, in T31950/R took over thirty hits and lost a track. The crew, some of whom were injured, baled out and made their way back on foot.

Major King was very lucky. A 50mm round came in through the open driver's visor and passed right through the crew area. Everyone quickly abandoned the tank but, once he could see that it was not going to burn, Major King and Lieutenant Howard climbed back in and brought the tank home. Over the next five days the crews got their tanks back into shape and Howard's tank was recovered. But the battle was moving away to the west and there was probably little point in trying to maintain five non-standard tanks which could not keep pace with the advance anyway. Kingforce was therefore ordered back to Alexandria, a journey that the tanks achieved on their own tracks, where it was split up.

A month after Montgomery launched his El Alamein offensive General Eisenhower commanded an Allied army that landed in Tunisia. There was no connection between the two events, the Tunisian landing had been planned well in advance, but it did not result in an immediate collapse of Axis forces. Granted Rommel's Afrika Korps was on the defensive and retreating west but the Germans and Italians in Tunisia were well established and the mountainous country conspired to assist them. The Allied offensive showed signs of bogging down and there was a call for reinforcements. The enemy was establishing strong positions in the hills, which could only be taken by direct assault, and this kind of work would suit the Churchills.

It has to be recognised that some Churchill regiments had been training continually in Britain for the best part of two years and, difficult as it is to imagine, most of them would welcome the chance to put it to the test in action. We have already established that, in the British scheme of things, infantry tanks were designed for a different role to that of the cruisers. Thus where cruisers tank regiments were formed into armoured brigades, normally in armoured divisions, infantry tank regiments (or battalions as they preferred to think

Reworked Mark III tanks of 25th Army Tank Brigade training in Britain before leaving for North Africa.

of themselves) were grouped in independent army tank brigades with no divisional associations. A contemporary document described the role of a tank brigade as 'a supporting arm to assist infantry on to a position'. The brigade would consist of three regiments (or battalions) with the usual sprinkling of supporting arms, which would be attached to infantry brigades for specific operations and Tunisia would be the first theatre in which this was properly put to the test with Churchill tanks.

The decision to send a tank brigade seems to have been taken in about December 1942 and the selected formation was 25th Army Tank Brigade which would consist of three Churchill regiments; the North Irish Horse, 51st Royal Tank Regiment and 142 Regiment Royal Armoured Corps. They sailed from Britain in separate convoys during January and all were established at Bone in Tunisia by the second week in February. It would be impossible to chronicle all their doings over the next four months so the practice - which will be followed in every case from now on - will be to draw highlights from their various experiences which illustrate the nature of the work they had to do and record anything particularly relevant to the performance of their tanks. They were all equipped with Churchills of course.

The North Irish Horse, raised in Northern Ireland as its name implies, was formed during the First World War as a militia regiment. It spent the years between the wars in suspended animation and was reformed in 1939 as a constituent part of the Royal Armoured Corps. With its tanks and men scattered on various ships which left the ports of Birkenhead, Cardiff and Swansea between the 4th and 7th of January it passed through the Straits of Gibraltar on the 24th and after touching at Algiers landed at Phillipeville on 10 February. A and B squadrons were at Le Kef on the 21st when news of the fight at Kasserine Pass came through and they received orders to create a defensive force. On the night of

A pair of Mark III Churchills and a despatch rider's motorcycle at a railway crossing near Hunt's Gap in Tunisia, the unit has not been identified but is probably the North Irish Horse.

26/27 February they moved to Beja where they were soon in their first action. A Squadron only had eleven tanks while B Squadron had just seven. One tank was knocked out by an 88mm round through the engine compartment while another was hit by artillery fire and broke a track. On the other hand, later in the day, Lieutenant Hern took a snap shot at a German tank which suddenly appeared about 500 yards away and discovered afterwards that he had knocked out a Tiger.

This defensive fighting went on through most of March and there is more than one entry in the regimental war diary to the effect that the tanks were being used as anti-tank pillboxes. But there are also references to squadrons being sent out with infantry on reconnaissance patrols, which seems to be unusual work for infantry tanks. By April the Allies were well into their stride and on the offensive. But the Germans had turned some of the hill top locations into defensive positions of immense strength which they believed to be tank proof. One such was known by the British as Longstop Hill and the North Irish Horse were instructed to support the 36th Infantry Brigade in their assault. It proved to be a costly battle for the British infantry. The Germans had all the lower slopes well covered by mortars and there were snipers operating everywhere. The tanks did their best to help but they had mines to contend with and, being unable to fire a respectable high explosive round, they could not achieve a great deal. Fighting continued around Longstop until the end of the month. Gradually, as the tank crews gained in experience and the infantry got used to working with them, their actions became more concerted. The tanks climbed higher up the slopes and soon they were supporting the infantry as they moved from ridge to ridge. Where ever possible they developed the practice of working around the flanks of a hill and then advancing to the crest from both sides. They destroyed machine-gun posts that were holding up the infantry and on more than one occasion a tank would find itself on a high spur, looking down on an anti-tank gun that was covering a valley below. At such times there was little hope for the enemy gun. Their first encounter with enemy tanks at Longstop took place around midday on the 30th. The German tanks were Panzer IVs. Three were engaged and driven off but another, with a long 75mm gun, appeared around the side of a hill, knocking out one Churchill and damaging another. The crew of the first baled out but the second was taken out of action by the co-driver, operating the reserve steering mechanism from his seat. They made it down the hill but there was a gully at the bottom which the Churchill hit and rolled over.

The second regiment in 25th Army Tank Brigade was 51st Royal Tank Regiment. It sailed from Britain on 20 January and its war diary gives a precise breakdown of its structure, which may be taken as typical. It consisted of 52 Churchills Marks III and IV, six Mark I and two Churchill armoured recovery vehicles. For reconnaissance it had nine Daimler scout cars, four White scout cars and ten Universal Carriers. It had seven Loyd tracked personnel carriers and two Loyds fitted out with extra batteries for starting recalcitrant tanks along with five Humber staff cars and 19 BSA M20 solo motorcycles. Its transport echelon consisted of three Austin utility pick-up trucks, 25 3 ton Bedford lorries and 10 Austin 3 tonners; ten Bedford 15 cwt trucks, two Bedford water tankers and a Morris-Commercial mobile office. The Light Aid Detachment (LAD) which took care of maintenance had a Scammell recovery tractor, an Austin stores lorry, two 15 cwt Bedfords

Churchills of the North Irish Horse, led by a Mark I, at the time of the battle for Longstop Hill.

one Austin utility and two BSA motorcycles. The signals section had one Bedford 3 tonner, three 15 cwts, two signals trucks and another BSA. Manpower was 41 officers, nine warrant officers, five Staff Sergeants, 40 Sergeants, ten Lance Sergeants, 62 Corporals, 58 Lance Corporals and 533 troopers.

They arrived at Bone on 2 February having lost one ship with nineteen tanks on the way. They were part of the defensive force at Le Kef but their first action, which took place on 28 February, is regarded as a classic. It involved A Squadron at a site known as Steamroller Farm near El Aroussa. The farm stood at the head of a valley which was heavily defended by German troops that included two battalions of the Hermann Goering Regiment and elements of a Panzer Grenadier regiment well provided with anti-tank guns. Thirteen Churchills were involved, supporting two companies of 2nd Battalion the Coldstream Guards, but the infantry were soon pinned down by enemy fire. The tanks were taking punishment as well, when orders came through to clear the pass at all costs. Two Churchills, commanded by Captain E D Hollands and Lieutenant J G Renton made a 1,500 yard dash across an exposed causeway which was covered by an 88mm gun. Firing at a range of about 200 yards this gun got off two rounds, and missed on both occasions, before it was charged by the tanks and the crew ran away.

Now the tanks were faced with a difficult climb up a steep and winding track more suited to goats. It was slow going with tight corners and a rocky surface but they finally arrived at the top to see, over the crest, the target of a lifetime. Firing down from their lofty position the two Churchills took out two more 88mm guns, two 75mms, two 50mms and many lesser weapons along with 25 assorted wheeled vehicles, two mortars and two Panzer III tanks. All this was achieved unsupported, because neither the infantry nor the remaining tanks could join them, and soon they had to withdraw. However they claim later to have intercepted a message from the German commander claiming that he had been attacked by a mad British tank battalion which had scaled impossible heights and forced him to withdraw. Upon such tales was the Churchill's reputation made.

Notes compiled shortly after the action recorded the damage sustained by ten Churchills at Steamroller Farm. It seems that 75mm rounds were penetrating on a regular basis, especially through the turrets, and there is evidence of internal scabs of armour breaking off and injuring the crew. However one major problem highlighted by the battle was that of close support. Since the new 95mm gun was not ready a number of Mark I tanks had been supplied to each regiment for this purpose. However at Steamroller Farm the close support tanks were unable to help because they did not have the range. The 51st RTR also pointed out that with the howitzer low down at the front it was impossible for the tank to go into a hull down position since it masked the gun. They recommended that the guns were swapped over, the 3 inch howitzer being placed in the turret and the 2-pounder in the hull. This was certainly done, but just how many tanks were involved is not clear. The regiment also put in a strong plea for high explosive rounds for their 6-pounder guns, pointing out that the majority of their targets were anti-tank guns or infantry which were difficult to hit with armour piercing rounds. Indeed a question still hangs over the logic of fitting tanks that were intended to support infantry with a gun only suited to combating tanks. Their wish was granted. A note in the war diary, dated 13 April 1943, notes that two lorry loads of 6-pounder HE had arrived so in future all tanks would carry 20 HE and just four High Velocity Armour Piercing in the turntable bins.

B Squadron, meanwhile, had been attached to the French 19 Corps while A and C Squadrons remained in reserve at Djelida for most of March; A Squadron being involved with the recovery of damaged tanks from El Arroussa. At the end of the month A Squadron was involved in one of those events that can irritate fighting troops out of all proportion. They were at Bou Arada on the 29th when a fleet of twenty tank transporters arrived with orders to carry eighteen of their tanks to Gafour, some twenty miles away. They loaded up and set out, but after five miles they came to a bridge which, upon investigation, proved too weak to take tanks. Thus they all unloaded, took the tanks on a route that avoided the bridge and then carried on, still on their tracks, for the remaining fifteen miles to Gafour. The tank transporters turned around and went home. The irritation is self evident from the war diary entry 'no doubt a satisfactory explanation is forthcoming as to why 20 transporters had been despatched from Guardienson, a distance of 70 miles from Gafour, in order to transport 18 tanks a distance of five miles only'. One wonders if it was.

That same month saw B Squadron at Siliana with the French. It was not a happy time. The landscape was bare, virtually devoid of cover, and the best they could do was to scatter the tanks across some farmland in what they christened Prickly Pear Plantation. While they waited for orders officers from the squadron began to reconnoitre the area. The ground generally seemed to be impassable to tanks while the roads were mere tracks, in very poor condition. Liaison with the French appears to have been poor. On 4 March the squadron moved off from Sidi Marchett, followed by their transport with petrol, ammunition and water. Mist was coming down which soon turned to heavy rain and the track, which ran along the edge of the mountain, was very poor. However they decided not to turn back and soon came to regret it. The scout cars found it hard going but the heavy tanks made it even worse. One tank broke a track, the tank behind it could not get past and the remaining sixteen kept going. The journey involved climbing to over 3,000 feet and the average speed was calculated at 3 mph. The transport soon became stuck. Yet the tanks got through in the end and the drivers were congratulated.

On the 9th the tanks were ordered into action. The British officers regarded the French scheme as unworkable and protested. They were joined in this by Captain Leenan of the US Army who was commanding a company of M5 light tanks in

Churchills of 145 RAC in Tunisia. They arrived in an overall coat of green but this regiment broke up the outline of their tanks with a pattern of sand paint. Some sources claim that it was a coating of local mud but it seems too smooth and even for that.

the same force. The squadron war diary paints a confused picture of the next few days. The French General Welvert seemed to be pushing for impossible results which the British and American tank men resisted, along with some of the French officers. Protests were made to higher authority which enraged the General but in the end a plan was made which involved French infantry riding on the Churchills. The action, described as a reconnaissance in force, took place on the 11th, despite a last minute signal from the British General Anderson telling B Squadron not to get involved. All the tanks returned but many minor mechanical troubles were reported, as a result of all the heavy driving of the previous days.

Even so on the very next day the squadron was ordered on another trip of 50 miles. There were two options. Set off at once with the tanks running on their tracks or wait 24 hours for transporters. The British tanks men favoured the latter but General Welvert said he needed the tanks urgently so they got under way. Meanwhile the squadron commander, Major Dow, went on ahead by car to prepare for the tanks. Reporting to General Welvert at Kesra they heard him say that the order to move the tanks was a mistake, and Dow promptly set off to try and turn the tanks back to Siliana.

For the next ten days the weather got steadily worse, the rain got heavier, turned to hail and made life a misery. Attempts to find routes for the tanks, or even better harbouring sites, only went to prove that the whole area was quite unsuitable for tanks which would only make matters worse if they were used. From a report dated 21 March it is clear that B Squadron was also blessed with Rota Trailers. They were a diabolical invention, a two-wheeled unsprung trailer designed to be towed by a tank. The body was a steel box intended to carry ammunition while the wheels were rotating, hollow tanks to carry fuel. On the march, along the wet and slippery track, Churchill drivers found that the heavy trailers would swing off the road, dragging round the back of the tank and forcing it off the road and into a ditch or against a rock. One tank broke a track while doing this. Perhaps the soldiers would have put up with this inconvenience if the Rota Trailers had proved their worth as carriers but they did not. Fuel leaked out from the wheel hubs and even after a short journey the stowed ammunition had been bounced around so much it was unfit for use and where the rain got in, as it did all the time, the rounds became soaked and ruined. The trouble was that even conventional transport could not be used easily to maintain supplies. The war diary records that tracked carriers were constantly in demand to haul the lorries out of trouble.

On 29 March, after yet another terrible journey, the surviving tanks of B Squadron prepared for action in the area Sidi Saidou Said. The attack began in the afternoon but it was a strange one. The French infantry had suffered casualties in the morning and reported nine enemy tanks in the area. Captain Leenan announced that he was unwilling to have his M5 tanks forward in the reconnaissance role so Major Dow agreed to a change. Thirteen Churchill tanks were waiting on the start line with their engines running at 14.30 hours when it was announced that the French infantry were not ready. They finally moved off at 15.10. The Churchills leading the way as if they were light tanks, followed by the French infantry and American M5s in support. There was no opposition and they reached their objective in about ninety minutes. German forces were seen preparing to counter attack but they went away when the French artillery opened fire.

B Squadron remained with the French for most of April, taking part in some vigorous actions, but they were back with the regiment by May and noted a change in enemy morale as more and more prisoners came in to give themselves up.

The third regiment in 25th Tank Brigade was 142 RAC. This somewhat stark title was now borne by the men of 7th Battalion the Suffolk Regiment. In 1941 the War Office had sanctioned the conversion of twenty Territorial Army infantry battalions to tank regiments, followed by a further fourteen in 1942. It was a drastic move, and not universally popular, but it was essential if the Royal Armoured Corps was to keep pace with the increased demand for armour. Many, but by no means all, of these regiments were trained on infantry tanks and ended up with Churchills. No doubt on the basis that these erstwhile infantrymen would have greater empathy with the troops they were supporting. The 142nd landed at Bone on 9 February. One of their first tasks was to replace another tank regiment at Sbiba, which raises an interesting point. The regiment in question was the 16th/5th Lancers which formed part of the 6th Armoured Division. It had arrived in Tunisia with a mixed establishment of Crusaders and Valentines but was now due to convert to Shermans.

As an armoured regiment in an armoured division the 16th/5th had little in common with an infantry tank regiment but it says much about the nature of tank warfare in Tunisia that they could duplicate one another's roles to some extent. Either that or it proves that senior commanders in the theatre were not aware of the difference. In late February the regiment, which had been fighting in support of US 34th Infantry Division, came under command of US II Corps. Later changes saw C Squadron move back under British command and five tanks from B Squadron go to join the French, but the remainder stayed with the Americans. By the end of the month the entire regiment had reverted to British command and was involved in action near Medjez el Bab. Undoubtedly the fighting around Medjez el Bab, which began on 21 April and lasted until the end of the month, was some of the most intense to be recorded in any of the war diaries for this period. Seen from regimental level it was an endless cycle of confusion with tanks attacking ridges in support of infantry and then retiring to leaguer for the night on one hour's notice. Often they got less than that. With German tanks around the infantry were very jumpy and tended to exaggerate enemy strengths. One entry in the regiment's diary mentions a report concerning an incredible number of German tanks which, inevitably, turned out to be a false alarm. But this is only learned after the call for help had gone out and every tanks that could move was taking up a hull down position on some exposed ridge.

As this action came to an end A Squadron was given a new task in the same area. The object was to take some high ground covering an enemy line of advance known as the Gab Gab gap and establish a defensive position for infantry and anti-tank guns. Each tank was to carry rolls of barbed wire and four of them would be towing 6-pounder anti-tank guns. Zero hour was 1700 hours on 5 May and the fourteen Churchills moved off in line ahead with the infantry marching alongside. At the entrance to the valley enemy fire came down and the infantry ducked into a dry water course, or nullah, so that they could advance under cover. The anti-tank gunners, who had been riding on the tanks, jumped down and tried to walk close alongside the Churchills but they provided little protection and before long the gunners had taken fifty per cent casualties. Although a tank path up the hill had been discovered on the previous night it now proved impossible to find again so the tanks began to explore every likely route on a trial and error basis. As night began to fall eleven tanks and two anti-tank guns had struggled onto a plateau near the top of the hill where they formed a defended leaguer for the night. The infantry were unable to follow, having taken too many casualties during the advance, so the tank men provided their own infantry protection during the dark hours.

By dawn they were joined by infantry and the tanks moved up to the crest where they commanded the enemy approach routes. All this took place under constant fire from machine-guns, mortars and *Nebelwerfers* but the tanks at least had the luxury of effective communications and when a column of 24 enemy tanks was reported to be approaching they called down artillery fire which drove them off. More tanks arrived during the day and by a leapfrog process they carried the advance over two further ridges. The action ended at 1900 hours on 6 May when the tanks withdrew. For their final contribution to the campaign 142nd RAC became involved in the surrender negotiations. A patrol of the Royal Sussex Regiment captured a German headquarters and Colonel Robertson, the Commanding Officer of the 142nd, offered the communications facilities in his tank which was netted to British headquarters.

In March 1943 the War Office decided to send a second tank brigade to Tunisia. This was the 21st Army Tank Brigade which sailed as an entity from the Clyde and arrived at Bone on 25 March 1943.

The senior regiment in the brigade was 12th RTR. Their first action was codenamed Operation Vulcan which began on 24 April. It was a combined tank and infantry attack against German positions at Sidi Medienne, Sidi Salem and Sidi Abdallah, all of which were known to be well protected by minefields. Since Sappers could not be found to examine the extent of the minefields in front of Sidi Medienne, which C Squadron was scheduled to attack, the commanding officer was instructed to use his Universal Carriers instead. They had already been loaded with ammunition for the battle so their reconnaissance of the minefields was hastily done and not very well organised, but there was no time to do a more thorough job and the attack went in. The tanks and infantry advanced through a cornfield until two leading Churchills blew up on mines, the next pair moved off to the right to find a way through followed by two more working round to the left. All four tanks ran onto mines. Without tank support the infantry had a difficult time and by dusk had run out of ammunition, but the Germans still held their positions. As the infantry withdrew they passed the six wrecked tanks, all with broken tracks, but with their crews still in position and capable of fighting. Fearing a German counter attack a decision was made to withdraw the tank crews overnight but there was no attack. Instead the Germans moved into the minefield and thoroughly destroyed all the damaged tanks.

On the next day six Scorpion flail tanks were brought up to deal with the minefield but, due to a mix up over signalling, a troop of Churchills from B Squadron also ran onto a mines with the inevitable result. This time, however, infantry went forward to cover the crews while they repaired their tanks, but the infantry pulled back before the work was completed and once they had gone the Germans put in a counter attack. On this occasion the Churchill crews stayed with their tanks and fought back so they were able to complete repairs in the morning and get away.

On the third day A Squadron took over and they managed to avoid the minefields. Instead their infantry left them and the tanks were decimated by anti-tank gunfire. These were joined by mortars and dive bombers until nine of the squadron's tanks had been knocked out. On the final day B and C Squadrons took over again. They had more luck and managed to get the weary infantry onto Sidi Abdallah but it had been a hard fight marred by bad planning and poor cooperation between tanks, infantry and artillery.

The 48th Royal Tank Regiment was the second senior regiment in 21st Army Tank Brigade. Its first major action involved A and B Squadrons on 21 April 1943 near a location christened with the homely British name of Handley Cross. The objective was a hill feature with two peaks named Jebel el Mehirigar and Jebel Jaffa, and 48th RTR was to support a battalion of the East Surrey Regiment. The first stage was entirely successful. A Squadron went in with the East Surreys while B Squadron got itself into a hull down position behind a nearby ridge and gave fire support. Stage two involved B Squadron passing through and sending two troops ahead to the next ridge, which is where the excitement began. German tanks were encountered at less than 200 yards range and a fierce battle ensued. Two Churchills were knocked out, one commanded by Lieutenant Peter Gudgin taking an 88mm round through the hull machine-gun position which ended up in the engine compartment. When one considers how crowded it is with five men inside a Churchill hull it is quite astounding to realise that all five men were able to evacuate the tank after it was hit. Another Churchill was struck in the turret by a Tiger at about 150 yards range, whereupon it caught fire. The crew prudently baled out but two of them then went back and tried to fight the fire with extinguishers. Soon the fumes

British troops gather around the Tiger knocked out by 48th RTR on the Djebel Djaffa. Whether any of the Churchills in the background are its victims is unclear.

Still attracting the curious the abandoned Tiger sits on the hillside while a Churchill thunders by in the background, accompanied by a Daimler Dingo scout car.

got so bad that they had to give up but the officer, Lieutenant Harvey, dropped into the driver's seat, released the hand brake and let the tank roll backwards behind the ridge. They were fearful of starting the engine in case it aggravated the fire but a short time later the driver and gunner requested permission to go back to their tank. This being granted they climbed in, started up and drove it safely out of action. Inspecting the battle site the next day the tank men counted three Panzer III, one Panzer IV and a Tiger knocked out. Apart from one Panzer III which was burned out the others showed very little sign of damage and the Tiger, which was little more than scratched, was later returned to Britain for evaluation and now resides in the Tank Museum.

As an example of the strong construction of the Churchill the 48th RTR war diary contains a brief summary of six damaged Churchills recovered on 26/27 April, following the action described above. All were apparently Mark IV variants and most had mine damage with broken tracks and suspension units. However two showed signs of being hit by enemy weapons and in one case the lower part of the hull was bulged inwards under the driver's seat by a double mine explosion. However this and three of the other tanks were all driven away after their tracks were fixed. Armoured Recovery Vehicles being used only for the other two.

Although it has nothing to do with Churchill tanks it would be a pity to ignore a small action involving the regiment's reconnaissance troop in Universal Carriers which took place as late as the 11 May. Reports had been received of an occupied farm from which a German soldier would emerge with a white flag, but from which fire was opened whenever British troops approached. The three Carriers got as far as the farm's driveway when the surrendering soldier appeared, claiming to be alone. This did not convince anyone so two of the Carriers mounted an attack, Churchill tank style. This resulted in the surrender of nine Germans, including a General; they attacked again and another 31 appeared. The next attack, delivered as the light was fading, netted a further 144 German troops along with the senior German naval officer in Tunisia. Still not satisfied the tank men kept the farm staked out until one o'clock next morning, by which time they had added another 30 to their bag. Happy that the place was now unoccupied they withdrew along with their prisoners.

Eighth Battalion the Duke of Wellington's Regiment provided the personnel for 145 Regiment Royal Armoured Corps when they converted to tanks in November 1941. They formed the junior regiment of 21st Army Tank Brigade. Their first taste of action was on the Jebel Jaffa alongside 48th RTR. They claim to have engaged a Tiger and knocked out two Panzer IVs having lost two of their own tanks and one ARV in return. During further fighting on the 23rd, against what the war diary describes as 'considerable opposition', they lost ten tanks with many wounded but just two men killed. They got their first Tiger on Guerriat Ridge on the afternoon of 24 April but the circumstances are not described. In most respects their actions mirror those of the other regiments already mentioned. They were attached to the Guards Brigade for most of the time which can never be regarded as a sinecure, but much of their work seems to have been of a defensive nature. Early in May, with A Squadron 48th Royal Tanks under command they were part of the force covering the Gab Gab gap where a heavy German attack was anticipated. It came late in the afternoon of the 4th with about twenty enemy tanks supported by anti-tank fire. The regiment lost two tanks but in return, assisted by its own anti-tank battery, managed to destroy two Tigers, nine other tanks which might have been Panzer III or IV, and what they described as a mortar carrier.

Two days later, while supporting 5th Indian Brigade, a tank from A Squadron ran onto an abandoned British mine dump and was blown to pieces. The debris damaged two other tanks and the regiment blamed the Royal Engineers for not marking the site. Even so the attack was a success, earning the regiment a special commendation from the divisional commander.

Attempts to reach any conclusions about the Churchill following its brief spell of action in Tunisia were limited by the amount of evidence available. As early as March 1943 the 25th Army Tank Brigade issued a report, based one imagines on very little experience, which nevertheless makes interesting reading. Among the points noted is the comment that not all of the tanks shipped out to Tunisia had been given the full rework treatment, which suggests either that the programme at home was falling behind or those responsible for shipping them were unable to identify all the modifications. Another matter, repeated more than once in the brief report, concerns the protection of stowed ammunition. As early as 1942 trials in the desert had proved what most crews already knew; that most fires in tanks were caused by incoming projectiles igniting the ammunition and not, as most theorists believed, by ignition of fuel tanks. The brigade was now calling for armoured ammunition bins; 4mm thick if made from armour plate or 6mm if mild steel was used.

Another interesting remark concerns the visibility for both driver and front hull gunner. Their inability to see to the sides of the tank left them feeling very isolated and there was an additional disadvantage for those of less than average stature. Short men had such difficulty looking through the periscopes at all that they preferred to drive with their front visors open and at least two tanks had been destroyed by rounds coming in this way. There appeared to be no easy way of curing this except by selecting taller drivers, but the view to the sides could be improved by extending the upper sections of the periscopes. Generally, however, crews were said to have immense confidence in their tanks. When it came to mobility the brigade advised that in order to get the best results from Churchill tanks movement in daylight should not exceed 12 miles in two hours for eight hours, with one hour's maintenance every four hours. This gave a figure of 42 miles per day, but the report noted that on one occasion nine tanks had moved 100 miles in 24 hours and were still perfectly serviceable at the end of the run, although their total lives had been shortened. A study of all three regiments in the brigade, up to the date of the report, revealed that most of the tanks had completed up to 400 miles. On average each tank had completed 340 miles and one belonging to 51st RTR had somehow managed to travel between 700 and 800 miles on its tracks.

On the gunnery side a case was made for more close support tanks. The argument was that, if an enemy tank was knocked out with an armour piercing shot it was nice to have a high explosive round ready to kill the crew when they baled out. It was recognised that the location of the 3 inch howitzer, in the front of the hull, was not ideal and the idea of exchanging it with the 2-pounder was mentioned again. However it was also suggested that the howitzer might be mounted in the hull of some 6-pounder tanks, accepting a reduction in 6-pounder ammunition stowage in the process. Where the 6-pounder was concerned crews agreed that geared elevation (or fixed elevation to use the contemporary term) was preferable to free elevation but it was noted that the elevating gear was badly located. It was also suggested that the turret BESA machine-gun should be transferred to the right side of the main gun. This would not only make the feed system easier to operate but relieved the gunner of any responsibility for attending to the machine-gun, which might distract him from his telescope. Of the telescope itself, the Number 39, it was agreed that magnification was insufficient for long range shooting and the cross wires were so thick that they totally obscured the target at 1,000 yards or above.

The different types of track were discussed, suggesting that all three types were in use in Tunisia. The light cast type was least popular but of the other two, the heavy cast and manganese, the latter was said to give better running and resistance to enemy fire but it had the unpleasant habit of breaking idler or sprocket teeth and jumping off altogether. This was particularly noted when the tanks were working over rocky ground. But mud was the biggest problem of all, and Tunisia was notoriously muddy. It worked its way under the track guards and, if left to harden, caused a lot of damage. The brigade report included a sketch showing their solutions; extra slots above the side doors and a hinged section, three feet long, either side of the turret which could be swung open for cleaning.

Finally there were some interesting comments on armour. Protection was required for the turret ring, something which British tanks were notorious for, but the most interesting thing is that the fabricated turret (the welded version of the Mark III) was regarded as tougher than the cast type. Most of the hatches were in need of a redesign to prevent small arms rounds from being deflected through the hinges and there were problems with the side doors. These were tapered where

Although photographed in Britain this appears to be an example of the conversion first mentioned in Tunisia. A Mark I in which the 2-pounder and 3 inch howitzer have exchanged places; the so-called Churchill Mark I Close Support.

they fitted into the hull but the brigade discovered that when a door was struck it was pushed in and wedged itself shut so that the crew could not open it to escape. They suggested an internal rim, such as they had noted on German tanks, which should prevent this.

In September 1943 the School of Tank Technology produced a summary based on a survey of battle damage. A total of 41 damaged tanks were considered, of which 20 were victims of mine damage. Basing their study on strikes from projectiles the STT report recorded 81 hits in all, effective and ineffective. More than 25 per cent of these came from 88mm weapons with 75mm the next most common. Of the 21 tanks not mined, 10 were definitely taken out by 88mm shots but whether from tanks or anti-tank weapons was obviously impossible to say. Part of the survey, which the report admitted was of dubious accuracy, claimed that of 20 strikes attributed to the 88mm, only 13 were effective while the 75mm scored 8 effectives from 16 strikes. They regarded this as encouraging, pointing out that it proved the Churchill's armour was not yet outclassed by the latest projectiles. They calculated that an extra inch of armour would make a great deal of difference, but that would depend upon where it was placed. The 25th Tank Brigades also records the loss of one tank, T68384 of 51st RTR, which was attacked by a dive bomber. One bomb broke the nearside track while the other tore off the engine cover and shattered it into three or four pieces before smashing in part of the engine. This in turn started a petrol fire which burned the tank out. It would seem difficult enough for a dive bomber to get one accurate hit upon a tank. Two would have to be quite remarkable.

Looking next at where the projectiles hit the report revealed that it was shared equally between the front, sides and all other areas. The area that concerned them most was the side. They estimated (why they could not be sure is unclear) that the sides of a Churchill were about .75 inch (19mm) thinner than the front and they believed that the addition of an extra 18mm of armour all round could have halved the casualties from shell fire. However the ratio of effective hits on the sides of tanks was far greater than those on the front and this gave rise to further speculation. There is no mention of the fact that the side presents a much larger target but the report concludes that, since the sides were vulnerable, it would make little difference if they were a bit thinner. Weight saved here could then be applied as extra frontal armour. These were only tentative conclusions based on those tanks that been knocked out. The report recognised that other tanks must have been hit without being knocked out and without details of them the sample was not truly balanced. Whether it would affect future designs remained to be seen.

9

The New Model Churchill

Whatever rigours the British Army and its Allies suffered during the Libyan and Tunisian campaigns they pale into relative insignificance when compared with what was going on in Russia. When Germany invaded the Soviet Union in June 1940 the Red Army had something in the region of 24,000 tanks. Not all of them could be mobilised at once but even so they outnumbered those of the invaders on a prodigious scale. Despite these odds the Germans began to decimate the Soviet armour and, in a matter of months, had virtually annihilated it. By December 1941 the Red Army was down to less than 5,000 tanks, nearly half of which were on their eastern borders and unavailable to meet the attack. Worse still the rapid German advance threatened to over run all the main Soviet centres of production so the government took the drastic step of evacuating the majority of their manufacturing plants well to the east, in the Urals.

The hiatus in production created by this massive industrial migration was made good, to a larger extent than the Russians were ever happy to admit, by supplies of Lend-Lease tanks from Britain, Canada and the United States. Shipments of British tanks, Matildas and Valentines to begin with, started in October 1941, the majority travelling in the gruelling Russian convoys. The arrangements made with Russia under what was known as the Protocol Commitment called for the delivery of 250 tanks per month but this required convoys to continue through the summer months when perpetual daylight rendered them particularly vulnerable. Thus it was one thing to promise tanks but quite another to deliver them. Convoy PQ16, which sailed from Britain at the end of May 1942 carried 468 tanks but, under constant attack from enemy aircraft and U-Boats seven merchant ships were lost, taking with them to the bottom 147 tanks. Worse was to follow. Convoy PQ17 which sailed on 27 June, comprised 36 merchant ships, carrying between them 594 tanks and other essential items. Hounded continually by aircraft and U-Boats the convoy and its escorts put up a stout fight until, on the evening of 4 July its escorting cruisers were withdrawn and the ships were ordered to scatter.

Naval intelligence in London, calculating from various reports, concluded that the battleship *Tirpitz,* and the so-called pocket battleship *Admiral Scheer* were preparing to attack the convoy. The order to scatter was standard procedure in such circumstances but it immediately left the merchantmen easy prey to air and submarine attack. PQ17 was devastated, only eleven of its ships finally made it to Russia where they delivered just 164 tanks. It is not clear how many of these were Churchills. Statistics merely show that up to and including PQ17 127 Churchill tanks had been shipped from Britain. All of those sent are known to have been of the 6-pounder type so these would almost certainly have been Mark IIIs. Russian records suggest that they received 301 Churchills, Marks III and IV. Forecasts issued in Britain suggest that the total, up to June 1943, should have been nearly 500 Churchills in all, but military forecasts are notoriously optimistic.

It may be that some 200 Churchills ended up on the bottom of the Barents Sea but it is more likely that the Russians simply stopped asking for them. They regarded the Churchill as a heavy tank like their own KV-1, but the two types could hardly be compared. The Soviet tank outclassed its British counterpart in almost every respect but above all else it was designed for Russian conditions. Soviet manned Churchills are know to have taken part in some of the great tank battles around Kursk in the summer of 1943 but details of their use are not readily available. There are good reasons to suppose that some of the Soviet Churchills fell into German hands but few to suggest that they were employed operationally by the Wermacht. Invariably, when a captured vehicle was adopted by the Germans it was issued with a new identification, and no such details are known for the Churchill.

Generally it seems to have been Russian policy to operate Western tanks on their southern fronts where winter conditions were not so harsh. The Churchill, having been designed for service in a temperate climate, later proved itself in much warmer climes but, well before that, two Churchills had been tested in very cold conditions. Between 2 March and 2 April 1942 T31108, a Mark I and T31113 a Mark II, both built by Vauxhall Motors, were tested in Kapuskasing, Ontario, Canada in very deep snow. The Mark I had the later pattern air cleaners and full track guards while the Mark II was in original condition. For two days, up to 11 March, both tanks lay out on the prairie while a blizzard raged and there was little to show but their turrets when crew members skied out to inspect them. Even so the report suggests that temperatures rarely dropped below freezing at this time so the greatest problem was snow. As with sand in the desert, snow seemed to get in everywhere. Both tanks suffered track breakages but the Mark I was further handicapped by the track guards. Snow packed in underneath and ended up bulging the tinwork which was later removed. Canadian built Ram and Valentine tanks also took part in the trials and if the photographs are any guide the Churchills seem to have spent much of their time dragging the other tanks out of ditches and snow drifts. Yet the impression given is that the trials were inconclusive and it is worth noting that this all took place some months before the Canadians took Churchills into action at Dieppe.

After a night on the Canadian prairies the first trick was to find your tank, then dig it out and prepare for the day's trials.

One result of these trials was a booklet entitled *Instructions to guide the preparation of Churchill III, IV, V, VI, VII and VIII for sub-zero operation (to minus 15 deg. F)* the first edition of which appeared in February 1944. Most of the precautions described involve the use of special covers and anti-freeze grease but there are one or two particular points to note. For instance the compressed air system had to be disconnected and certain parts removed, otherwise condensed moisture could freeze in the pipes and damage the equipment. Special batteries were provided and, of course, the engine cooling system had to be drained when the tank was not in use. At times, when a tank had to be left standing for long periods, special catalytic heaters were provided. These would be placed inside and then the tank would be covered completely by a huge sheet of canvas material, draped loosely over the entire vehicle. This would maintain a temperature of at least zero degrees but crews were advised not to climb inside unless it was essential because the fumes could kill.

Towards the end of 1940, Babcock & Wilcox Ltd., of Renfrew considered the construction of a welded hull for the Churchill tank. This caused some interest at the Ministry of Supply and an order for one, placed with the Scottish company in November 1940, was followed by a second in Janu-

Many pictures from the cold weather trials in Canada show the Churchills rescuing other vehicles. This is a case in point.

ary 1941. Work on the hulls coincided with the programme to produced welded turrets for the 6-pounder gun and this enabled the firm to solve some of the anticipated problems on a smaller structure. Welding armour was regarded with a good deal of suspicion in Britain, despite the fact that much experimental work had been done at Woolwich Arsenal before the war. One fear was that heating armour to effect the weld could spoil the integrity of the plate. Another suggested that the weld itself would be vulnerable to direct hits from anti-tank rounds. Many British companies engaged in tank production emphasised these fears and also pointed to the great shortage of skilled welders since they employed traditional riveting techniques in most of their plants.

The attractions of welding were many. It saved production time and weight. It ensured a watertight hull and did not require such accurate preparation of plate. But concern about the degradation of armour continued so Babcock & Wilcox initiated a procedure known as postheating. This involved giving the assembled hull a second application of the heat treatment after welding, but there were those who expressed doubts about this. In order to settle the matter some turrets were given the postheating treatment and others were not. Firing trials suggested that postheating made no difference and there was even some evidence to suggest that it did indeed cause more harm than good. The trials were therefore extended to the experimental welded hulls. The first one was given the treatment and the second was not. Both hulls were then completed as tanks and subject to a 1,500 mile running test. When this was completed, in July 1942, the unpostheated tank was stripped down and the hull placed out on the ranges to be shot at. It survived remarkably well and took the punishment even better than a comparable riveted hull, so plans were made to change over to welding entirely.

The results were studied by the Welding and Gas Cutting Branch of the Department of Tank Design who announced that welding would save about 20 per cent in man-hours, 4 per cent in weight but require at least 50 per cent more floor space for assembly. This was balanced out to some extent by a reduction in the demand for space where the plate was cut, because it did not now have to be drilled to anything like the same extent. Since it would create a good deal of upheaval in production there was no immediate attempt to change over to welding and indeed. when it was introduced many other changes were incorporated.

On 6 April 1943 the Secretary of State for War and the Minister of Supply issued a Joint Memorandum on the subject of Tank Supply Policy for consideration by the Defence Committee of the War Cabinet. It contained a telling sentence. "A substantial proportion of the British 1943 tank requirement will be met, provided shipping is available, from assignments of American tanks". Whatever this implied for the future of British tank production it did at least mean that, to some extent, the pressure was off. Since the majority of tanks on order from the United States would be Shermans this would go a long way to correcting the imbalance between infantry and cruiser tank production which had proved impossible to remedy since the preference in favour of cruisers had first been announced in 1940. Things might have been different if Vauxhall Motors had shown more enthusiasm for the proposed changeover from Churchill to Cromwell production but they fought a long, hard rearguard action in respect of the Churchill, and they were about to win.

While crew members repair a track on their Mark III one can study the panels of appliqué armour fitted to the sides of the hull and the side door.

Looking at the current production programme for 1943 the Joint Memorandum noted that 1,300 Churchills remained to be built from the agreed final total of 4,000 and that these should all be completed by early 1944, when production would cease. Yet, the report noted, "as a result of continuous development, the Churchill type has been greatly improved in reliability and fighting performance". It went on to say that in a recent telegram to the Prime Minister, General Alexander had emphasised the value of the Churchill in the fighting in Tunisia. "Its relatively heavy armour enables it to discharge tasks for which no other British tank at present seems equally fitted". The report went on to recommend that Churchill production should continue, at a reduced rate, throughout 1944 and that a further 1,000 should be ordered, bringing the total up to 5,000 in all. A footnote to the report includes a suggestion for the improvement of reliability and, if possible, armour protection for the Churchill and it concludes; "The War Defence Committee is invited to endorse the policy as stated above". It must have been music to the Prime Minister's ears.

Despite the comforting assurance of Shermans from the United States the War Cabinet Defence Committee made it clear that they were not prepared to rely entirely on America for tanks. It was, to some extent, a matter of national pride, although there were industrial implications to consider as well. Thus the Defence Committee announced that Cromwell would continue to be developed, as would the Churchill 'for the time being'. The implication was that something better should be produced and the Defence Committee made it clear that they would be satisfied with a much more heavily armoured tank. Indeed they said that speed might be sacrificed to increased protection and that they would not mind if the speed came down to as low as 6 mph in such a case. This was only marginally better than British tanks of the First World War. In the meantime it was agreed that between 200 and 400 Churchill tanks should be up-armoured with as much extra plate as possible and the Ministry of Supply was invited to comment on this. They responded on 14 May with the rather disheartening information that it would be impossible to increase the turret armour of existing tanks, because the turret ring would not accept the weight. And they estimated that no more than half a ton of extra armour should be applied to the hull because the suspension would not take it. This modest increase would bring the top speed down to 11 mph. In June the Department of Tank Design produced a drawing to show how the extra armour should be fitted to a Churchill Mark III but went on to point out that this extra modification could not be done as part of the rework programme. If it was done at all it would have to be done in the field.

We have already noted comments from Churchill users in Tunisia concerning the need for high explosive ammunition but it was not a new demand. Unfortunately the HE round supplied for the 6-pounder gun was not capable of holding a significant explosive charge. Its destructive effect was hardly worth the effort of firing it and it compared unfavourably with the HE round fired by the American 75mm gun in the Grant or Sherman. This failing was felt strongly by Churchill crews in North Africa and one result of this need will be studied later.

Meanwhile in Britain others were thinking along the same lines. The general background to subsequent developments has nothing specifically to do with the Churchill, and it is covered elsewhere*, but the ultimate result affected the Churchill in a major way. A report, delivered to the Tank Board in December 1942, revealed that Vickers-Armstrong was investigating the possibility of adapting the 6-pounder to fire 75mm ammunition. This involved enlarging and relining the barrel and altering the breech end to accept American 75mm rounds. The result could hardly be regarded as a major advance in the science of tank gunnery but it had two distinct advan-

* Fletcher, *The Universal Tank,* HMSO London 1993.

Churchills on exercise, probably with 9th RTR. A Loyd Carrier is bringing up the rear.

***Bradford*, a well stowed Mark IV taking part in a winter exercise. Two locals have come along to see the mess the tank is making of the road and their fence.**

tages. It put British tanks on a par with their American counterparts, especially with the ability to fire a respectable high explosive round, and it meant that any tank (or armoured car come to that) which could take the 6-pounder could now be adapted to take the 75mm weapon; and this adaptation could be achieved by doing little more than changing the gun.

Of course there was more to it than that, the new gun would require new sights to match and the internal ammunition stowage would also have to be altered, but the main problem was elevation. The modified gun had a greater effective range than the 6-pounder when it came to firing high explosive shells and it was no longer possible to trust to free elevation and the gunner's skill for accuracy. It was now imperative that some form of geared elevation should be introduced and, since Vickers-Armstrong developed the revised gun and its mounting this became their problem. However it also poses a problem for anyone attempting to produce an accurate history of the Churchill tank.

The problem is this. Did the 6-pounder Churchill have geared elevation or not? It seems to be a simple enough question but one is now forced to rely, almost entirely, on written evidence and that is incredibly contradictory. It would be tedious to review the subject in detail, just to prove that the researcher had done his homework, but the following might at least provide some insight into the problem. A diagram from the Illustrated Parts List for Churchill tanks Mark III and IV, published in July 1942 (long before the 75mm weapon was considered) clearly shows what is described as the elevating and firing gear in which a handwheel operates a threaded shaft within a sleeve to activate the elevation while the handle of the wheel includes a firing trigger. Next a handbook on 6-pounder co-axial mountings, published in January 1943 quotes the Churchill III and IV as having 'mounting 6-pr and BESA Med. M.G. No. 1 Mk I or MkII' and states unequivocally *these mountings are of the geared elevation type*.

A report from Lulworth, dated 10 February 1942, throws a bit more light on the subject. It concerns firing trials of a Mark III Churchill and refers to elevating gear of Mark I and Mark II types. The former, which was designed originally for the 6-pounder Crusader tank and AEC armoured car, is given a very poor report and even the improved Mark II does not seem much better. The pistol grip and trigger are so difficult to handle that it is impossible for the gunner to fire accurately when the tank is moving or even when tracking a moving target from a stationary tank. The lack of a shoulder piece is also said to make the gunner's task more difficult.

In January 1944 Vauxhall Motors published a heavyweight tome entitled *Weight Analysis of the Churchill Tank* which covers all types of Churchill from Mark IV to Mark VII and illustrates each version with a beautifully executed side elevation section drawing. That for the 6-pounder Mark IV shows, quite distinctly, a horseshoe shaped shoulder piece which clearly indicates free elevation, along with a foot firing pedal which again is typical of this system. A Service Instruction Book for the Churchill III and IV, published in October 1944 refers to four mountings for the 6-pounder and 75mm gun which, it claims, divide into two types; those that have cable firing gear and those with electrical firing gear with an auxiliary foot firing pedal, but it then goes on to state *the mountings are of the "free" elevation type, that is they are elevated or depressed by the gunner exerting pressure on a shoulderpiece fitted to the mounting,* and again this appears to be quite unequivocal.

Reference to the previous chapter will reveal that reports from Tunisia already mentioned geared elevation on 6-pounder Churchills and the only logical conclusion is that both types existed, but if that were the case one would expect it to be mentioned in contemporary publications, especially service handbooks and parts lists. The closest one can find dates from September 1945 in the form of the First edition of a handbook for the Churchills Mark IX, X and XI. These will be treated in more detail later but suffice it to say for now that the Mark IX would carry a 6-pounder, the Mark X a 75mm and the Mark XI the 95mm howitzer. This handbook has nothing specific to say concerning elevation systems at all, except as an aside when discussing the construction of the 6-pounder gun. In describing the muzzle counterweight the book notes that there are different styles for those guns that have geared elevation and those with free elevation. It is a small enough matter but a curious one, made the more so by the obvious confusion in contemporary publications and the almost total lack of an explanation.

Unfortunately one cannot quite leave it there. As already explained, when Vickers-Armstrong developed the new gun and mounting they introduced a geared elevation system of their own design. This was subsequently the subject of much criticism for its poor quality. It was claimed that the only way to ensure accuracy was to wind the gun up to the desired angle and then turn the handwheel backwards in order to take up the free play which would otherwise cause the gun to jump, on firing, and miss the target. Yet we have seen from the above that a form of geared elevation already existed for the 6- pounder mounting as early as July 1942 and one has to assume that this was a more reliable product than the Vickers version. If so, why was it not adapted to the new gun instead?

A crewman demonstrates how one man can lift the engine out of his tank with the aid of an Atherton jack, fitted to the turret.

Churchill tanks of an unidentified regiment, supported by a platoon of Carriers and, for some reason some motorcycle despatch riders. In the distance another regiment from the brigade awaits orders to move. It is very unusual to see a complete regiment of Churchills in one photograph.

The history of the adoption of the British 75mm is another lengthy saga of muddle and misunderstanding, some of which has already been noted. In February 1943 the Armoured Fighting Vehicle Liaison Committee announced that the priority order for converting tanks to the new gun was Churchill first, then Cromwell, then Centaur, and finally Valentine. On the 23rd of that month the Chairman of Armoured Fighting Vehicles at the Ministry of Supply announced that the 75mm mounting had passed endurance tests and the Liaison Committee stated that the 75mm Churchill with geared elevation would be ready to enter production by August, subject to the number of 6-pounder tanks still required. The Committee also agreed to increase the number of Close Support (95mm) Churchills from 250 to 300. It was at this point that the Prime Minister elected to interfere once again in the fortunes of his namesake.

On 14 May he decided to comment on the Joint Memorandum referred to earlier. He concurred with their views on up-armouring, even at the expense of speed, but had his own ideas on armament and in this instance he was commenting on tanks in general. Believing, as he did, that a new, high velocity 75mm gun was still a possibility he questioned the widespread adoption of the medium velocity weapon, the Vickers-Armstrong conversion. In his view, pending availability of the HV 75mm, only 30 per cent of production tanks should have the MV 75mm; the proportion of close support tanks with the 95mm howitzer should be increased from 10 to 20 per cent and the remaining 50 per cent should retain their 6-pounder guns. The War Office bit back sharply. In a note destined to be read before the Defence Committee on 15 June 1943 they state;

Tank armament policy cannot be decided on purely arbitrary percentages ... but must, in the end, be governed by an organisation, within the unit or formation, at once tactically sound and administratively practicable.

Tank policy, here and in America, is clearly directed to the production - apart from a relatively small number of specialised equipments - of standard types of medium tanks mounting, as main armament a developed high-velocity 3 inch gun, effective both as an AP and HE weapon ...

... which was as good as telling the Prime Minister to shut up. The War Office was quite prepared to admit that percentage figures were necessary for production planning, with the unstated proviso that it was up to them to produce such figures. They then went on to expand upon the subject.

They noted that, during a recent visit to the Mediterranean theatre of war, the Deputy Chief of the Imperial General Staff encountered a remarkable degree of unanimity among tank users that what they required as an all-purpose tank with dual-purpose gun. Achieving this, they explained, was a long term policy which could not become fully effective before the second half of 1944. In the meantime they would have to make the best use of what was available. Of Churchills all they said was that there should be sufficient to equip all Army Tank Brigades, which was probably just as well because they finally stated that its predecessor, the Valentine, was no longer acceptable in the field. On the subject of guns they explained that British tank crews were now so convinced of the superiority of the 75mm gun in the Sherman that they would accept nothing less for their own tanks, but the War Office experts had other ideas for the Churchill.

A Churchill Mark III*, fitted with the 75mm gun, possibly photographed in France.

To this extent they appear to be agreeing with the Prime Minister when they say that, in tank battalions of army tank brigades equipped with Churchills, the percentage of 6-pounder tanks might be higher, up to 50 or even 66 per cent. The reason for this, as explained in their note, provides a useful insight into the whole subject of infantry tank armament. Their reasoning is the old one, that infantry tanks work, primarily, with unarmoured troops who might be defenceless against enemy tanks. Thus it would be important for them to be supported by tanks which could drive enemy armour off until the infantry had their own anti-tank defences consolidated. This goes right back to the reason for equipping the Matilda with a 2-pounder gun and one might have thought that there was enough cumulative experience from the last three years to show that what the infantry really needed was a tank with a good high explosive firing capability. Yet on this point the War Office is quite firm;

There is no evidence at present that more than the normal proportion (about 10 per cent of unit equipment) of 95mm howitzer armed close support tanks are required.

In its final comments the War Office report states that from September 1943 all Churchill gun tanks will come off the production line equipped with a mounting that would take either the 6-pounder or 75mm gun, and with ammunition bins adjustable to take either type of round. In the case of the

A Churchill Mark IV with 75mm gun or a Mark VI. Normally one would expect the latter to carry an All Round Vision Cupola. The marking on the turret side, a black cannon on a yellow and red diamond, represents the Gunnery School at Lulworth Camp, Dorset.

95mm it had been found impossible to make the guns interchangeable so close support tanks would have to be built specifically as such. Any attempt to rework 6-pounder/75mm gun tanks to 95mm would have to be undertaken by the manufacturer.

In April 1943 a pilot model Churchill Mark IV with the 95mm howitzer had been tested at Lulworth where problems had been identified with the traversing gear, depression limits and sighting arrangements. Shortly afterwards the Defence Committee said that they would like to review the relative merits of the 75mm and 95mm guns as dual purpose weapons and asked for a demonstration to be laid on at Lulworth. They were told that the 95mm armed tank was not in a fit state to take part in such a show but it is interesting to see that the 95mm howitzer was even being considered as anything but a high explosive firing gun at this stage. A high explosive anti-tank (HEAT) round was certainly developed for the gun, although one authoritative source suggests that it was not very effective and probably never issued for service. Nevertheless some official figures claim that this could penetrate 110mm of armour at ranges up to 2,500 yards which was better than the British or American 75mm guns could do at any range, and even better than the 6-pounder, firing APDS, at anything over 1,000 yards.

Returning to the war diaries one can observe more regiments as they come to terms with their new tanks. The old 10th Battalion, the King's Own Royal Regiment, now serving in Churchills as 151 Regiment, Royal Armoured Corps was stationed at Prudhoe on Tyne in February 1942 when its first Churchill arrived. But they did not follow in a rush. They had twelve tanks by the end of March and just one more in April. At the end of May they supplied two tanks for a trial on the River Tyne where they were ferried across on a raft composed of pontoons supporting a length of Bailey Bridge, a system that would be employed later on the mighty River Rhine.

The regiment had 21 tanks by the end of that month but they make no mention of marks until the 21st tank is delivered, it being described as fitted with a 6-pounder, as is the 22nd. However when their next tank arrives, on 28 June, it is described as a Mark II converted to Mark III which is an odd expression since it would have been fitted with an entirely different turret and internal stowage. It is difficult to understand how a regiment could know that a tank had changed mark unless it was one of their original stock, sent back for rework. However there is no evidence at all of this practice. Normally tanks sent away for rework were simply replaced by new machines.

By September 1942 151RAC had moved to Westgate-on-Sea near Margate in Kent and by the end of that month it had 45 Churchills. The last delivery consisted of two Mark I, six Mark II and one Mark III. By December the regiment had moved around the south coast to Worthing in Sussex where it took part in an exercise code-named Advent III. It is fascinating to note how the War Office conjured up names for its many exercises. To call one Advent around Christmas might reflect the spirit of the festive season but it hardly suggests an attempt to maintain secrecy. In March the regiment took part in the massive, twelve day exercise Spartan which occupied a large portion of the British Army in south east England that spring. It was all part of the long term build-up for the forthcoming invasion of France. Further evidence of this is recorded at the end of May when the regiment's REME officer, along with a regimental officer and seven men travel to

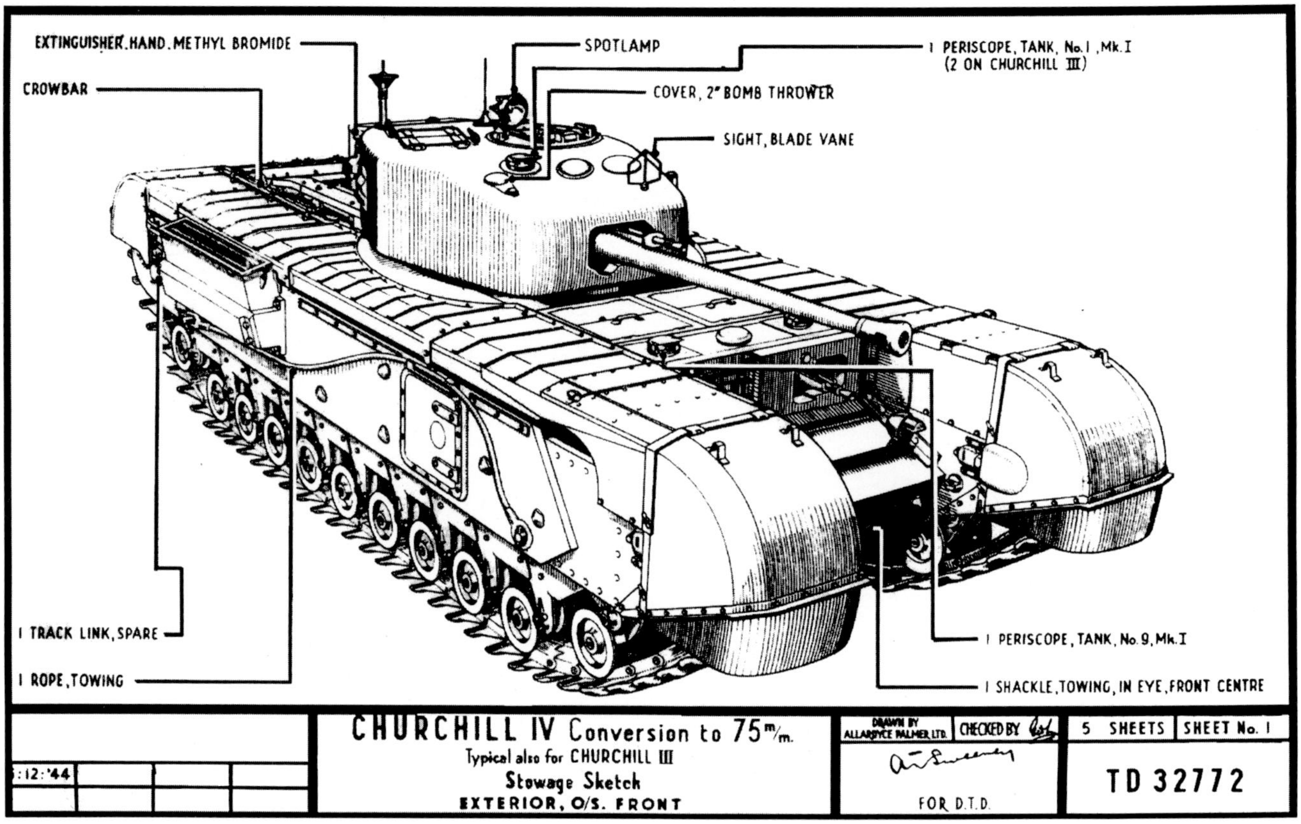

Stowage diagrams for the Churchill Mark IV with 75mm gun.

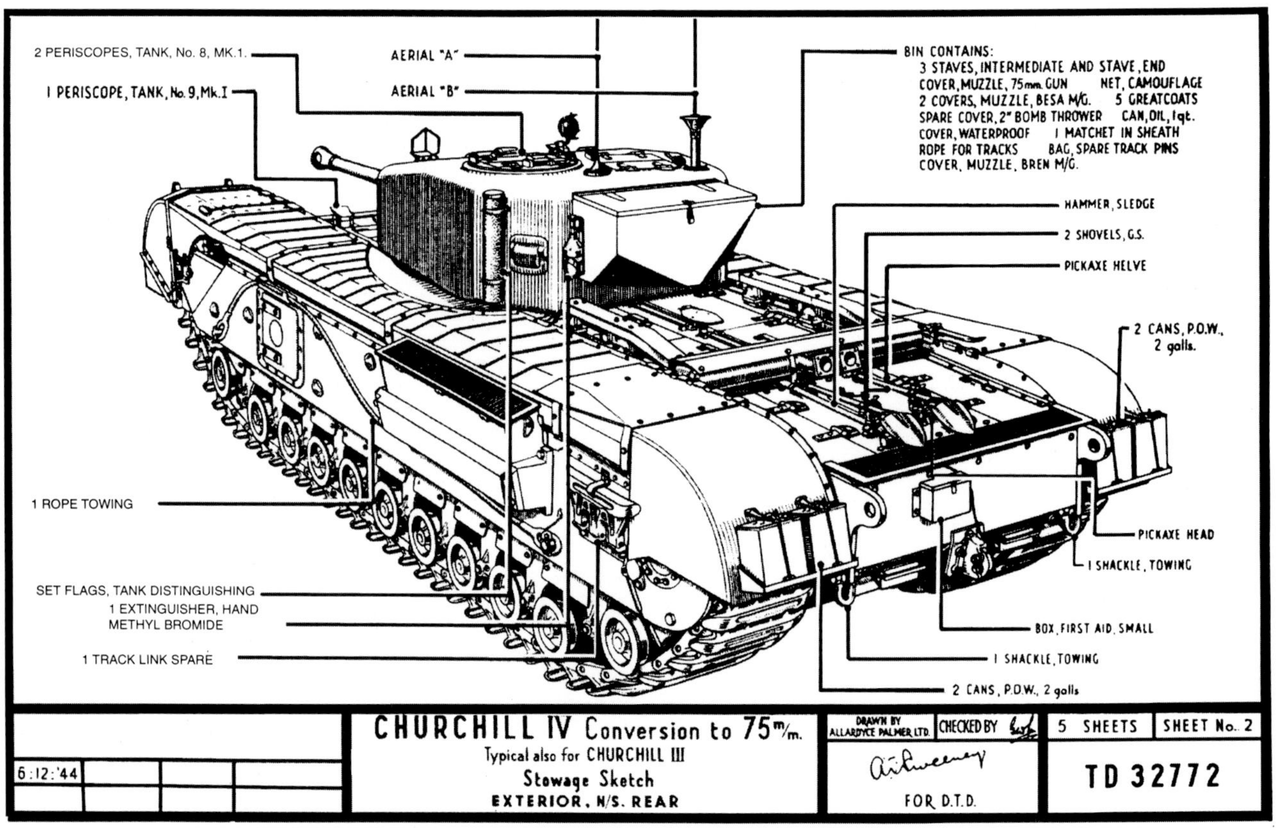

A cross section drawing of a Mark VI Churchill showing the 75mm gun mounting.

Burry Port in Carmarthenshire, Wales where they assist in experimental trials on the waterproofing of Churchills.

Their next exercise, codenamed Harlequin took them to Otterden Park, a large private estate near Charing in Kent and ended with a most unusual expedition. According to the war diary, the regiment was ordered to move from Otterden to Broome Park, another stately pile in the area which had once been the home of Lord Kitchener. Measured as the crow flies Broome Park is about fifteen miles from Otterden and even a Churchill could have covered the distance comfortably in three hours, barring breakdowns. Yet if the war diary is to be believed the tanks were moved by sea. Unfortunately the war diary does not say where they were shipped from, but common sense argues that it must have been somewhere on the north Kent coast. The nearest logical point being Whitstable which, by the direct route, is also about fifteen miles from Otterden. One hopes their point of departure was not Westgate-on-sea since the route there from Otterden would take them via Canterbury, and Broome Park is just outside Canterbury, on the Dover Road. Anyway, according to the war diary parts of the regiment sailed in two shiploads, round to Dover. The journey by sea must be about 50 miles and it would take the ships eastwards, around the North Foreland and down into the English Channel. It was a notorious stretch of water in both wars, threats from mines, U-Boats and German coastal forces are only part of it. The entire route is well within the range of German aircraft operating from airfields in Holland, Belgium or France and for the last few miles into Dover they are also within range of large guns on the French coast which regularly fired at Dover. Just how much of the regiment travelled in this way is not recorded, but assuming the point of the exercise to be a practice in moving tanks then the vessels must have been large merchant ships, rather than landing craft which could only carry five or six tanks each at best.

By November the regiment was back on the coast at Folkestone where they participated in Exercise Canute II which, one might suspect, had a marine theme. In fact it is described in the war diary as an advance and landing upon an imaginary beach. Having survived this and their brief coastal cruise in safety the regiment now comes under fire. On 4 December 1943 the war diary reports that two Universal Carriers, six of the regiment's trucks and two of the despised Rotatrailers were destroyed by long range gun fire, presumably from one of the big railway guns on the French coast which amused themselves by firing random shots into Kent around this time. The year ends with the regiment involved in wading trials and then learning that it is to be disbanded; its fate is now tied up with that of 107 Regiment RAC.

In fact 107RAC was another battalion of the King's Own that had converted to tanks in 1941. Its role changed in February 1942 when it was designated an Advanced Training Regiment for Churchill tanks. What this meant in practice was that it took men who had completed individual training at a Royal Armoured Corps training regiment and formed them into tank crews with their own Churchill to maintain and operate. When this stage was complete they would learn to operate with other tanks as a troop, and then the troops amalgamated to undergo squadron training, after which the men would be released to operational regiments. The 107th was based at Leyburn in North Yorkshire, not far from Catterick where our story began. A typical weekly training programme, as published in the war diary, shows how the system worked. There were four training areas around Leyburn; Rigton Moor, Lindley Moor, Stainburn Moor and Denton Moor. And on each weekday one of the three squadrons was on one of the areas doing a specific task. For instance on Thursdays A Squadron was on Denton Moor after a road march where they practised harbouring for the night. Meanwhile B Squadron was on Rigton Moor training on minefield drill and C Squadron, on Stainburn and Rigton Moors learned to throw hand grenades from tanks, and at them. Saturdays was set aside for maintenance and inspection while Sunday was the inevitable day of rest. There were diversions. In June C Squadron, work-

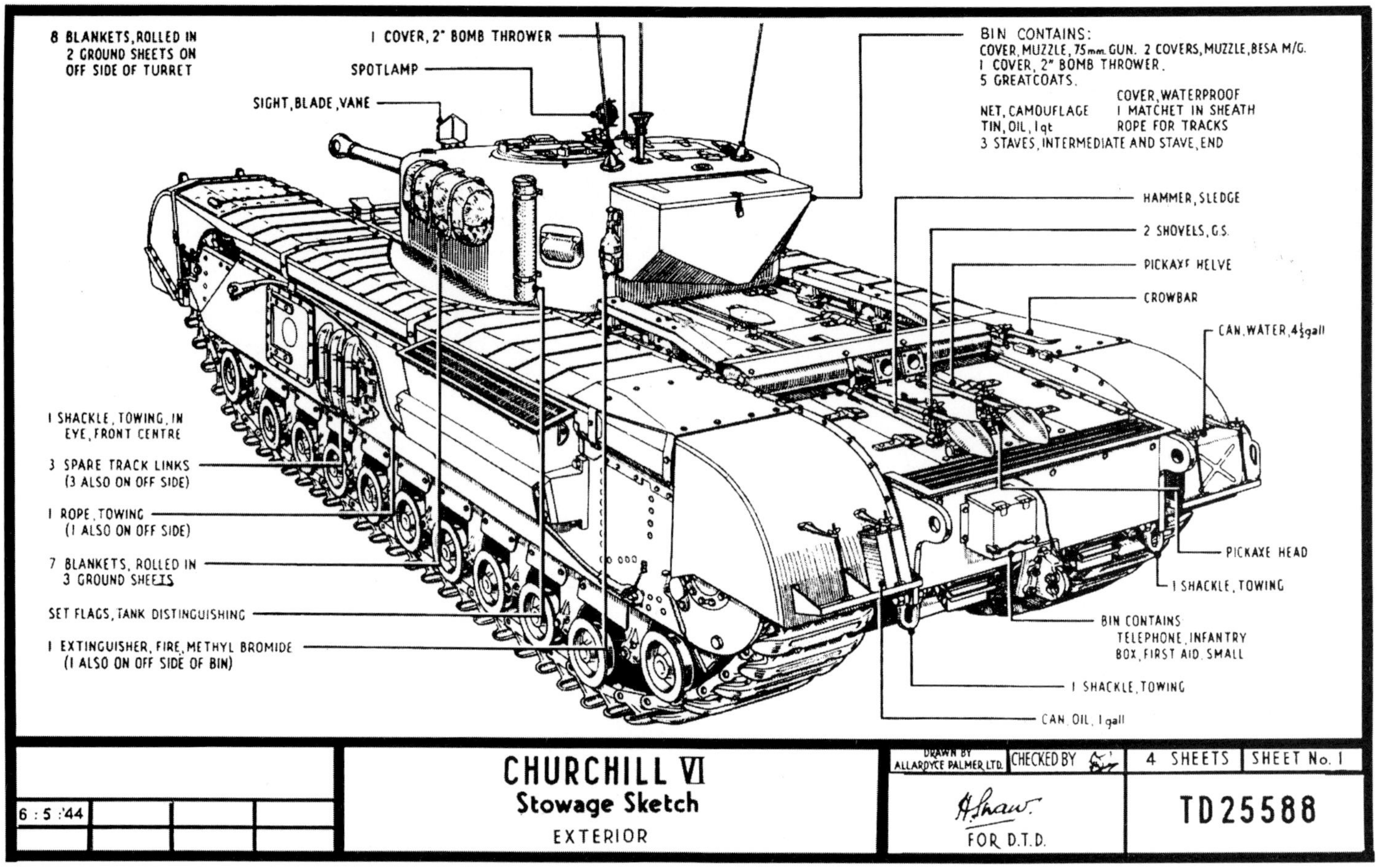

Stowage diagrams for the Churchill Mark VI.

ing with the local Home Guard, carried out a practice attack on what is described as a Blacker Bombard position. The Blacker Bombard was a crude, dangerous but effective form of spigot mortar supplied for use with the Home Guard. It was not much of a threat to a tank, and none at all to a Churchill, but no doubt it was useful experience for the Home Guard.

In October 1943, after the regiment had moved to Otley in West Yorkshire, it found itself back on Denton Moor involved in what is described as a tank versus tank exercise with live ammunition, which sounds highly dangerous. Yet there are no recorded casualties, to tanks or men, and a month later the regiment moves down to Boston in Lincolnshire. During this period the war diary continually reports men and equipment being drafted away from the regiment, concluding on the 21st and 22nd of the month when all the tanks due for rework were loaded onto trains at Boston for despatch to Broom & Wade. Immediately afterwards the regiment is disbanded and amalgamated with 151RAC but retains its title.

In January 1944 the amalgamated regiment is at Folkestone. It starts to receive M3A1 Stuart light tanks which will form its reconnaissance troop. Its days as a training regiment are over, now it is girding up for war. Its first Churchill V arrives on 15 January, its second on the 19th. On 3 February the regiment is addressed by General Sir Bernard Montgomery, a clear sign that it is being prepared for something and in March it travels way up north to Warcop Range in Cumberland where Regimental Headquarters and A Squadron practices firing 95mm HE while B Squadron fires 6-pounder HE and C Squadron learns to control the fire of 25-pounder guns of the Royal Artillery from its tanks. In May it is involved in Exercise Quack, which had to do with medical matters, presumably evacuating casualties from tanks. On the 16th of that month it receives the significant one-word order "CONCENTRATE". This is not an injunction to pay attention but the instruction to start work on sealing and waterproofing its tanks.

But we have got well ahead of ourselves. Returning now to the north of England in 1941 we find the 1st/5th Battalion, the Lancashire Fusiliers, converting to 108 Regiment RAC when the 125th Infantry Brigade, of which they are a part, becomes 10th Armoured Brigade virtually overnight. They are stationed at Barnard Castle in County Durham in November where they are visited by General Martel and the senior NCOs spend a day at the Vickers-Armstrong tank factory in Newcastle to see what they have let themselves in for. It is a generalisation, and probably quite unfair, but anyone who understands British regimental loyalty will know that the task of changing the hearts and minds of the senior members of the Sergeants' Mess is the key to any successful change. Convince them that it is a wise move and everything else follows smoothly, more or less. On 8 December the regiment is told that it must adopt the Royal Armoured Corps beret and as if that is not bad enough their first tank, which arrives just before Christmas, is a Covenanter. By May 1942 the regiment has 25 Covenanter but at this point it learns that it is to become an Army Tank Battalion. Three Churchills arrive within a few days and the Covenanters are given to the Polish Armoured Division. By June they have sixteen Churchills, seven Valentines and four Universal Carriers. They are also shown a film on the Churchill tank. In August they move to Rufford Abbey in Lancashire where, among other things, they are treated to a demonstration of the Atherton turret hoist. This is a device which can be mounted on a tank's turret and used to lift components on or off the tank. The sort of thing which is issued during training, practised with diligently for months, and probably never used again once the regiment is on active service. From November tanks start going back for rework, the first three being despatched to Vauxhalls. In January 1942 the regiment supplies a Churchill tank and crew to the Ministry of Supply to tour factories in the Midlands which are producing tank components. Unfortunately the officer commanding the tank is injured a few days later and ends up in hospital. Then, in March 1943, they learn that they will be changing back to an armoured regiment once again. Their Churchills are sent away to be replaced by an amazing collection of worn out cruiser tanks of very type from Mark I to Covenanters, so they are right back where they started. Their first Shermans arrive in April. They have 49 of them by June but are disbanded, as an armoured regiment, in December.

This story can be repeated, almost word for word, in the case of their sister battalion which became 109 RAC in November 1941. The only curiosity, which has nothing directly to do with our subject, involves their brigade commander who, on 4 December 1941, informs the regiment that it should ignore the War Office directive and continue to call itself 1st/6th Lancashire Fusiliers. It will continue to wear regimental buttons and badges instead of the brash new RAC embellishments and the soldiers will be address as Fusilier, instead of the RAC rank of Trooper. It is a delightful example of that ingrained British habit of rejecting change by pretending that it has not happened and it contrasts particularly with the 1st/5th Battalion which accepted the change so readily. This reactionary attitude seems to infect the entire regiment if their war diary is any guide. It hardly mentions tanks at all beyond admitting that it had some Covenanters. It continues in this idiosyncratic way through until May 1942, when it becomes an Army Tank Battalion, after which tanks are never mentioned again and then right around to August 1943 when the RAC Record Office informs them that they are 109 RAC, like it or not, and must use that title. However in the end the War Office bows to the inevitable and 109RAC is disbanded as of 1 January 1944 without, apparently, reverting to being an armoured regiment.

A few details will suffice to bring out the flavour of some other Churchill regiments in Britain at this time. The 110th Regiment RAC, 5th Battalion the Border Regiment as was, starts converting from Covenanters to Churchills in May 1942 but then becomes a Training, Holding and Drafting regiment in 77 (Reserve) Division at Catterick and is subsequently disbanded. Meanwhile 111 RAC, the erstwhile 5th Battalion, the Manchester Regiment, converted to Churchills in June 1942. Men from the regiment attended a lecture on Jungle Warfare at Catterick in September but go no further than Wensley in North Yorkshire. From there, in January 1943, it sends ten Churchills by rail to Castlemartin in South Wales where, according to the war diary, guns are fitted to their tanks for firing practice. Two-pounder and 6-pounder guns are fired on Castlemartin ranges but this regiment is also disbanded early in 1943.

At the seaside resort of Worthing in Sussex 141RAC (7th Battalion The Buffs) carries out anti-aircraft practice on the sea front, firing machine-guns at drogue targets pulled by aircraft, and gas filled balloons, out to sea. It is a regiment with an amazing propensity for losing things and holding Courts of Enquiry to find out where they have gone. First it is some morphine tablets from the first aid kit in a tank, then a soldier's greatcoat. This is followed by a spate of tools being lost from tanks which seems to become endemic, and then a complete set of batteries vanishes from a Churchill tank. This would be remarkable for weight alone, and might stand as some kind of record if they had not managed to lose a complete Bedford truck in January 1943! They spent that winter at Ashford in Kent and in February took part in an interesting exercise with number 184 Squadron, Royal Air Force, which is described as a Tank Buster squadron. The mock attack by these aircraft, which are not identified by type, is reported as very effective but the regiment then puts up a smoke screen with each tank firing its two inch smoke mortar and the regimental close support tank helping out with its 3 inch howitzer. The war diary describes this smoke experiment as "something quite new" and it seems to work very well. In February 1944 the regiment is told that it will be changing to become a Crocodile flamethrower unit.

Finally 144RAC, otherwise 8th Battalion the East Lancashire Regiment, finds itself at Rufford Abbey in Nottinghamshire in January 1942, equipped with 24 Churchills Mark I and II. In March one of its troops of Churchills goes to nearby Thoresby Park where they join up with Churchill troops from 148 RAC and 48 RTR and companies of infantry to take part in a demonstration of tank, infantry and aircraft co-operation with Tomahawk fighters of the RAF. They move to Berkhamstead in Hertfordshire and in October 1942 enjoy a spell of live firing on Studland ranges in Dorset. This evokes an interesting comment from the war diary which states that for the first time tank crews are provided with a liberal amount of ammunition. They fire 3 inch howitzer, 2-pounder and BESA machine-gun and we are told that each gunner was allowed to fire fifteen rounds of 2-pounder which was obviously something quite special. By January 1943 they are at Shorncliffe Camp near Folkestone, the place at which Sir John Moore introduced his revolutionary method of training which proved so useful both to Moore and his successor Wellington during the Peninsular campaign of the Napoleonic Wars. Such are the shadows of history.

The regiment changed over to Shermans, for what was described as a special assault role, but a memorandum, compiled by their commanding officer while they were at Ashford in November 1943, speaks for all tank regiments in Britain at this time. Lieutenant Colonel Waddell is clearly writing direct from the heart and his complaints probably tell us more about the ruling situation, and the muddle it creates, than any official report.

His first point concerns the rework policy and therefore is highly relevant to Churchill regiments:

> *There is no relation of any kind between the tank policy (i.e. the withdrawal of tanks from the regiment for rework and the allocation of new tanks) and the requirement of the regiment for tank crew training.*
>
> *No tank regiment should be stationed in a county where the training grounds are so few, so small and in such tremendous demand. Tanks require big training areas and plenty of them.*
>
> *To kill Germans we must be able to kill quickly. This requires plenty of shooting practice on the move with snap targets. We cannot have too much. We get far too little; only twice in two years. Most present gunners and tank commanders only once now.*
>
> *We are now GHQ troops but have not got clear of an infantry division yet. The necessity for training virtually untrained infantry is preventing us from getting nearly enough tank training.*

A brief note dated 7 August 1943, claims that the Churchill Mark VI was about to begin its acceptance trials. This was the designation given to a reworked version of the cast turreted Mark IV when fitted with the 75mm medium velocity gun; the modified 6-pounder. However a report, issued in July 1945, confuses as much as it clarifies this issue. It differentiates between a Mark IV fitted with a 75mm gun and the Mark VI which has the same weapon with geared elevation. This report claims that 242 Mark VI were issued while 820 Mark IV Churchills were fitted with the 75mm gun. This latter may well refer to the otherwise unknown Churchill Mark IVB.

An overhead view of a brand new Churchill Mark VII. Notice in particular two features of the turret; the welded top panel and thickened lip around the base.

Acceptance trials for the Mark VI were carried out at Lulworth in September 1943 and among other things they reveal a fault apparently common to the 6-pounder and the bored out 75mm guns. Both were described as requiring high firing effort and reference to this has already been noted with the Mark III trigger mechanism. On the Mark VI this had been replaced by a foot pedal, but one that still required considerable physical effort to operate. Colonel O'Rorke, the officer commanding the Experimental Wing of the Gunnery School said that this problem would be cured once solenoid operated firing gear was introduced but he said that he found it hard to understand why it was taking so long to produce this equipment. He noted that a firm requirement had been established more than a year earlier.

The Mark III, when it was fitted with the 75mm, simply became the Mark III*. Again the figure of 242 is given for the number of Mark III Churchills converted to take the new weapon. When the first edition of the Service Instruction Book was issued for these tanks in October 1944 it simply referred to Churchill III (75mm) and Churchill IV (75mm) and presumably precedes the appearance of the Mark VI. The new gun was known as the Ordnance, Quick Firing, 75mm Mark V or Mark VA. The barrel was described as externally similar to the Ordnance, Q. F. 6 pdr. 7 cwt., Mark V, that is the longer barrelled version of the 6-pounder. It was also noted that with the exception of the extractor levers and firing pin the components of the breech assembly were identical, and indeed interchangeable with the 6-pounder. The difference between the 75mm Mark V and VA was simply in the number of threads on the end of the barrel which were used to secure the single baffle muzzle brake. The Mark VA had twice as many. As already noted above it is this instruction book which claims that the 75mm mounting in these tanks is of the free elevation type, and goes on to describe the shoulder piece in some detail.

It is worth noting that the 1945 report refers to the 75mm Marks III and IV as follows; 'A Field Conversion scheme was introduced for replacing the 6-pounder in Churchill III and IV by the 75mm gun. This scheme affected stowage, sighting arrangements etc., and was carried out by units in order to get sufficient 75mm gun tanks into the hands of the troops, pending the appearance of Churchill VI and VII'. The implication clearly being that the conversion of a Mark IV to a Mark

(text continues on page 134)

This Mark VII, viewed from the rear, has its air intakes stowed on the track guards, which suggests that it is ready to travel by rail. The bolts on the back of the hull show where the coupling for a flame fuel trailer would be attached.

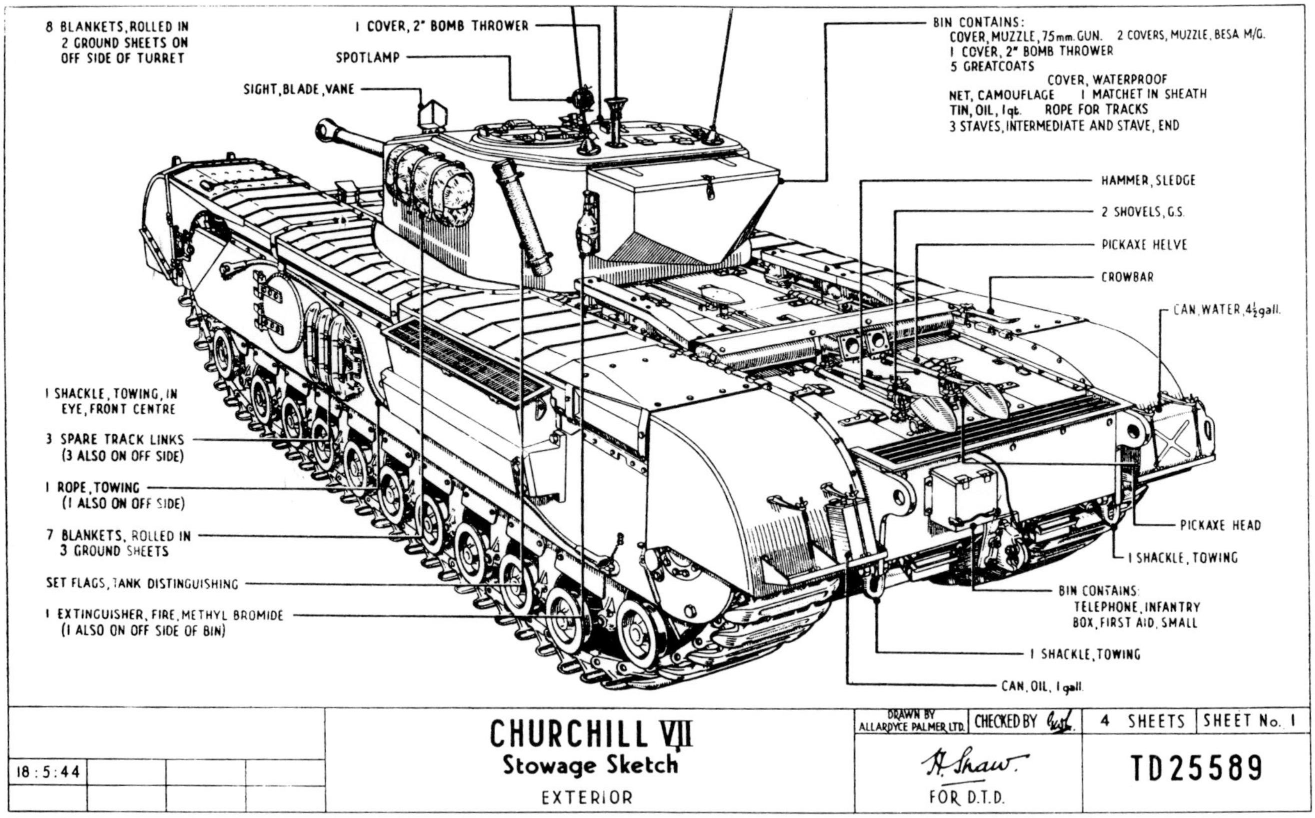

Pages 130-133: Stowage, armour thickness and other diagrams for the Churchill Mark VII.

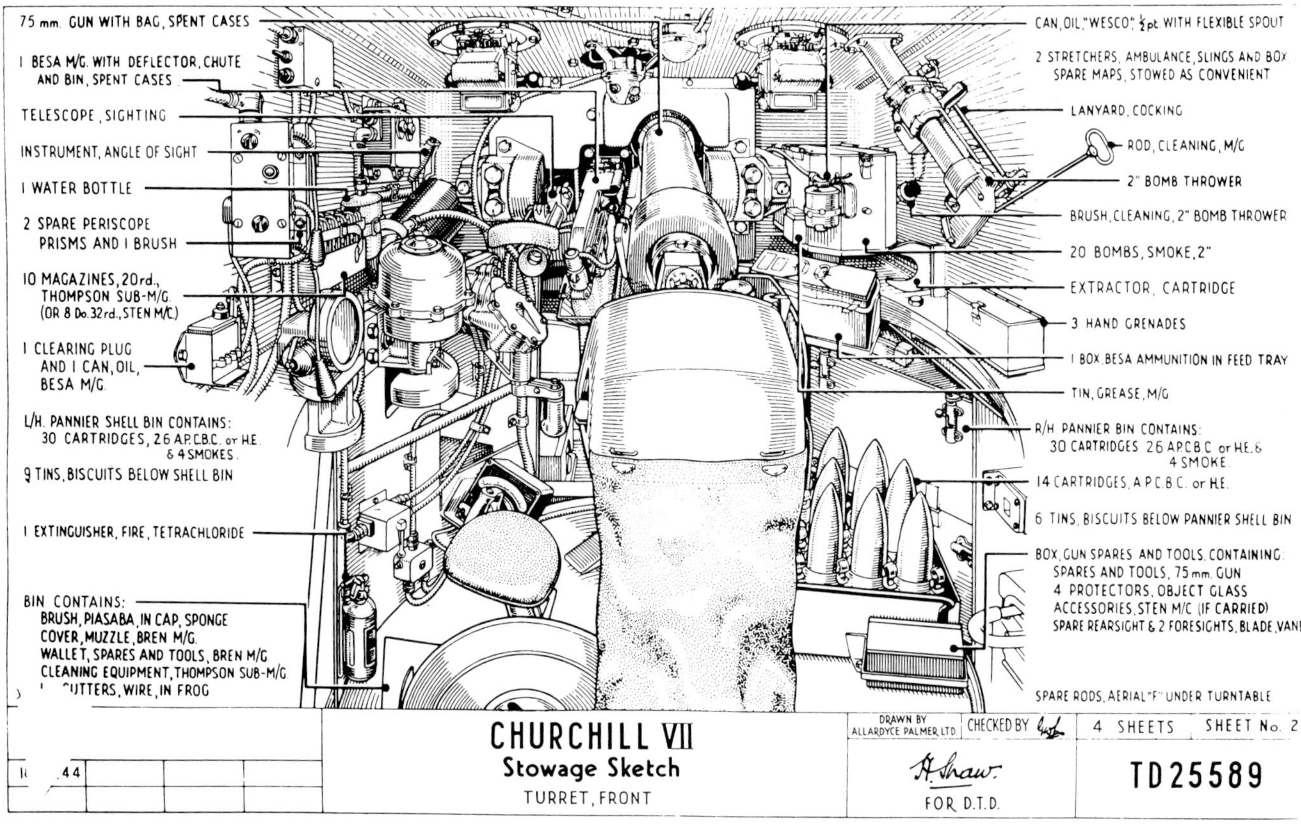

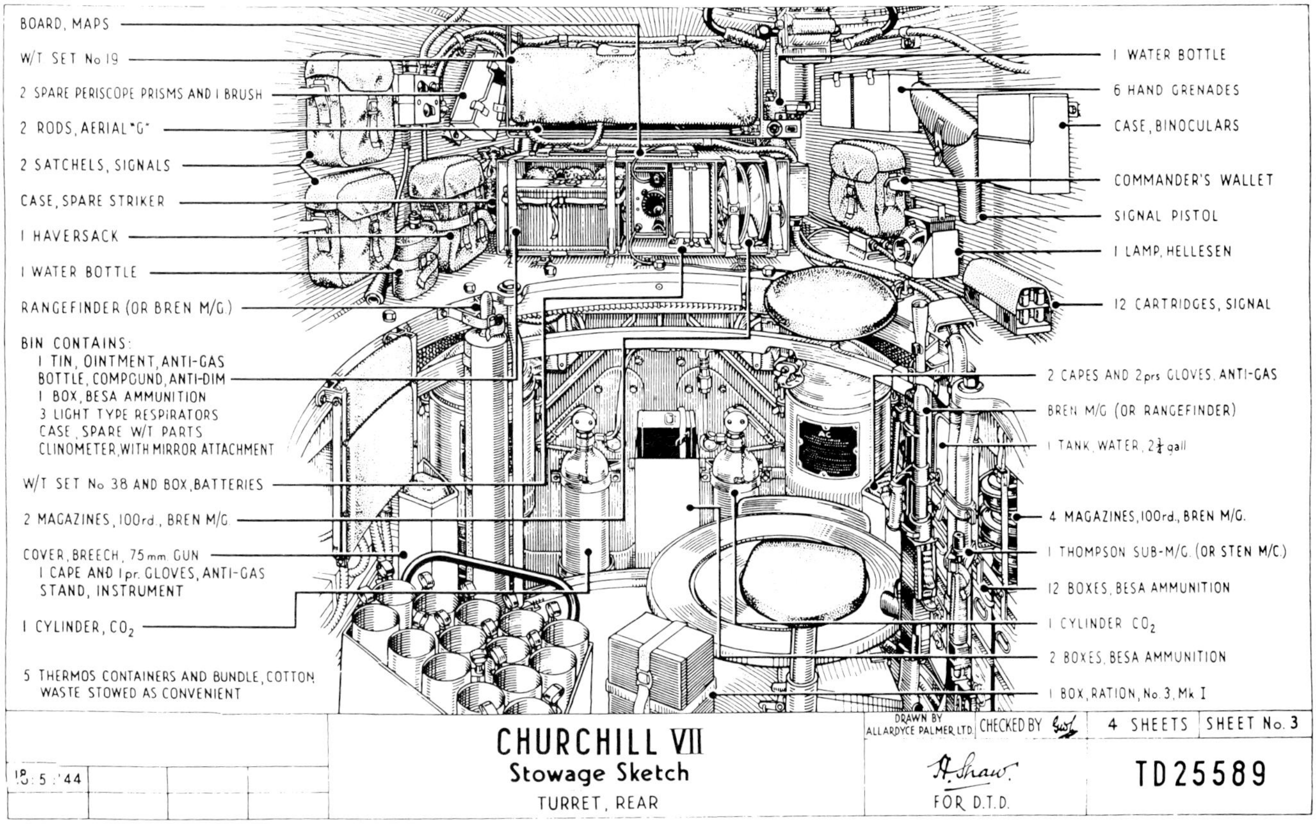
BOARD, MAPS
W/T SET No 19
2 SPARE PERISCOPE PRISMS AND I BRUSH
2 RODS, AERIAL "G"
2 SATCHELS, SIGNALS
CASE, SPARE STRIKER
I HAVERSACK
I WATER BOTTLE
RANGEFINDER (OR BREN M/G.)
BIN CONTAINS: I TIN, OINTMENT, ANTI-GAS BOTTLE, COMPOUND, ANTI-DIM I BOX, BESA AMMUNITION 3 LIGHT TYPE RESPIRATORS CASE, SPARE W/T PARTS CLINOMETER, WITH MIRROR ATTACHMENT
W/T SET No 38 AND BOX, BATTERIES
2 MAGAZINES, 100rd, BREN M/G
COVER, BREECH, 75mm GUN
I CAPE AND I pr. GLOVES, ANTI-GAS
STAND, INSTRUMENT
I CYLINDER, CO_2
5 THERMOS CONTAINERS AND BUNDLE, COTTON WASTE STOWED AS CONVENIENT
I WATER BOTTLE
6 HAND GRENADES
CASE, BINOCULARS
COMMANDER'S WALLET
SIGNAL PISTOL
I LAMP, HELLESEN
12 CARTRIDGES, SIGNAL
2 CAPES AND 2prs GLOVES, ANTI-GAS
BREN M/G (OR RANGEFINDER)
I TANK, WATER, 2¼ gall
4 MAGAZINES, 100rd, BREN M/G.
I THOMPSON SUB-M/G. (OR STEN M/C.)
12 BOXES, BESA AMMUNITION
I CYLINDER CO_2
2 BOXES, BESA AMMUNITION
I BOX, RATION, No 3, Mk I
DRAWN BY ALLARDYCE PALMER LTD.
CHECKED BY
4 SHEETS
SHEET No. 3
18.5.'44
CHURCHILL VII
Stowage Sketch
TURRET, REAR
FOR D.T.D.
TD25589

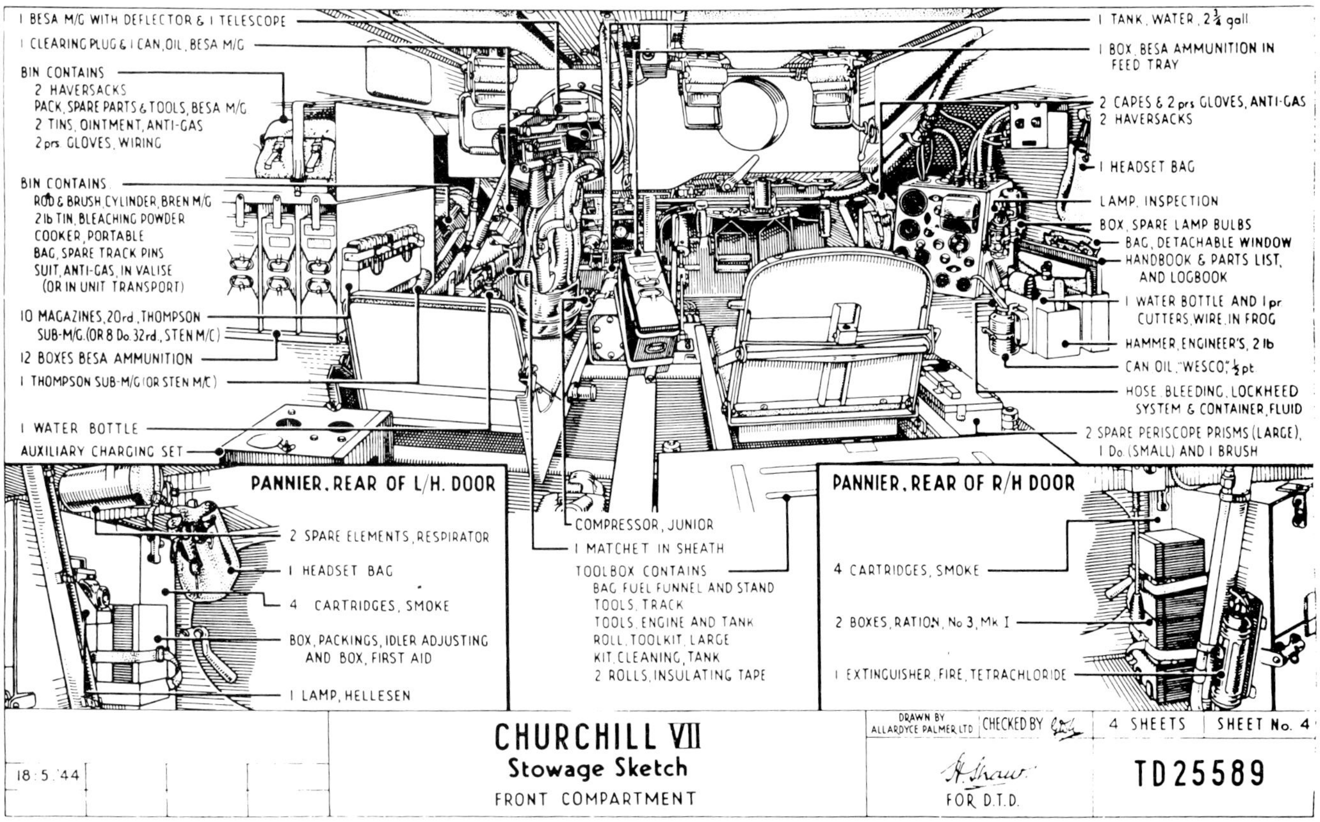
I BESA M/G WITH DEFLECTOR & I TELESCOPE
I CLEARING PLUG & I CAN, OIL, BESA M/G
BIN CONTAINS 2 HAVERSACKS PACK, SPARE PARTS & TOOLS, BESA M/G 2 TINS, OINTMENT, ANTI-GAS 2prs GLOVES, WIRING
BIN CONTAINS ROD & BRUSH, CYLINDER, BREN M/G 2lb TIN, BLEACHING POWDER COOKER, PORTABLE BAG, SPARE TRACK PINS SUIT, ANTI-GAS, IN VALISE (OR IN UNIT TRANSPORT)
10 MAGAZINES, 20rd, THOMPSON SUB-M/G. (OR 8 Do 32rd, STEN M/C)
12 BOXES BESA AMMUNITION
I THOMPSON SUB-M/G (OR STEN M/C)
I WATER BOTTLE
AUXILIARY CHARGING SET
I TANK, WATER, 2¼ gall
I BOX, BESA AMMUNITION IN FEED TRAY
2 CAPES & 2prs GLOVES, ANTI-GAS
2 HAVERSACKS
I HEADSET BAG
LAMP, INSPECTION
BOX, SPARE LAMP BULBS
BAG, DETACHABLE WINDOW
HANDBOOK & PARTS LIST, AND LOGBOOK
I WATER BOTTLE AND I pr CUTTERS, WIRE, IN FROG
HAMMER, ENGINEER'S, 2 lb
CAN OIL, "WESCO," ½ pt
HOSE, BLEEDING, LOCKHEED SYSTEM & CONTAINER, FLUID
2 SPARE PERISCOPE PRISMS (LARGE), I Do (SMALL) AND I BRUSH
PANNIER, REAR OF L/H. DOOR
2 SPARE ELEMENTS, RESPIRATOR
I HEADSET BAG
4 CARTRIDGES, SMOKE
BOX, PACKINGS, IDLER ADJUSTING AND BOX, FIRST AID
I LAMP, HELLESEN
COMPRESSOR, JUNIOR
I MATCHET IN SHEATH
TOOLBOX CONTAINS BAG FUEL FUNNEL AND STAND TOOLS, TRACK TOOLS, ENGINE AND TANK ROLL, TOOLKIT, LARGE KIT, CLEANING, TANK 2 ROLLS, INSULATING TAPE
PANNIER, REAR OF R/H DOOR
4 CARTRIDGES, SMOKE
2 BOXES, RATION, No 3, Mk I
I EXTINGUISHER, FIRE, TETRACHLORIDE
DRAWN BY ALLARDYCE PALMER LTD.
CHECKED BY
4 SHEETS
SHEET No. 4
18.5.'44
CHURCHILL VII
Stowage Sketch
FRONT COMPARTMENT
FOR D.T.D.
TD25589

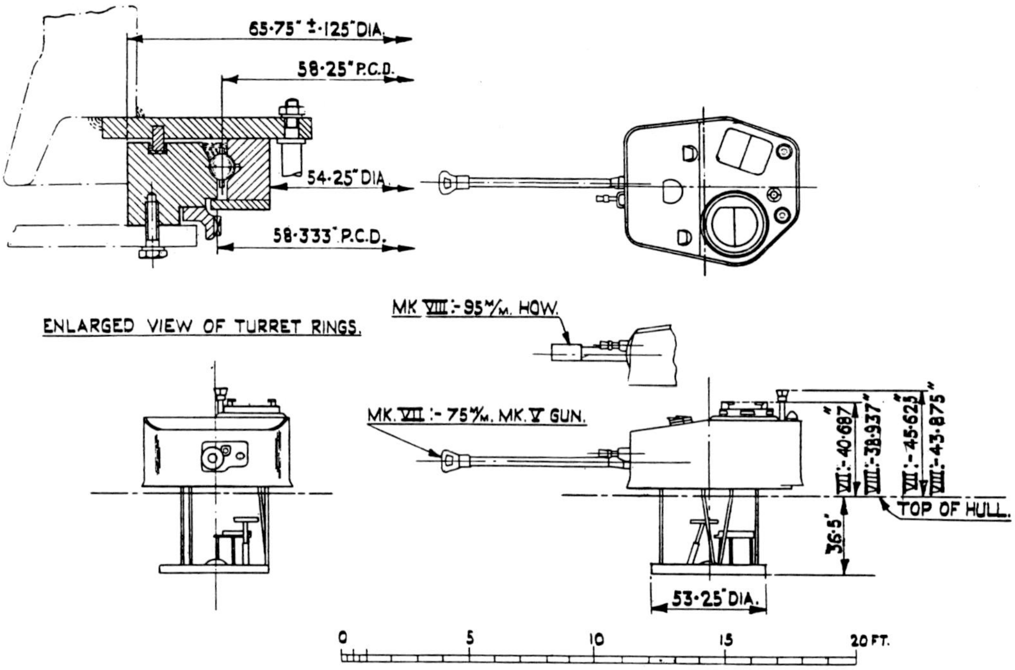

CHURCHILL VII & VIII.
TURRET.

CHURCHILL VII & VIII

CHURCHILL
scale: 1:48

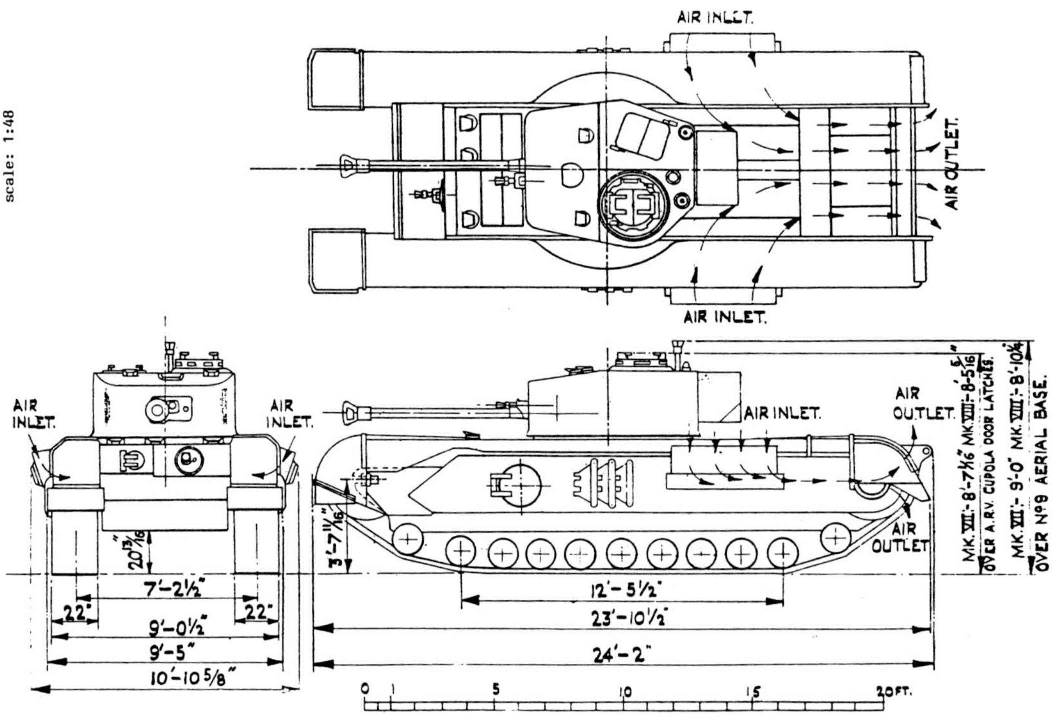

CHURCHILL VII. (TYPICAL FOR VIII.)
OUTLINE.

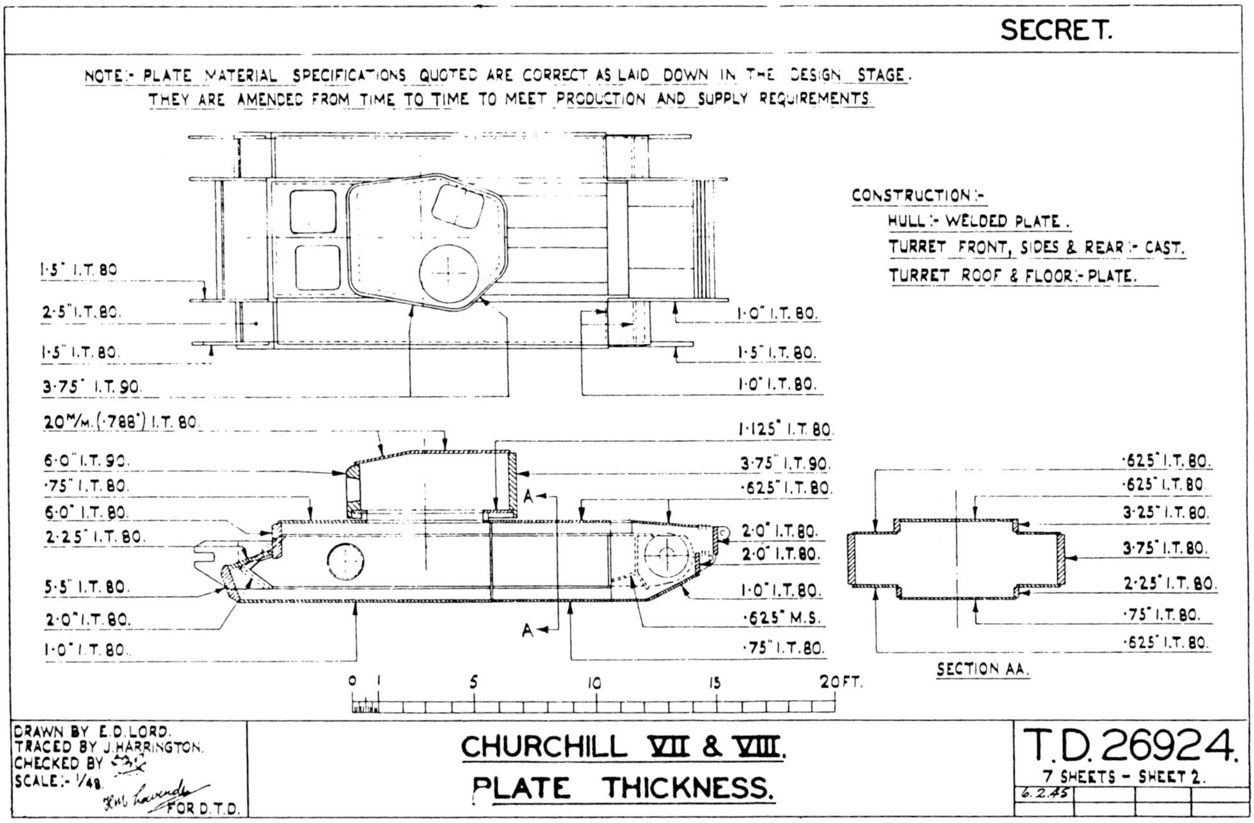

TURRET—CHURCHILL VIII AND XI.

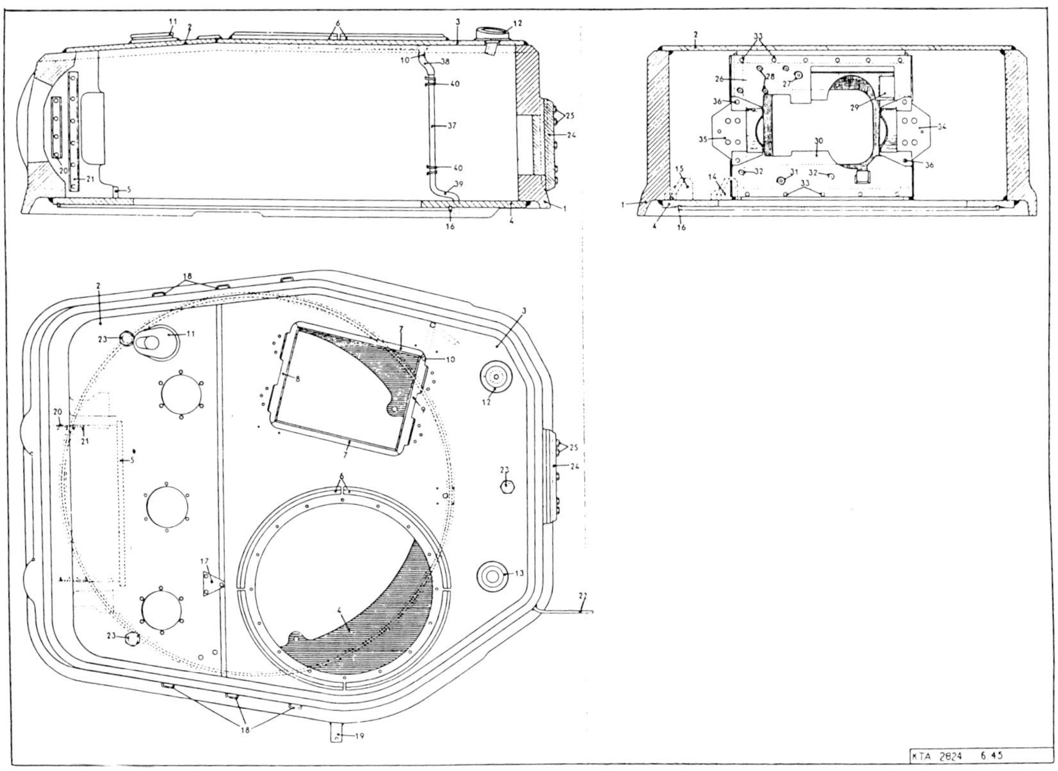

VI was something far more significant than simply swapping over the gun and a few other internal details.

Now that the rework programme was well under way and many of the existing tanks were also being fitted with 75mm guns there was time to consider the design of the next batch. Clearly everything that could be done to the present design had been done, so if the Churchill was to be further improved it could only be achieved by a substantial redesign. The new version was originally described as the Heavy Churchill although, in the end, it only amounted to a difference of one ton. The specific designations were Mark VII for the 75mm gun version while the close support model, with the 95mm howitzer, was classified Mark VIII.

When the much-quoted Joint Memorandum appeared in May 1943 it discussed the so-called Heavy Churchill but was rather vague about details. A month later the Director, Royal Armoured Corps offered some suggestions, including the need for the floor of the tank to be proof against mine blast, thicker armour for the hatches over the driving compartment and a stronger turret ring to take heavier turrets. Design work was aided by the fact that certain major aspects of the new version had been anticipated. Back in August 1942, when Babcock & Wilcox issued their report on the original welded hull, they put forward the suggestion that in future hulls the doors at the sides, and the driver's front visor should be circular instead of square. It was a sensible suggestion because it reduced the risk of weak points developing near the corners on square door apertures. This was noted but not acted upon at the time, partly because it interfered with production but mainly since the future of the Churchill tank was then in doubt.

The new turret design had also been anticipated but it is not entirely clear whether this was another idea generated by Babcock & Wilcox, or the Department of Tank Design or indeed Vauxhall Motors themselves. The problem had been to create a new and stronger design but it was generally agreed that neither welding nor casting would be ideal, although both had merits. A cast turret was integrally stronger but it was difficult to control the relative thickness of armour on all surfaces of a one-piece casting, with the result that in most cases one ended up with a roof much thicker than it needed to be, and hence an unnecessary weight penalty. With welded structures, on the other hand, it was quite simple to have a different thickness of plate on each surface but the thicker such plate got the more difficult it was to produce a satisfactory joint by welding alone. The ultimate answer appeared to be a composite turret in which the four vertical faces were created as a single casting while the thinner top and bottom plates were welded in place. This arrangement gave another advantage for it proved possible to recess the top plate onto a lip inside the turret shell so that no welded edge was exposed to enemy fire. There were instances on record where turrets that had an upper plate welded or bolted would come away at the top when struck a glancing blow on the joint. There are few things any tank crewman likes less than the sight of daylight shining in through an unauthorised hole in his armour. It immediately creates a sense of vulnerability which is bound to affect the performance of the crew unless it is remedied at once. The concept of a composite turret was not entirely new, it already existed on the Matilda and the Valentine although in that case the top panel was bolted on, but the new Churchill turret was much heavier with an even thicker cast skirt around the base which gave added protection to the turret ring.

Vauxhall Motors had produced a very similar turret earlier in 1943 for the new version of Cromwell tank that they were expected to build when the Churchill programme came to an end. A prototype had been produced and was undergoing firing-at trials at Shoeburyness by April 1943.

There were clearly those at the War Office who regarded the heavy Churchill as a new tank and there is evidence that at one stage it was allocated the General Staff number A42, rather than the more logical A22F which became the accepted designation. However at a meeting of the Director, Royal Armoured Corps' Advisory Committee meeting held at the Vauxhall plant on 2 December 1943 the Deputy DRAC, Brigadier C A L Dunphie, stated that the Churchill VII was regarded as a new type of existing tank. On that basis the committee members, who would normally work through a lot of detail concerning the general fightability of the tank, confined themselves to matters of stowage. A fully stowed tank was made available for them to study.

In fact, to the untrained eye it is difficult to identify the Mark VII from other Churchills and one does best to rely on the turret shape and various rounded apertures to be certain. There were other distinctive features, of course, but they are less obvious. At the rear of the hull the air outlet was now deeper, eight inches instead of five, as foreshadowed in the original rework programme. Pistol ports were also eliminated from the turret and the hull pannier doors since they had proved almost completely useless. Inside was another change. The auxiliary steering bar for the hull machine-gunner was removed, probably because it got in the way and was never used.

Mechanically there were no major differences beyond a change in gear ratios to allow for the extra weight, the new gearbox being designated type H41. Other developments that had taken place in respect of engine and transmission design were across the board and could be applied to any marks, in theory. The various changes were distinguished by suffix designations applied to the engine number. Suffix R, part of the original rework programme, signified improved oil feed to the distributor gear and the adoption of 37mm carburettors. R1 marked the adoption of an Amal, positive drive fuel pump but included the original rework. R2 denoted the use of bolts on the connecting rod big ends instead of studs and naturally included modifications R and R1. And so it went on through to R6, covering a change to mechanical, adjustable tappets, instead of hydraulic tappets, strengthening of the crankshaft centre main bearing cap and various other details along the way.

Precisely the same system applied to the H4 gearbox. R, in this case, indicated strengthened selector forks, among

The hull of an A22F, held in a rotating jig, while two workmen position the front roof section. The jig can be used to turn the hull right over to make the welder's task easier.

Working on the casting of an A22F turret. The workman appears to be grinding off excess metal from the interior before the roof plate is fitted.

other things. R2 recorded a change from double helical to spur gears in third and top along with improvements to the power take-off while R4 showed that the gear change mechanism had been revised. In this case modifications went through to R8 which meant that the splines on the layshaft had been modified along with all the foregoing changes, and others. The H41 gearbox, adopted for A22F was only issued as H41B, indicating the first production box and H41C which was the same modification as R8 on the H4 box.

The reason for this change in transmission had to do with the increased weight, although this was theoretically restricted to one ton. A Churchill Mark IV, V or VI, with all modifications, fully stowed and complete with crew weighed 39 tons. A Mark VII or VIII Churchill in the same state weighed 40 tons. Yet armour on the new model had increased from 4.5 to 6 inches at the front, by an inch to 3.75 on the sides and even by a fraction on the hull floor and roof. Only the rear remained the same. The turret likewise was now 6 inches at the front and 3.5 on sides and rear while the top remained the same at one inch. Savings would have been made by welding of course, in particular on the inner faces of the hull panniers, but it is still remarkable that the designers were able to increase frontal protection to the point were it exceeded that of the celebrated German Tiger tank without adding more than one ton to the overall weight. However in order to allow for this increase the gearbox ratios were modified. It was not a dramatic change but it reduced the maximum speed from 15.5mph to 13.5mph.

Modifications to the suspension had also been on-going and there is a bewildering list of bogie units and their relevant distinguishing features in the Churchill Workshop Manual but for the Marks VII and VIII a new type, with much stronger springs, was introduced. Since this was not interchangeable with the earlier type an identification disc was applied to each bogie which the fitter was supposed to check. This was easy enough on the right side of the tank, because the discs were readily visible, but on the left side they were on the inner face so in theory at least the individual would have to crawl underneath the tank to make sure.

Another major change which featured on Marks VII and VIII but was applied retrospectively to earlier marks concerned the commander's cupola on top of the turret. Most British tanks produced early in the war carried huge, drum shaped cupolas but they quickly went out of favour. The rumour went around that cupolas could be shot off, beheading the commander in the process. Throughout the middle years of the war most British tanks and the majority of American types dispensed with cupolas altogether. Rather, as on the Churchills built at this time, the commander had a simple two-piece hatch with one or two periscopes, set into a rotating collar.

The return to European conditions with the landings in Sicily, and the German habit of employing snipers to pick off Allied tank commanders, revived the practice of closing the tank down in action and immediately there was a cry for improved visibility. The result was a low profile device known as the All Round Vision (ARV) cupola. The ARV cupola featured the normal two-piece hatch surrounded by eight episcopes, one of which could be elevated. The entire cupola could still be rotated but under normal circumstances it was set with a group of three episcopes at the front, and the tilting one at the back. It came in two versions. On cupola No. 1 Mark I the two hatches remained vertical in the open position while on the No. 1 Mk. II they could fold flat. Newer tanks were also fitted with the complicated Twin Blade Vane Sight No. 2 which lined up with the cupola and enabled the commander to get an accurate lay on a target even from the turret down position.

Firing trials of a new Churchill Mark VII took place at Lulworth early in 1944 although it would appear, from reading the report, that this was not the first such trial. A number of improvements are listed, from the elevating gear to the

Turret meets hull on the A22F production line at Vauxhall Motor's Luton plant. Notice the two lines of rails on the floor, along which the bogies run; also the way the tanks are chained together.

Cross section drawing of a Mark VII showing how the redesigned turret gave more room than the old Mark IV casting.

padding on the commander's seat and the comment that the cupola still leaks is indicative. The report refers to Mark III elevating gear, which must be an improved version. It seems to work satisfactorily with just seven minutes backlash but a more alarming fault is turret lift. Colonel O'Rorke said that when turret rotation was reversed, on power traverse, the turret jumped and it was most disconcerting. He suggested either that the turret ring was oval, or that it dipped at the 6 and 12 o'clock positions and he had been promised by Vauxhall Motors that it would not occur in later models. However he felt that it should be watched in case the situation got so bad that it affected the accuracy of the gun.

Trying to establish the precise number of heavy Churchills built is complicated, as usual, by the way in which the contract cards have been compiled. Two very vaguely dated contracts with Vauxhall Motors include Churchills Mark IV to VIII. Of these contract M5701 for 550 tanks includes 58 type A22F which carried WD numbers T173142 to T173199. Contract M7454 covers 1,000 tanks of which T251505 to T251647 are A22F, that is 143 tanks. Contracts specifically related to A22F are the undated M14283 for 500 tanks (450 Mark VII and 50 Mark VIII) and M10577, dated 11 November 1943 for the same quantity in the same proportions. All of these contracts were placed with Vauxhall Motor as the parent company and it seems clear from photographic evidence that they were now back producing new tanks, which suggests that the rework programme was coming to its end. However these last two contracts are complicated by the fact that no WD numbers are given for M10577 and those for M14283 are contradictory. On the front of the card the numbers given are T341209 to T341708 but it is endorsed on the reverse with the information that T341409 to T341508 were issued to the

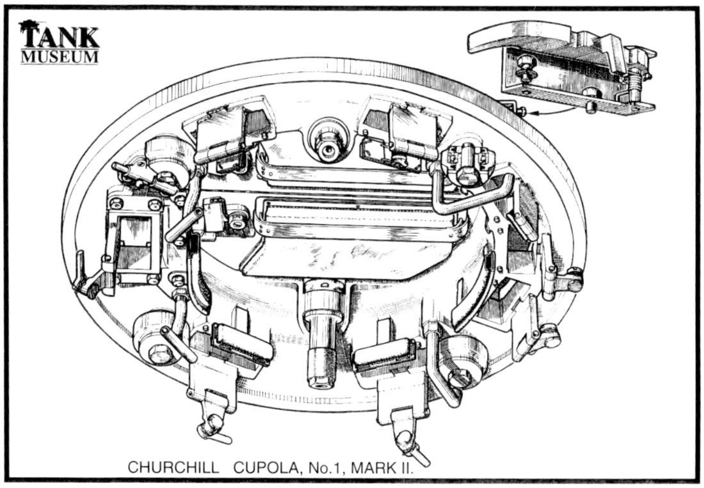

Drawing of the all-round vision cupola.

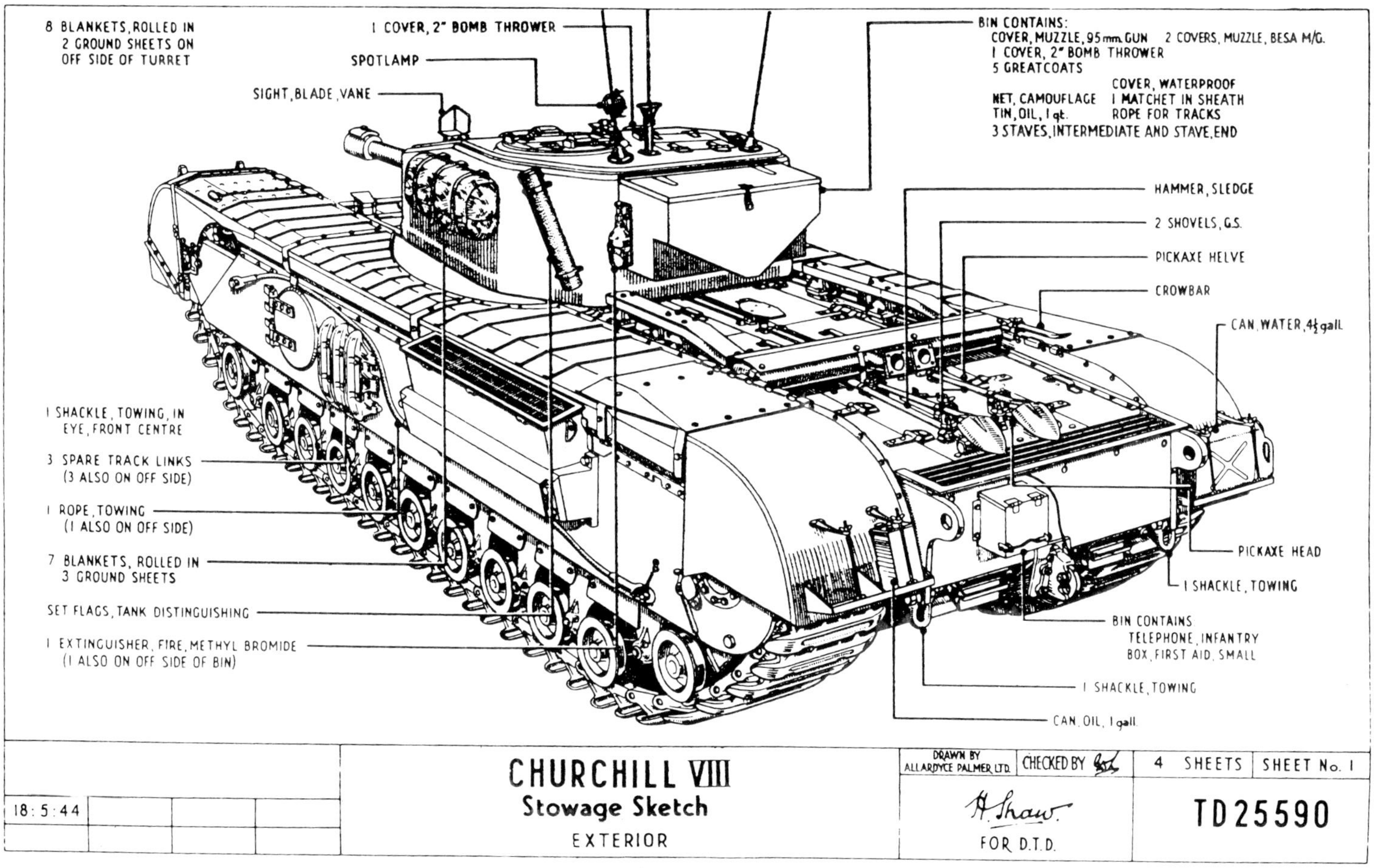

Stowage diagrams for the Churchill Mark VIII.

Gloucester Railway Carriage Company while below that entry it states that T347659 to T348158 were also issued. Whether these relate to M10577 is not clear. Two other contract cards appear to have complementary information. M10521, dated 9 November 1943, covers the supply of 95 Mark IV Heavy Welded Hulls to Vauxhall Motors Ltd. specifications - this with the Gloucester Railway Carriage and Wagon Co. While M12574, dated 11 March 1944 with Charles Roberts and Co. refers to the assembly of 120 Churchill tanks with welded hulls.

Clearly there is more to it than this, and there may well be more contracts for which no records survive. A booklet published by the Ministry of Supply in 1946 claims that in order to expand production the contract was extended to various other firms including Newton Chambers Ltd., of Sheffield, the Gloucester Railway Carriage & Wagon Co., and the Whessoe Foundry & Engineering Company in Darlington. It is interesting to note that when production of heavy Churchill hulls began Babcock & Wilcox, the pioneers, were employing a horizontal-vertical welding technique known as flat-bead welding while the Whessoe Foundry introduced a gravity welding technique which was also taken up by GCR&W. Most other contractors adopted the Babcock & Wilcox system although, as they gained experience, some firms appear to have introduced variations of their own which, presumably, should permit an expert in the science of welding to identify the manufacturer by inspecting the tank.

The technicalities of welding, not to mention the broader subject of tank construction, could easily fill a book but it might be a very boring one. Most firms appear to have adopted the practice of assembling tanks in situ, bearing in mind that the majority were put together from components which arrived from a variety of factories. However in their war history *An Account of our Stewardship*, written by W J Seymour, Vauxhall Motors give some specific details of their own technique. Taking advantage of the double flanged wheeled bogies on the Churchill the plant at Luton was laid out with double, narrow gauge tracks like a pair of little railways running the length of the production line. Once the new hulls were fitted with suspension units they were placed on these rails, one behind the other. The tanks were then moved down the production line in sequence. At first the company used a winch and then chained the line of tanks together. As each completed tank was winched forward onto its tracks it drew all the others one tank length behind it. Having practised this for some time it occurred to someone at the factory that the winch was superfluous. The next stage involved starting up the leading tank, once its tracks had been fitted, and having it pull all the others forward in its wake.

Finally it should be noted that the new hull, whether it ended up as a Mark VII or VIII, was completed in such a way that the tank could be quickly adapted to receive all the fittings necessary to convert the tank into a Crocodile flamethrower. Development work had been undertaken on an earlier model but all production Crocodiles were of the Mark VII type with at least one Mark VIII, but whether this was true of every heavy Churchill hull it is impossible to say. If not all then it certainly applied to the majority.

Early in 1944 a new Mark VII Churchill was subjected to firing-at trials which revealed a number of detail faults. Modi-

Head on view of a Mark VIII with the 95mm howitzer and All Round Vision Cupola. If surviving photographic evidence is any guide then the Mark VIII saw very little active service.

Probably training for D-Day a column of Churchills on the move. The leading tank has the number of its appointed landing craft chalked on the track guard.

fications were put in hand, both to remedy these and incorporate other improvements suggested by Vauxhall Motors themselves and another tank, altered to these standards, was despatched to Shoeburyness to serve as a target. The trial took place over a period of five days starting on 31 May 1944 and the results were published in July. Among the changes listed were improvements to the hull side doors both to reduce the ingress of splash and strengthen the hinges; to the driver's vision port which was also letting in splash and proved weak around the locking handle and a stiffening girder running crosswise beneath the roof, just behind the driver's top hatch and another running fore and aft between the two hatches. Other work was also carried out, mainly to strengthen internal fittings which, presumably, were being dislodged when the tank was struck by a projectile.

The tank became a target for various weapons including medium and heavy machine guns, a 17-pounder firing armour piercing ammunition and a 25-pounder firing high explosive. A 4.2 inch mortar was also used. The trial report is a substantial document, far too detailed to review here, but some of the more serious faults can be mentioned. Under attack from armour piercing or high explosive rounds many welded joints appear to have fractured and the air inlet louvres were badly damaged. The side doors still let in splash and they tended to jam shut when struck by heavy rounds. Worse still the roof hatches above the driver and hull machine-gunner also jammed so that, under certain conditions, it would be difficult for all crew members to escape in a hurry. More modifications were recommended but how many of them actually were done, and how successful they were is not revealed. In any case many of the tanks were already in service and the invasion was imminent so these improvements would only appear on later versions and one could only hope that crews survived long enough to appreciate them.

In fact the first two Churchill regiments ashore had a rough time of it. Seventh Royal Tank Regiment, which had been recreated from the erstwhile 10th RTR, along with 9th RTR formed the armoured element of 31st Army Tank Brigade. They spent their last months in England at locations around Kent and Sussex where the 7th did some range practice at Lydd in March and then went to the South Downs training area in April. Later that month they moved to their concentration area, at Blackdown in Hampshire where they received a number of new Churchills from 6th Guards Tank Brigade. At the end of April they began to waterproof their tanks and strict mail censorship was imposed. In effect what this meant was that they ceased to have a proper postal address; virtually ceased to exist because all mail had to be directed simply to 7th RTR, Army Post Office, England which, as far as their families were concerned, could have been anywhere. General Eisenhower came and spoke to them on 25 May and three days later they were told that they were on six days notice for movement. On 11 June they moved to their assembly area, Fort Gomer at Gosport and on the 18th they moved down to the docks to embark. The next day their landing craft were hove-to off Courselles-sur-Mer in steadily worsening weather. Some elements came ashore in France on the 20th but the rest remained afloat, no doubt in increasing misery, until the 22nd.

Clearly a posed view but it embodies the period of waiting for embarkation. The partly camouflaged tank sits at the roadside while its crew pretend to be spotting for hostile aircraft.

In reality there is a lot of hard work to do. Here a crew from 9th Royal Tanks, assisted by REME fitters, prepare a tank for wading in a street at Gosport near Portsmouth.

During that time one LCT was mined; they lost two men and six Churchills.

Ninth RTR were at Hove when postal censorship was imposed and they went to their concentration area at Farnborough on 8 April. Sir Charles Broad, one of the great Royal Tank Corps pioneers of the inter-war years, visited the regiment on 4 May and on the 13th they took delivery of 14 US half-tracks. During the period 18 to 25 May they were busy converting 17 Mark III Churchills from 6-pounder to 75mm, which seems to have been done by their own REME personnel assisting a team of travelling fitters. On the 31st they took delivery of six Crusader anti-aircraft tanks, fitted with twin 20mm Oerlikon guns, which would form their AA Troop. On the 12 and 13 June they moved down to Gosport and sailed at 2300 hours on the 18th. Their arrival off the French coast coincided with the height of the great storm, which effectively wrecked a large proportion of the artificial Mulberry Harbours. It had been building up for days, blowing force 6 or 7 from North North West, veering North East during the night 19/20 June and turning the entire Bay of Seine into a lee shore. Thick, low cloud was racing overhead and the seas building up, even within the protected area created by the blockships. Many craft started to drag their anchors and some were forced to beach in order to save themselves. The landing craft carrying regimental headquarters lost its anchor, followed by two kedge anchors and then the barrage balloon that it was flying. Then one engine failed and it was pure luck that it was not wrecked. Most of the men were seasick but the regiment came ashore intact on the 22nd.

The 107 RAC, which has already been mentioned in this chapter, was also addressed by Monty, at Dover on 3 February but its war diary for the months up to D-Day seems to be obsessed by the weather. On 28 May it transfers to its concentration area where it remains at six hours notice to move. It must have been a frustrating time. Nearly a month goes by. D-Day comes and is gone. The regiment moves into its marshalling area on 21 June and then sits out another long wait. Finally, on 3 July, they move down to Gosport for embarkation, and war. They came ashore at Graye-sur-Mer on Juno Beach at 2130 hours that same night.

Here they joined with 147 RAC, which had landed at Colombieres-sur-Seulles on the previous day and the third regiment of 34th Army Tank Brigade, 153 RAC. Their war diary chronicles many more details of the preparations for this great event. They were at Broome Park near Canterbury in February 1944 and a month later were still firing 6-pounder armour piercing rounds on Seaford ranges in Sussex. On 14 April they returned 23 Rota Trailers to stores and no doubt heaved a sigh of relief that they did not have to take them to France. Theirs is the only war diary inspected for this book to mention Mark VII Churchills at this time, of which they received three on 24 April. More came over the next few days but they also acquired a similar number of Mark III* Churchills, those mounting the 75mm guns. The men were issued with evasion kits on 25 May which, presumably, were intended to help them escape if they found themselves behind German lines and on 20 June they were issued with French currency. They embarked at Gosport on 2 July and came ashore the next day in France.

In our study of the Churchill tank and its gradual acceptance by the British Army we have, so far, not mentioned three important regiments. These are 4th Battalion the Grenadier Guards, 4th Battalion the Coldstream Guards and 3rd Battalion the Scots Guards which, together, formed 6th Guards Tank Brigade. A Guards Armoured Division had been formed in 1941 but when the composition of armoured divisions was revised in 1942 an independent armoured brigade was formed from the redundant regiments and it was equipped with Covenanter tanks. It converted to an army tank brigade early in 1943 and was equipped with Churchills but its future was never secure. Twice between then and the invasion of Europe it was threatened with being broken up and the men sent as reinforcements to Guards infantry battalions in Italy. But each time an appeal to the Prime Minister saved the day.

The brigade spent most of 1943 and 1944 in the north, training with various infantry divisions in the art of tank/infantry co-operation. As we have already seen the brigade was obliged to hand over its new 75mm gun Mark VI Churchills to 31st Army Tank Brigade in April 1944. This was a direct consequence of the second attempt by some authorities to return the brigade to an infantry role and before they finally embarked for France the 6-pounder tanks they had received instead had to be converted to 75mm by brigade workshops. They moved to Kent in April and one gets the impression that the county was simply crawling with Churchill tanks at this time. However they were still there on Sunday, 24 June, when a V1 Flying Bomb, intercepted by an RAF fighter on its way to London, came down on the brigade workshops killing 51 men and injuring a further 40. The loss of so many skilled men just a few weeks before going to war was a serious blow. Coincidentally the brigade held a Church Parade in Canterbury Cathedral one Sunday in June when the service was taken by the Archbishop. One young tank commander in the Scots Guards, Lieutenant R A K Runcie, who may well have attended that service, subsequently became Archbishop of Canterbury in 1980.

The brigade sailed from Southampton Water on 19 July and came ashore on Gold and Juno beaches the following day. According to their history the regiments found that the Military Police guides who had been instructed to meet them did not turn up and it took some time for the various tanks to find their way to the assembly area near Bayeux.

10

Churchills in Normandy

The Churchill tank played such a significant part in the battle to free Europe, and so many photographs can be found, that one is tempted to assume that there were dozens of regiments of them over there, supporting the Allied infantry in all their operations. Yet if we disregard the various specialised units, and concentrate on conventional regiments it is perhaps surprising to discover that there were never more than eight, and from August 1944, almost until the end of the war in Europe only seven. These were the three regiments, or battalions as they called themselves, of 6th Guards Tank Brigade. The two (7th and 9th RTR) of 31st Army Tank Brigade and the three (107, 147 & 153 RAC) of 34th Army Tank Brigade. It was 153 RAC which was disbanded that August. This means that for most of this period there were never more than 350 Churchill gun tanks with fighting regiments at any one time. Their effect on the campaign, at certain crucial times, would appear to belie this figure.

One reason for this was the degree of mutual confidence that grew up between infantry battalions and the tanks they worked with; another is neatly summed up in a comment made by the war diarist of 9th RTR; *the Churchill stands up well to anti-tank fire. If set on fire it burns slowly and the crews have a good chance to bale out - Panthers, on the other hand, tend to blow up.* The Churchill was a survivor and this was just as true of its resilience to anti-tank mines. Or look at it another way; 4th Grenadier Guards, which saw as much hard fighting as any Churchill unit in north west Europe, only suffered 54 fatalities - five officers and 49 other ranks, during the entire campaign up to VE Day.

On 25 June the Churchills of 7th and 9th RTR moved from their assembly area within the beachead towards the village of Cheux, west of Caen. It was the start of an offensive codenamed Epsom. Caen had been a prime British objective for D-Day itself but for a variety of reasons this proved unattainable. Fully aware of its key location, German troops quickly poured in to defend the city and before long there were five Panzer divisions and one infantry division in the area. Since a head-on assault now appeared costly Montgomery planned to encircle the city, going west about, and Epsom was planned to that end with an initial aim of crossing the Odon river and creating a bridgehead on the southern side. The great storm delayed plans for the operation by four days but even this hardly gave the two Churchill regiments time to get over the effects of sea sickness and clear the waterproofing from their tanks before they were catapulted into action.

Things had changed a good deal since the fighting in Tunisia and it was now common practice to allocate one squadron of tanks to each battalion of an infantry brigade and in this instance 7th RTR was supporting 46th Highland Brigade in 15th (Scottish) Division. Moving out to cross the start line at 0730 on the morning of 26 June B Squadron was accompanying 2nd Battalion the Glasgow Highlanders while C Squadron worked with 9th Battalion the Cameronians. A Squadron, with 7th Battalion the Seaforth Highlanders acted as reserve.

It is probably only fair at this stage to point out that there was a lot more to this, and most subsequent battles, than a couple of regiments of Churchill tanks and their infantry. Epsom was a substantial operation involving one armoured division, an independent armoured brigade and two other infantry divisions but it is clear why the two Churchill regiments were placed under command of 15th Division (9th RTR was supporting their 44th Brigade) since this was the central punch of the entire attack. Their task was to break through the German defences and hold the bridges over the Odon river that would enable 11th Armoured Division to cross. The historian Sir Basil Liddell Hart has described Epsom as a strategic failure but a tactical success and there is no doubt that it was a hard fight. Not the ideal situation for men in relatively untried tanks to begin their combat careers. Both leading squadrons of 7th RTR ran onto minefields early on, C Squadron lost five tanks here and B Squadron four but they were in Cheux by 1100 hours. The village only falling after some grim hand to hand fighting by the Glasgow Highlanders who took over 200

An instructive picture of Churchill tanks of 147 RAC at rest. Most of the crews have acquired some spare track for added protection but they have all fitted it in different ways. The Mark V, foreground left, has its turret and nose covered while the tank closest to the camera has some covering the side door, more over the driver's visor and an unusual arrangement on the front of the turret.

casualties during their very first action. A Squadron, following up, was engaged by enemy tanks on the right flank which resulted in the deaths of one officer and ten other ranks.

On the following day C Squadron, supporting the Highland Light Infantry, assisted in an attack to gain a river crossing at Gavrus, south of Grainville-sur-Odon, during which they got into a fire fight with some Panthers and Panzer IVs. They claim to have knocked out three enemy tanks. At 1800 hours on the 29th the bridgehead was subjected to a powerful counterattack by 1st SS Panzer Division which cost the British twelve Churchills. C Squadron remained in a defensive position while B Squadron pulled back to rally.

Meanwhile 9th RTR was in action nearby. On 26 June C Squadron attacked the village of Colleville, on the road from Cheux to the other river crossing at Tourmanville. The Churchills of 9th RTR claim one Panzer IV and a Panther for nine of their own but they were unable to hold the village in the face of strong opposition. The Odon was crossed by the Shermans of 23rd Hussars on the 27th but the bridgehead was attracting German armour from a wide area, including Tiger tanks of 1st SS Panzer Corps. The Ninth's war diary records that on the evening of the 28th they were strengthened by the arrival of five new Churchills and ten extra crewmen from 269 Field Delivery Squadron which is rare recognition for one of the many invaluable support units which kept the front line battalions up to strength.

South of the Odon there is high ground which overlooks the next valley, that of the river Orne. This high ground is dominated by a feature known to the British as Hill 112 from its highest contour. British forces were established on its northern slopes by the 28th but they got no further for the present. All the while Caen and the high ground to the west was held by German forces they could pose a threat to the flanks and indeed counterattacks continued on the north side of the Odon for some time.

By the evening of 9 July British and Canadian troops had effectively liberated Caen and attention was again turned to the range of hills south west of the city and in particular Hill 112. Once again 31st Tank Brigade was called forward, this time with 349 Battery, Royal Artillery under command, equipped with American built M10 tank destroyers. For this attack 7th RTR was to operate with A Squadron supporting 4th Battalion the Wiltshire Regiment and C Squadron working with 4th Battalion the Somerset Light Infantry while B Squadron was held in reserve with 7th Battalion the Hampshire Regiment. Panther tanks, backed up by dug-in Tigers, presented formidable resistance and neither Churchill regiment made very much progress once they arrived on the southern slopes of the hill.

A Churchill of 7th RTR during the Epsom operation approaches the edge of a field with infantry in close support.

A Churchill IV, probably of 4th Battalion Grenadier Guards, passes a ruined church and the battered remains of a Panzer IV during its advance.

In their war diary 9th RTR remark that "the closeness of the country around Maltot (a village on the southern side of Hill112) made it almost impossible to locate these tanks until too late" but they were learning. There was little to be gained in trying to slog it out with Tigers, or even Panthers so the new practice was to call down medium artillery fire upon them. After a certain amount of this treatment the German tanks would withdraw into adjacent woods or orchards and the process began all over again. The Seventh seem to have had a relatively easy time of it. They claim nine Panthers and Tigers knocked out for eight of their own Churchills with four officers and one trooper killed. Ninth, on the other hand, suffered badly. They make no claims for enemy tanks but lost sixteen of their own along with 1 officer and four other ranks killed but 29 other ranks are given as missing.

The German defenders were told that Hill 112 was the pivotal point of the whole position and in no circumstances may it be surrendered, but the Duke of Cornwall's Light Infantry held it for most of the night of 11/12 July and the Germans were never able to recover it entirely. The fighting here would be resumed in a few days. Having created a salient to the west of Caen Montgomery launched Operation Goodwood to achieve the same thing in the east. It was the responsibility of British VII Corps and featured three armoured divisions but none of the regular Churchill regiments. Goodwood began on 18 July but three days earlier the Churchills were operating again around Hill 112. British forces in this area now swung to face roughly south west and 31st Tank Brigade was joined by the three Churchill regiments of 34th Army Tank Brigade. They had come forward from their assembly area earlier in the month in case they were required for the previous action but now it was time for the real thing.

For 107 RAC it started at once. At the forming up point, where they met up with their infantry on the 15th, heavy artillery fire and mortars meant that the tanks stayed closed down and there was little communication with the infantry. Moving off at H-Hour the tanks found that they were virtually blind. Operating closed down, in a cloud of smoke and dust kicked up by the barrage, they could do very little but, fortunately, the infantry attack was a success. Meanwhile 147 RAC, supporting 7th Battalion the Royal Welch Fusiliers on 17 July, had a different experience. With A Squadron leading they started to move forward through waving cornfields. The tanks found that every time they halted the infantry went to ground behind them and it was a problem to get them started again. In due course the tanks were told to advance without halting and they soon got well ahead of the infantry. German troops soon started popping up and surrendering in large numbers

A long column of Grenadier Guards' Churchills in the Normandy countryside. All of the tanks have local camouflage applied to their turrets and the rearmost tank has its turret reversed.

Proof, if it were needed, of the confused nature of the fighting in Normandy. Churchills of Guards Army Tank Brigade, going one way, pass Shermans of Guards Armoured Division moving the other way. On the Continent they observe French rules of the road. The location is near Vassy.

but without infantry the tanks were unable to handle them. Behind them B and C Squadrons were called up to keep company with the infantry while A Squadron, now unsupported, took most of the punishment. They lost eleven tanks but suffered no casualties and as darkness fell were pulled back. On 16 July, while they were preparing to move off, 153 RAC was attacked by a dozen Focke-Wulf 190 fighter-bombers which swept the area with machine-gun fire and bombs. The regiment suffered no casualties but they claim to have got one of the aircraft. During the subsequent action five tanks from A Squadron were seen to pass over the crest of a hill and lost to sight. A few days later all five tanks were found, burnt out north of Evercy. One of these tanks, named *Egypt* contained the regimental Intelligence Officer who appears to have had the war diary with him so the surviving document is a reconstruction, compiled from memory.

On the morning of 17 July tanks of 9th RTR carried out an indirect shoot using 75mm and 95mm guns against German positions which they claim to have produced modest results. Trying the same trick in the afternoon brought swift retaliation from a battery of *Nebelwerfer* mortars. From then until the end of the month the regiments were in and out of action almost on a daily basis. On the 22nd, 7th RTR took part in Operation Express with the Sherman Crab flail tanks of C Squadron, 1st Lothian & Border Horse under command and the next day the 7th lost eight Churchills from C Squadron when operating west of Maltot. Pulling back into the village they were able to claim one Tiger. On 29 July their Anti-Aircraft Troop of six Crusader AA tanks was disbanded. The flat wheatfields north of the Odon had now given way to the picturesque bocage, so typical of Normandy. From the air it looked like a random patchwork of tiny fields surrounding small farms, but on the ground it was a maze. The narrow country lanes ran between high banks and these would also be around all the fields. Each field was bordered by a high earth bank, strengthened by stones lifted from the fields and bound by shrubbery and small trees in the top. Many of the fields contained orchards which increased the available cover for defenders and helped to fill the tanks with hundreds of small, hard apples if they drove through with turret hatches open.

Indeed it was country designed for the defender. Infantry had a very hard time moving from one field to another and it was even worse for tanks. At least Shermans and Cromwells found it difficult and preferred to call in Royal Engineer bulldozers to break the banks down, or use the improvised blade developed by US Sergeant Curtis Culin which the tank used

Fourth Battalion Grenadier Guards again and a Churchill leading infantry through standing corn. The turret crew are clearly not expecting immediate action.

to cut its own way through the bank. None of this really bothered the Churchills very much. They found it quite easy to climb the banks, as we shall see, so instances of them fitted with Culin's prong device are rare; the only problem was what might be waiting for them on the other side. The Germans knew the country well and intended to fight for every field, but the Allies were learning fast.

Planning with their infantry 9th RTR worked out a routine for handling bocage fields. Infantry would lead the way, clearing the far side of the bank for the tanks which would then come crashing through. Here they would pause and lay down high explosive and machine-gun fire on the next hedge line while the infantry worked their way down the sides of the field. They would then clear the defenders from the far side, call the tanks through and start the process again. One problem they could not solve was the follow up, especially the infantry battalion's anti-tank guns. These could not deal with the banks at all and by the time they got through their infantry could easily be under counter attack. The tanks were unable to leave them in such situations and there were times when they might have to keep the infantry company all night.

Reports from 34th Army Tank Brigade contain very little that is new from this period. During the move towards Gavrus on the 28th, 147 RAC explain that they made their tanks travel

One of a famous series of photographs taken during the advance towards Caumont. Churchills of 6th Guards Tank Brigade carrying infantry of the Royal Scots Fusiliers along a taped route which has been swept clear of mines.

slowly and in small groups to keep down the great clouds of dust that might otherwise give their position away. At Gavrus each tank moved into a special, bulldozed space where it would wait until summoned in the event of a counter attack. On the 28th, at Maltot, 153RAC has a tank hit on the glacis plate, at which point the plate broke away and dropped into the driver and hull gunner's laps, injuring both men. The same complaint can be found in other accounts of Churchill fighting which suggests either that it was endemic or, like the old cupola horror story, one of those exaggerated tales that seem to entertain the morbid soldiery. It would be interesting to know whether this fault, if it was a serious one, applied to the Mark VII or earlier types. If so then it suggests a problem with the welding, which might not be surprising, bearing in mind the thickness of plate being supported at this point.

The horrors of the bocage. A Churchill rears up to cross a field bank during the advance from Caumont to Hill 309, exposing its vulnerable underside to the enemy.

Considering the number of Churchill regiments in France their losses, in terms of tanks and men, were not great, but it was still necessary to keep them up to strength. This duty fell to the forward delivery squadrons (FDS) of the Royal Armoured Corps and, since very little has ever been written about them a mention here might be appropriate. Number 269 FDS has already been mentioned and, from a study of its war diary it would appear to have been dedicated more or less specifically to 31st Army Tank Brigade. Certainly from early 1944 it appears to specialise in Churchills and even joins the fighting regiments on exercises. Even so it was not just there to deliver tanks. Its tank crews were fully trained in the various combat roles and ready to replace those lost from fighting regiments as required. The FDS move to Farnborough when the tank regiments moved to their concentration areas but crossed over to France somewhat later.

They boarded the landing craft at Gosport and Lee-on-Solent on 19 June but spent a few days moored up in the Solent while the great storm blew itself out. Their convoy, named Starlight 37, sailed on the evening of 22 June. It consisted of 35 landing craft escorted by two motor torpedo boats and a French destroyer. All arrived safely off the Normandy coast despite an air attack in mid-Channel and landed near Courselles-sur-Mer about noon on the 23rd. They arrived with 26 Churchill tanks (some adapted to take Crocodile equipment for 141 RAC which formed part of the brigade at this time) four Stuarts, a Loyd Slave Battery Carrier and sixteen B vehicles. Once off the beach they followed the route to their concentration area which was marked with the green diablo sign of 31st Army Tank Brigade. From their base at St Gabriel they sent seven tanks up to 7th RTR on the 25th while the remainder, with their crews, were held at two hours notice to join whichever regiment might need them.

Their war diary shows that on the 28th they supplied six Churchill Mark VI, along with 3 officers and 27 other ranks to

A panoramic view of wheatfields between Cheux and Saint Manvieu on 26 June 1944. Churchills of 7th RTR in the distance with infantry from 6th Battalion, Royal Scots Fusiliers.

Viewed from the other side the Churchill crashes down into the next field while a file of infantry prepares to follow.

7th RTR. Three Churchill Mark VI and one Mark V with 15 other ranks to 9th RTR and two Churchill Observation Post (OP) tanks and eight other ranks to brigade headquarters. Their commanding officer visited the Armoured Replacement Group from which they drew their Churchills and trained men on the 29th, indicating a further link in the chain that replaced battle casualties in both machines and men. On 3 July they were at a new base near Secqueville-en-Bessin where they were joined by a battery of 5.5 inch guns. Confident that what they were doing was useful, the men of 269 FDS came to the conclusion that the only thing the artillery battery could make was noise. The Crocodile regiment, 141RAC, left 31st Army Tank Brigade on 4 July and the Field Delivery Squadron's war diary notes that all first line refits, presumably meaning Churchill Crocodiles which the FDS was holding were to be handed over to the regiment.

On 6 July eleven tanks with full crews were sent to 7th RTR and ten went to 9th RTR. However this regiment sent most of its new crews back, which the FDS interpreted as proof that many crews were surviving, and remaining fit, even though their tanks were knocked out. Indeed crews without tanks, which the war diary refers to as 'unhorsed', are soon turning up at the FDS to collect new machines. On 10 July ten Mark VII Churchills, with crews, arrive from 258 Corps Delivery Squadron and these are forwarded to 9th RTR on the 11th. It is an indication of how closely the FDS identified with its brigade that their war diary carries brief details of the actions that their fighting regiments were engaged in. They move to Putot-en-Bessin on the 12th, to a field full of half buried cattle which they first have to dispose of. On the 14th they note that it had been decide to form a troop of OP tanks at brigade headquarters, since casualties to these tanks and their crews was so heavy. Apparently the original scheme was to hold OP tanks at the FDS and send them out as required.

They also suffered from a fair amount of attention by enemy aircraft. They seem to be visited most nights and on the 16th remark that strafing hedges appears to be the enemy's

In September 1944 B Squadron 141 RAC was rushed to Brest to assist American forces to reduce the fortress there. This was a Crocodile regiment but it included conventional gun tanks on its strength. Two of which are shown here preparing to move forward. These crews have gone to some lengths to camouflage their turrets.

latest fad. However on the 17th they are told that air attacks may not be dealt with by small arms because there have been instances of mistakes in aircraft identification which 269 FDS blames on 'some other unit'. By 23 July they note that it is now trained men who are becoming scarce. On the 25th they have to admit that scarcity of tank crew refits coming from 2nd Armoured Replacement Group prevented full reinforcing of brigade casualties in both tanks and men. Up to the 26th they learned that 43 officers,414 other ranks and 94 tanks had been sent forward to the brigade but as far as more men were concerned 'reinforcement channels remain dry'. Two days later the OP Tank Troop is detached from Brigade HQ and sent back to the FDS so clearly that experiment did not work. On Sunday, 30 July, nine Churchill VIIs were removed from 9th RTR who received nine Churchill IVs and VIs instead. The Mark VIIs, according to the war diary, needed modifications. Could this have anything to do with the damaged glacis plate referred to earlier? Is the modification an attempt to strengthen it?

We last encountered 6th Guards Tank Brigade ashore near Bayeux on 20 July. Most of them had an uncomfortable crossing and these three regiments also pay tribute to the U. S. Navy landing craft crews who offered to share their excellent rations with the Scots tank men. Unfortunately nobody was in a fit state to take advantage of the offer and, indeed, one battered old LCT carrying Churchills of the Scots Guards had to turn back when some of its tanks broke loose half way across the English Channel. Once ashore, with time on their hands, the men of all three regiments started poking around the graveyard of wrecked tanks which was steadily building up behind the beaches. It rapidly became apparent that, well protected as the Churchill was, the Germans had weapons that could make holes in it. And it needed very little imagination, having peered inside some of these tanks, to appreciate what went on when a tank was hit. Thinking upon this the men wandered back to their tanks, and then back to the graveyard again where they started to collect up all the spare track plates they could find; Churchill, Cromwell, Sherman, it did not matter. These were dragged back to the tank park and welded all over the front and turret of the tanks until some of them were hardly recognisable.

An experienced British tank officer, having observed this practice in Italy, reported on it to the War Office. In his view the covering of old track plates made absolutely no difference at all, apart from increasing the tank's weight and slowing it down, but he noted that crews had unbounded faith in the practice. Some were happy to explain to him how the irregular surface of the track would deflect shot, when in fact it was just as likely to trap it, while others told him that they preferred to have the track hanging loose so that it swung with the incoming round and slowed it down. Nonsense it might have been but the crews believed that it helped them and as long as they were happy officialdom saw no reason to forbid it.

The stubborn fighting that characterised the campaign on Normandy took a new turn on 25 July when American forces launched Operation Cobra which broke through the German positions at St Lô and started a sweeping move designed to encircle the main enemy concentration facing the combined British, Canadian and Polish forces. Even so it was no joy ride and the advance down the valley of the Vire River, south east from St Lô, was hampered by German positions on the heights to the east, including the dominant Mont Pincon. The British were requested to neutralise this from the Caumont position which they had taken over from the Americans and the result was Operation Bluecoat. As usual it is not our intention to cover the entire battle but merely to deal with that part of it which involved Churchill tanks, and in this case it is a very particular one. This would be the battle debut of 6th Guards Tank Brigade.

The three Churchill regiments would move out from the hilltop village of Caumont and drive due south to take the objective known as Hill 309, about five miles away. *En route* the Scots Guards would swing to the left and take a ridge above the village of Les Loges in order to cover the eastern flank. The Guards would be working, in the main, with 15th Scottish Division with whom they had trained for the past twelve months in Britain. The understanding between these two units, coupled with the combat experience already gained by the Scottish infantry and the exceptional professionalism of the Guards augured well for the battle. Churchills had been selected for this part of the operation because of the nature of the country between Caumont and Hill 309. Within a triangle of roads whose apex was Caumont was a maze of tiny streams, orchards, small farms and hills, crossed north west to south east by one narrow lane. Keeping direction would be quite difficult enough, and the opportunities for ambush were extensive but to make matters worse the area had been sown with mines at various times by German, American and British troops when they were in the area and this had a serious effect on the progress of the infantry.

The start of the action, on 28 July, saw 4th Battalion the Grenadier Guards descending the slopes from Caumont and taking two important objectives within the first hour. These were Luvain Wood, at the foot of the slope and the village of Sept Vents on the Caumont to Vire road which was the western side of the triangle. Down behind them came 3rd Scots Guards who soon found their infantry lagging behind. Within an hour orders came through that they should go on alone and the Coldstreams on their right got the same message. It was a choice between waiting for the infantry or taking advantage of the artillery barrage which could not change its timetable. So the tanks went on, rearing up over each of the bocage field boundaries and then crashing down on the far side in what has been described as a sort of armoured steeplechase. In their regimental history the Scots Guards claim that no other Allied tank would have coped with the conditions but it meant a rough ride for all crews and a lot of punishment for the tanks. However they took their hilltop position and sat down to wait for the infantry. These were later reported to be in the village behind the ridge but their support weapons and transport were still miles back, snagged up in an enormous

A fine illustration of just how much dust a Churchill could throw up when moving fast along a dirt road.

traffic jam and the Scots Guards had to remain in front. About 6pm the Germans put down an artillery barrage on the ridge which forewarned of a counterattack. This came suddenly in the form of three new Jagdpanther 88mm self-propelled guns, the first of their type to be encountered by British troops. They appear to have come onto the ridge from the north west, one remaining in support while the other two, making good use of natural cover, raced through the British regiment destroying tanks as they went. The Scots Guards retained an infantry style by referring to their three Squadrons as Right Flank, S and Left Flank Squadrons; and it was S Squadron that bore the brunt of this attack. Within seconds eleven Churchills were blazing hulks, at least one had its turret blown clear off. The remaining tanks fired as best they could at the departing Germans but they all appear to have had high explosive rounds loaded. Even so, when the battlefield was examined a few days later, two of these big SP guns were found below the ridge with their tracks broken and one of these is attributed to a 75mm HE round fired by one of the Churchills.

The Coldstream Guards, having left their infantry behind, set off down the Vire road and made good progress until they arrived outside the village of La Morichesse les Mares at about 3pm. Here they learned that the village was infested with Panthers so the CO swung the regiment off the road and went, bald headed for Hill 309 by a cross-country route. The three squadrons (simply numbered 1, 2 and 3 as were the Grenadiers) spread out abreast and crashed through the orchards which effectively wiped off many wireless aerials. They approached the hill at a point where the Germans believed it to be quite impossible for tanks to climb, but the Germans did not know their Churchills. Changing down the tanks dug their tracks in and worked steadily up the slope. One rolled over but the rest made it to the top and deployed.

The Grenadier Guards moved out later in the day, carrying infantry of the Glasgow Highlanders on the tanks. They also got caught up in a major traffic jam and having reached an orchard below Hill 309 decided to drop the infantry off there. Throughout the 29th and 30th the Germans tried every

Church towers provided excellent observation platforms for German artillery observers so they were invariably destroyed. Two Churchills rumble past this one and the leading tank is very well covered with extra track.

means they knew to throw back the British armour. Unfortunately although the Churchills and their infantry had done exceptionally well the armoured and infantry division working on each flank, which in fact represented the main theme of Bluecoat, were seriously delayed. The Grenadier Guards were involved in fighting around the Bois du Hommes, a large wood to the east of Hill 309 where they reported finding some dummy Panther tanks made of wood. They regarded snipers as one of the worst threats to tanks in this close country and claim that the best antidote was a ten inch high armoured plate welded behind the cupola.

The Coldstream Guards caught up with some real Panthers a week later. On one occasion at the village of Estry a Coldstream Churchill and a Panther met face to face and the British tank fired first. Unfortunately it was loaded with high explosive which did not hurt the Panther. The German tank then replied and destroyed the Churchill. It is worth noting that here we have seen two cases of infantry tanks having high explosive rounds 'up the spout' when encountering enemy tanks. This raises interesting points. Clearly when working with infantry high explosive was essential to deal with enemy machine-gun and mortar positions, along with infantry. However it may be that the 75mm armour piercing round was considered so ineffective that they did not bother to use it very often. It is worth noting, for example, that when the Scots Guards approached the village of Chenedolle and were warned that it contained a Panther tank, they blocked all the exits from the village and summoned up some 6-pounder Churchills to deal with it. When they arrived the Panther crew drove their tank into a barn and set it alight. Of course by this time there was a discarding sabot (APDS) round for the 6-pounder which had a far better penetration figure than any projectile supplied for the 75mm, at least at medium range. Not that the sabot round was very much use against a Tiger. The comment by the Coldstreams, that even sabot rounds bounced off Tigers like ping-pong balls, has been quoted by many writers and they claimed that the only way to deal with Tigers was to call down a medium artillery barrage which, if it did not knock the German tank out, at least encouraged it to clear off.

While Operation Bluecoat was going on the other Churchill regiments were still fighting on the Orne but it is a repetitive story of attack, counterattack and waiting. One or two incidents from this period will suffice to give a flavour of events. Having crossed the Orne on 7 August 107 RAC covered their infantry (7th Bn North Staffordshire Regiment and 7th Bn Royal Norfolk Regiment) when a German counterattack came in. The infantry withdrew, leaving the Churchills to face the on-

Crew and fitters work on the track of a Churchill Mark VI while another crewman keeps watch from the turret. A Churchill Armoured Recovery Vehicle Mark I has arrived to help and the mugs reveal that tea has been provided for the workers.

slaught of Panther and Tiger tanks attacking their flanks, and many tanks were lost. Still they held on and next day B Squadron came up from reserve to relieve the remnants of A and C Squadrons for the next night. The following day they were relieved in turn by 147RAC.

On the 12th this regiment, 147RAC, was involved in an attack on the village of Thury-Harcourt. They lined the tanks up on a hill overlooking the village and pounded every door and window with high explosive shells and concentrated machine-gun fire. This went on undisturbed for 45 minutes. Under this tank generated barrage it was assumed that the infantry (2nd Battalion the Glosters) were doing well but when a signal came from the battalion CO for more help it revealed a different state of affairs. The Glosters were so depleted that they would not be able to support the tanks within the village, and in any case the route down from the hill was difficult to negotiate, so it was decided to withdraw. The tanks, still in their substitute artillery role, now laid down smoke shells to cover the infantry as they came out.

While on the subject of 34th Army Tank Brigade this is probably the place to record that on 18 August, at a conference called to consider manpower problems, it was decided to disband 153 RAC. Their war diary records that the last tank left the regiment on the 21st and the remaining personnel were posted away on the 28th.

Finally in this part a brief look at 7th RTR. On 2 August they were ordered to take part in what was described as an Armoured Drive against the Villers-Bocage to Caen road, following up an enemy withdrawal. They moved off at 16.30 hours with a squadron of flail tanks and a troop of Churchill AVREs under command. They met very little enemy resistance but mines were a problem to the extent that they lost seven Churchills and two flails along the way. On 17 August the regiment was transferred to 34th Army Tank Brigade and made what the war diary describes as 'a remarkable run' of 54 miles by road to the Falaise area. What made it so remarkable was the high standard of reliability. They were on the road from 0900 hours until midnight but suffered very few mechanical breakdowns. The regiment was now in for a lengthy period of rest and maintenance when, among other things, according to the war diary, they were busy lifting the turrets off their Stuart reconnaissance tanks.

Operation Astonia, the attack on Le Havre, was scheduled for 10 September 1944. It was to be the first of a series of assaults on German held ports along the Channel coast that had been by-passed by the main Allied forces which, even now, were sweeping forward into Belgium and Holland. The Churchill regiments that took part tend, in their histories and war diaries, to make light of the ensuing action although it was hardly an easy battle. The 107 RAC got through their part of the proceedings without incurring any casualties at all, and the same appeared to have been true of 9th RTR. The attack on Le Havre was co-ordinated by Canadian I Corps. It could only come in from the north and east because of the location of the city port, on the north side of the Seine estuary. It was first necessary to break through a fortified outpost line and then reduce the defences of the city itself. The three Churchill regiments were all under command of 49th Infantry Division which took the eastern sector in the attack and their main task seems to have been tackling occupied enemy pill boxes. They had help of course, from the flails and AVREs of 79th Armoured Division and the Churchill Crocodiles of the Buffs. These gave an excellent example of their powers at one stage when 9th RTR was dealing with a pill box. After a bout of shelling the enemy put up white flags, so an interpreter was sent forward to negotiate. He was told that a brief

The assault on Le Havre. Churchills of 7th Royal Tank Regiment at their forming up point in the evening. Notice that the Mark VII in the foreground has no track guards at all.

Crews from 7th Royal Tanks watch as Lancasters of the Royal Air Force batter Le Havre before the attack goes in.

With the fighting in Normandy over it is time to move east. *Bugler* is guided onto its transporter. The appliqué armour shows up well and notice that the centre section of the track guard has gone to avoid the risk of jamming the turret.

respite was required in order to deal with casualties and after waiting for a while hostilities were resumed. At this point a couple of Crocodiles came up and laid on a brief demonstration, at which point the white flags appeared again, and this time the British infantry just walked in.

Seventh Royal Tanks certainly had a tougher time of it. First they had a substantial minefield ahead of them and they had to wait while three lanes were cleared by flail tanks. Three of these were knocked out by anti-tank fire coming in from the right flank but when all was complete the Churchills of C Squadron started across, only to lose four of their number on mines. B Squadron, operating on the left, came through unscathed. By 21.30 hours all positions were taken and A Squadron had come through to provide cover for the night. B Squadron managed to rally but when C Squadron tried to go back there was so much traffic congestion that they got stuck. This meant that at dawn the wrong squadrons were in the wrong place so A Squadron had to improvise and when it ran low on supplies it was replenished from turretless Stuarts which managed to get through the minefields. Most of their work seems to have been blasting away at pillboxes with 75mm HE which usually appears to have had the desired effect. The battle for Le Havre effectively ended on 12 September and when it is remembered that British troops were in Brussels on the 3rd, and Antwerp on the 4th while First and Third U. S. Armies were even further east it illustrates how quickly things were moving.

Churchills had no place in this kind of warfare, even if they could catch up. It is interesting to note that such progress as they did make was almost entirely on their tracks. Transport and fuel was at a premium. Those tank transporters which were available were rushing Shermans forward to the fighting front so the Churchills would just have to manage. As an example on 17 September, the day the airborne operation to capture Arnhem was launched, 7th RTR made a 50 mile road run to a place called Belmesnil where it went into a reserve area for maintenance and training. Further eastward moves, starting on 29 September took them via Abbeville to a location south of Dieppe where they found themselves resting close by 9th RTR. At a dinner, given for Lieutenant Colonel A R Leakey, the commanding officer of 7th RTR, by the officers of 9th RTR, the Colonel was presented with a model representing a Churchill tank disabled in a minefield. It was intended to honour his prowess and leadership in discovering more minefields in France than any other unit commander and is described as being received with 'grave and dubious laughter', as well it might be.

11

Campaigning in Italy

When the Allies landed in Sicily, in July 1943, and subsequently on the Italian mainland in September they did not take any British built tanks with them at all. This was understandable in the case of the Crusader, and not unreasonable when one considers the Valentine, but why not the Churchill? Two more or less official answers can be traced. One concerns the nature of the country. Southern Italy was largely rural and underdeveloped compared with the rest of the country so the roads were simply incapable of handling heavy traffic. Bridges, and even under-road culverts, could easily be damaged by heavy vehicles and the difference in weight, something close to 10 tons, between the Sherman and Churchill, excluded the latter. However the men of the Churchill regiments in Tunisia believed it was the gun. The high explosive round supplied for the 6-pounder was a mere firecracker when compared with that fired by the American 75mm gun and since experience had shown that tanks supporting infantry needed an effective high explosive capability this was another possible reason for discarding the Churchill.

Other explanations have been offered. One being General Montgomery's belief that the difference between cruiser and infantry tanks was largely academic and that an all-purpose tank was preferable. It was also claimed that the two Churchill brigades had taken such a battering in Tunisia that they were in no fit state to take part in another campaign. However the underlying reason could easily be that British tank troops in the Mediterranean theatre had simply lost faith in British tanks and favoured the reliability of American machines.

Thus it was that the 21st and 25th Tank Brigades spent a frustrating twelve months in Algeria and Tunisia between the Victory Parade in Tunis and the time when they finally crossed over to Italy. Since it is obvious that it would not have taken a year to restore the two brigades to combat readiness one has to assume that the authorities in Italy were quite happy with their Shermans until they came up against the formidable German positions in the north of the country. They had a foretaste of this at Monte Cassino, but for now we must observe the regiments in North Africa.

Selected entries from the various war diaries give a flavour of this period. The 21st Tank Brigade was based for most of this time at Penthievre in Algeria. For 12th RTR this was a time for moving tanks about. Ten of their Churchills went to 145 RAC in August and in order to make up numbers they were equipped with a variety of other tanks. For instance their brigade headquarters, in September 1943, consisted of one Churchill III and four cruiser tanks, which we subsequently learn are Crusaders. Later that month they initiated Exercise Greaser, the purpose of which was to run in newly delivered Universal Carriers and solo motorcycles.

In October they give their strength as 48 Mark III and IV Churchills plus four Mark I and on the 20th they begin Operation Kismet which ended on the 22nd. Two days later a signal from brigade announced that, in order to conserve their tanks they would be limited to a total of 650 miles. This caused sixteen tanks to be frozen immediately while 24 more had only 100 miles each in hand. The remaining twelve were low mileage machines, and the regiment being six tanks below war establishment this seriously limited further training. Early in November a troop of 17-pounder anti-tank guns was attached to the regiment and shortly afterwards they were visited by a Mr Jack, the Vauxhall Motors representative in North Africa, of whom more later. By the end of the month their strength is given as 43 Marks III and IV, five Mark I, eight unidentified training tanks and three Stuarts; the Crusaders having been handed in. At the end of 1943 four senior NCOs were sent to El Guerrah on a course to learn the 75mm gun.

Early in the new year they learned that Honeys (US M3A1) would replace the Universal Carriers in their recce troop and eight more arrived. Their next report explained that 10 of their Churchills were at Brigade Workshops 'as per programme' which appears to include modifications to the cupola and what are described as '12 RTR BESA modifications to all tanks' but these are not explained. The regiment also noted that all new vehicles were painted in European camouflage colours. In March the men attended a lecture on the Italian campaign.

The 48th RTR were also at Penthievre and their activities match those of the others in the brigade. This long period of waiting provoked some people to experiment. We shall encounter an important case shortly but for the present we will look at the work of a Captain Olgin, the REME officer in charge of 48th RTR's Light Aid Detachment. Having listened to the complaints from members of the regiment concerning the limitations of the hull mounted three inch howitzer in the Churchill I, Olgin resolved to do something about it. He took a sample tank and installed a second howitzer in the turret, replacing the regular 2-pounder. His idea was to use the turret gun as the initial ranging weapon before bringing in the hull gun. With a bit of practice the turret crew discovered that they could drop a shell within a 50 yard square on the third round. The information was then passed to the hull gunner who could hit the same patch of ground with his first or second shot. The problem was that the turret gun, being still on a free-elevation mount, would occasionally jump so that the range accidentally decreased or increased by as much as 100 yards. Work now concentrated on developing geared elevation for the turret gun but the war diary fails to report further developments.

At the end of December 1943 returns show that the regiment had 49 Marks III and IV, three Mark I and what is described as a Churchill training tank which, presumably, had been rejected for combat service. In March 1944 two Sherman tanks, type M4A2, arrived, harbingers of another change.

The third regiment in the brigade, 145 RAC, found themselves back in Bone after the victory parade in Tunis. The very place they had landed on arrival from Britain. Their war diary is one long catalogue of training and maintenance but it is interesting to note that their reports on exercises Trumpet and Kismet are written as if reporting real actions. At Penthievere, in October 1943, a tank from C Squadron gave a demonstration of towing a 17-pounder gun belonging to 14th Anti-Tank Regiment. This practice had first been used by Valentines of 23rd Armoured Brigade during the advance into Tunisia and it is interesting to observe that it had not been forgotten. However at that time the anti-tank gun was the 6-pounder which weighed just over 1 ton whereas the 17-pounder was more than double that. Even so cross country towing trials by a Churchill were described as very satisfactory.

Since it was brigaded with two Royal Tank Regiments 145 RAC treated Cambrai Day (20 November and the RTR's regimental day) as a holiday but at all other times it was an endless cycle of maintenance and training. In February 1944 tank crews begin training with the new rear smoke emitters and in the middle of April they took over seven Shermans from 25th Army Tank Brigade.

Meanwhile 25th Tank Brigade was recording similar events. The North Irish Horse, based at Ain Mokra near Bone, has nothing of note to report for this period, and neither does 51st RTR beyond announcing that their commanding officer,

A demonstration of rear smoke emitters. These were ignited electrically and could either be dropped behind the tank or used to lay a smoke screen from a moving vehicle.

This appears to be one of the Mark I tanks converted in Tunisia to mount a pair of 3 inch howitzers; one in the turret and one in the hull. Presumably it belonged to 48th RTR and was mined in Italy.

Lieutenant Colonel R B Holden, flew across to Italy in March to study conditions over there. In fact this became standard procedure for all the Churchill regiment COs around this time for obvious reasons. Later that same month some Stuarts and Shermans arrived for training and the regiment prepared to move overseas. The junior regiment in this brigade, 142 RAC, has a similar story to tell. It went to Ain Mokra from Tunis and then to Nabeul in Tunisia in September 1943. By the end of the year they were at a place called Oued el Aneb. On 20 March 1944 they receive a warning order for movement to Italy and early in April the commanding officer and second in command visited the Royal Armoured Corps Schools in North Africa to examine a Churchill tank fitted with a Sherman's 75mm gun. They were then told that the regiment would receive 18 M4A1 Sherman tanks pending delivery of these up-gunned Churchills. Mixed regiments of cruiser and infantry tanks seem, on the face of it, to be a curious combination. But experience in North Africa had blurred the distinction and since these were the people who had seen action their views had to count when placed against the dogma being expounded in Britain.

That was the question, and we have already alluded to it. Was there really any difference, apart from the gun, between the Churchill and Sherman tanks when it came to operational use? Granted they had their differences, and neither of them was in any way comparable to the new generation of German tanks, but was it really enough to justify operating the two types, never mind all the other tanks then being developed in Britain and the United States? Each type had its proponents, naturally, but these were invariably predictable. If one had trained and fought in Churchills the general view was that they were the best. But if your tank was the Sherman then that was considered the better machine. Of course if one came down to specifics then there were difference, and these could be tested.

As we have seen, actions such as Steamroller Farm and Longstop Hill had convinced a lot of people that the Churchill had an ability to climb that was hardly inferior to a fly on a window. How would the Sherman do under similar conditions? The obvious answer was to find out. British 1st Armoured Division agreed to sponsor a trial. The site chosen was in hilly country south east of Bou Farik and the date for the trial was 6 December 1943. A course was laid out about one mile long over which each tank would have to climb a scrub covered slope of average gradient, traverse the side of a hill which was cut up by small wadis and then tackle a very steep hill. After that each tank had to demonstrate its ability to climb out of a wadi and finally cross some heavy plough land. It was a condition of the trial that the land should be dry, which it appears to have been on the day. The report does not explain how many tanks were used, neither does it identify the precise marks in either case, but we do know that the tanks ran in pairs, one Churchill and one Sherman, over the course and we can assume that there were at least two of each type. This is clear from the fact that the first pair both managed to break a track ' for no accountable reason' as the report puts it.

The results appear to have been inconclusive, at least if one were looking for a distinct winner, and the most that the report was prepared to say was that the Churchill was slightly faster than the Sherman over difficult country, otherwise; 'There was no obstacle which the Churchill surmounted that proved impossible to the Sherman on the day of the trial'. Both completed the course and both were defeated by the climb out of the steep wadi at the end. They managed to get out a bit further along the wadi, where the bank was not so steep and, depending upon how one reads it, there is a suggestion that both tanks tackled the worst part of the bank successfully later on. Clearly the results depended to some extent on the skill of the drivers and their experience. The second attempt at the wadi would appear to bear that out, but it did nothing to justify the champions of either tank. Indeed a turretless Honey was also put over the course and it got right through before failing at the last wadi when the tracks got wet and, being rubber padded, failed to grip on the obstacle. Wet conditions were another matter and, conveniently, it rained for 24 hours on 11 December so it was decided to repeat the trial.

This time the going was much tougher yet again there was no real difference to be noted. Neither the Churchill nor the Sherman managed to get more than one third of the way along the course and both broke tracks when mud got into the drive sprockets and under the tracks. A pair of turretless Honeys failed at the first ditch. Summing up the report agreed that there was little to choose between the British and American tanks although it was noted that on a side slope in the wet the Sherman was much more difficult to control. Yet the Churchill still had its supporters. The officer commanding the Churchill troop added some comments of his own to the report. In his view the Churchill was far better at climbing than the Sherman because of the higher set of its idlers and the longer length of track on the ground. He pointed out that the main reason why the Churchill failed on the wet plough land was because the right hand final drive spline stripped. He also explained that at one point the Churchill had to tow its competing Sherman out of the wadi and then go on to complete the course.

So if there was so little to choose between the two tanks in terms of mobility, what about the gun? A REME officer, writing after the war, explained that the 6-pounder, with a range of 800 yards firing armour piercing and no ability to fire high explosive was no match for the German 75mm and 88mm weapons. He also explained that the aperture in the front of the Churchill turret was vulnerable to machine-gun fire and, in bright light, cast a deep shadow for enemy gunners to aim at. Possibly the officer was loading his case, the anti-tank performance of the 6-pounder was not that bad and there is little evidence for the belief that even German gunners could aim at a point so small as the mantlet area of an Allied tank.

This officer, Captain P H Morrell, had good reason for playing down the value of the 6-pounder Churchill. He was, in 1943, commanding a detachment of 665 Tank Troops Workshop REME, attached to 21st Army Tank Brigade at Bone. Morrell's detachment was subsequently absorbed back into its parent workshop, stationed at Le Khroub, and he was appointed second in command. The task of this unit was the scrapping of tanks damaged in action and considered beyond economic repair. The workshop park was full of these wrecks, mostly Churchills, Valentines and Shermans, and Morrell soon started to notice that in many cases the 75mm guns on the Shermans he was cutting up were virtually new. Morrell was convinced that the reason why the Churchills were not being sent to Italy was due entirely to the poor performance of their guns. He felt that the men of the two Churchill brigades were becoming bored and frustrated by the long months of inaction and he suggests that this feeling had also permeated his unit. The war was going on without them. What they were doing might be useful but they were part of fighting unit and should be in action.

Whenever he had the time Morrell used to wander among the wrecked tanks, studying the different models. Then he started measuring and soon came to the conclusion, with a bit of work, he could fit Sherman guns into the cast turrets of Mark IV Churchill tanks and he started to prepare a letter to the Deputy Director of Mechanical Engineering in North Africa, a Colonel Green. Now Morrell shared a large table with his CO, Major Chris Bevan, and before the letter was finished they were visited by the said Colonel Green. Since he was not required Morrell gave up his chair to Green and left the office, but while the Colonel was talking to Bevan his eye lit upon the letter. Morrell was recalled and told that Green liked the idea but it seems that his CO did not. He advised Morrell to keep quiet, and other officers in the unit pointed out that if such a thing were possible someone would already have thought of it; ergo it was impossible.

However Morrell persevered and within two weeks of sending the letter he was visited by Mr J C Jack, the Vauxhall Motors representative in North Africa. Morrell and Jack had known one another in Britain and the civilian engineer was impressed with his friend's scheme. As he pointed out, in a letter to Army Force Headquarters, the Sherman gun would physically fit into the Churchill's turret and, if the many problems could be overcome it could be made to work. Furthermore Jack calculated that the turret ring assembly of the Churchill was even stronger than that of the Sherman, so it should be able to handle the strain. On the basis of Jack's recommendation Morrell was ordered to report to the Director of Mechanical Engineering, Major-General W S Tope who, like Morrell was a civilian engineer in uniform for the duration of the war.

For two days Tope questioned Morrell, probing for weaknesses in the idea and coming up with problems that the young REME Captain had not even thought of. It was an uncomfortable time. Tope did not mince his words but he clearly believed it could be done. Morrell was told that he could have a brand new Churchill Mark IV plus all the staff and facilities he might require to carry out the conversion. Tope told Morrell that he was sure it could be done but reminded the young officer that, so far, he had enjoyed a successful military career. However it hardly needed the General to point out that, if the project failed and the British taxpayer was deprived of a new and expensive tank, Morrell might consider that his career had gone about as far as it could go.

Fitting the Sherman 75mm gun into a modified Churchill Mark IV turret by REME personnel in Tunisia.

The finished article, Churchill turret and Sherman gun. The neighbouring turret has already had its frontal aperture enlarged to accept the American weapon.

The project was declared Top Secret and given the codename Whitehot. It would require too much space to detail every stage in the conversion but the first step was simply to prove that the gun could be made to fit. This being done Morrell said that all of his crew agreed that it looked right which, as many engineers can testify, often indicates that a thing is right. There were problems of course, notably when it came to arranging for the 75mm gun to be fired from the right instead of the left. Then there was the sight which, in these early Shermans, took the form of a combined telescopic periscope that had to be adapted to the British turret. It had also been agreed that the Sherman mounting should retain its coaxial Browning .30 calibre machine-gun instead of the Churchill's BESA. Morrell now discovered that with the gun on maximum elevation the back end of the machine-gun knocked against the turret traverse gear so he adapted the mounting to overcome this problem. During elevation the machine-gun kept in line with the 75mm until the last few degrees when it ceased to move, leaving the main gun to continue. In order to avoid confusion over ammunition stowage the hull machine-gun was also changed for a Browning.

In his memoir Morrell fails to give precise dates but it seems that the initial work took place around February or March 1944. The actual conversion of the prototype is reported to have taken just ten days and it is known that a report on firing trials at the RAC Ranges was certainly issued on 21 March, followed by another early in April. In general this was complimentary about the work, the only complaint being that the Churchill's turret fume extractor, which had been moved to accommodate the sight, did not clear enough fumes from the turret. Once the matter of the gun had been settled work began on the less glamorous task of sorting out ammunition stowage and Morrell discovered that he was still able to fit 81 rounds into the tank which matched the quantity carried by the 6-pounder version. Once this had been settled the workshop was transferred to Bone where a production line was installed and in due course 210 Mark IV Churchills were converted to take the Sherman gun. They entered service as the type NA75, the letters naturally standing for North Africa.

Of course, as we have already seen, 6-pounder Churchills in Britain had already been converted to accept a British 75mm gun but by this time the Italian theatre had become something of a poor cousin, in terms of supply, to northern Europe so the NA75 Churchills provided a useful stopgap. Bearing in mind comments made by some users on the British weapon it might even be fair to claim that the Churchill with the American gun was the better version. One might speculate as to why this solution never occurred to anyone in Britain. Spare Sherman guns must have been available from tanks which the British were happily converting into armoured recovery vehicles and other roles, much to the irritation of the Americans, or indeed those that were being adapted to fit the British 17-pounder.

Details of the service use of the NA75 Churchill will be considered later but it is pleasing to record here that Morrell was duly rewarded for his inventiveness and the moral courage he displayed in going ahead with it. He was promoted Major and taken onto General Tope's staff and was subsequently made a Member of the British Empire.

The 25th Army Tank Brigade departed for Italy around the middle of April 1944. Most of the men sailed direct to Naples while the tanks and transport were shipped to Taranto in the south and then brought up by rail and road transporters. They were joined by 21st Army Tank Brigade at the beginning of

A photograph of a completed Churchill Mark IV (NA75) signed by many of those who were involved. The signature of M J E Jack, the Vauxhall Motors representative, is visible bottom right.

Churchills and Shermans at a tank park in Italy. The former are all Mark IV, the latter all M4A1.

May. Precise figures are not available for all six regiments but to take 142 RAC as an example they now had 18 Shermans (M4A1) to replace a similar number of Churchills. It was said at the time that these tanks were provided for their 75mm guns, pending the arrival of 75mm gun Churchills. Many of the commanding officers had visited the RAC Schools before they left North Africa, to inspect Morrell's converted NA75. All of the regiments appear to have been organised on the basis of two troops of Churchills and two troops of Shermans per squadron. However as we shall see there appear to have been many occasions when these tanks operated independently.

In covering the activities of the tank brigades in the Italian campaign one is faced by the problem of separating actions involving Churchills or Shermans. Where the war diaries identify particular battles in which Churchills took part, or make any other relevant comments, these have been recorded, but the wider aspects of the Italian campaign have been covered elsewhere so there is no point in going into too much detail.

When it arrived in Italy 12th RTR immediately handed over about half of its Churchills to a tank delivery squadron and received, in its place, eleven Stuarts, ten Shermans two scout cars and six half-tracks. In July the regiment attended a demonstration on the use of fascines and the Snake, explosive mine clearing apparatus. They were also inspected by the King and enjoyed an audience with the Pope.

Twenty fifth Army Tank Brigade was selected to support the Canadians and spent the early part of its time in Italy training with 1st Canadian Division at Lucera, near Foggia. The high standard of teamwork created there was soon put to the test during the attack on the Hitler Line in the sector between Aquino and Pontecorvo. The line was said to be thinly held, although it was not unusual for assault troops to be fed this sort of information but it included, among other things, Panther tank turrets dug into the ground which were sited to cover obvious approach routes. They combined potent firepower with minimal target area and, when skilfully emplaced became difficult to detect and harder still to knock out.

An account of the attack on the Hitler Line, prepared by Lieutenant G L Simcox, throws an interesting light on the real meaning of infantry support and how the Churchills and Shermans worked together. Simcox commanded the Sherman troop in C Squadron of 51st RTR and it is worth noting that a summary of this battle indicates that over the eight day period 17 to 25 May 1944, following the regiment's arrival at the front, tank crew members snatched, on average, 24 hours of sleep. Detailed planning at troop level was entirely absent and, indeed, Simcox describes a briefing at which most of the troop leaders were asleep and that, on returning to his tanks he decided to let the men sleep for as long as possible rather than wake them up try to explain the plan, insofar as he understood it.

Thus the action was entirely a case of follow your leader. The infantry set off with an objective in mind, followed by the Churchills which in turn were supported by the Sherman troop. Starting out in darkness and thick mist, under a perpetual barrage from both sides, the tanks simply followed the troops

Damaged by Allied fire and abandoned by its crew an emplaced Panther turret no longer represents a threat on this Italian hillside.

Churchills on the move along an Italian road. Freedom of movement like this was the exception rather than the rule in Italy.

and reacted to each problem as it occurred. Simcox records that at one point he encountered uncut enemy wire, which he knew the Churchills had been ordered to flatten. He and his troop sergeant started to deal with it when a troop of Churchills appeared, from a totally unexpected direction, trailing masses of barbed wire behind them and endangering their own troops. The Shermans were used to tear the wire off the Churchills and then followed them on to the enemy positions. The action resolved itself into a static fire fight, with British and German tanks simply slogging it out, gun for gun, but a dug-in Panther turret was doing most of the damage until it was taken out by the Commanding Officer's Churchill.

During the action Simcox was called upon to deal with an enemy gun position which was troubling the Canadians. He found difficulty in manoeuvring his Sherman into a suitable position and remarks at one point that it would have been easier with a Churchill. However in the end he is attacked by a hand held anti-tank weapon and decides to finish off the German position on foot, with his Tommy gun. After repeated visits to the German dug-out he killed or captured all of the occupants and was subsequently awarded the Military Cross.

The 142 RAC took a particular hammering on the Hitler Line. They had actually taken part in preliminary operations on 16 May and B Squadron reported that their Churchills worked well with the infantry. However one tank was lost when, bumping over a ridge, the commander's head set fell off and, unable to communicate with his driver he was unable to stop him before the tank got bogged down. During operations on the 17th C Squadron lost four tanks to German *Faustpatrone* anti-tank weapons. On the Hitler Line, on the 22nd, C Squadron had seven Churchills damaged by mines, three disabled with track trouble, four Shermans mined and two more halted with electrical problems.

With the Hitler Line finally subdued and the two Churchill brigades becoming firmly established as part of the army in Italy there was a distinct swing in favour of the infantry tanks. During their time at Lucera in June 145 RAC personnel were told. "Following the recent operations in which Churchill tanks have taken part, and the success gained, it is clear that the Churchill has made its name, and higher authority are now fully in favour of the tank and its capabilities". Quite who this meaningless message was meant to impress is unclear but the average Churchill crew man would probably interpret it as saying "you are about to be sent into action in a selection of obsolete old tanks with inadequate guns but, never mind, your superiors have every faith in your ability to achieve results anyway".

By mid-June 142 RAC had formed a composite squadron of its Shermans while its Churchills were kept out of action. This may have something to do with mud. For example 51st RTR had reported roads so bad during that month that none of their tanks could be moved and a plan to send the Shermans up first had to be abandoned. But there were reports of mud causing serious damage to the Churchills. It was heavy, glutinous stuff which worked its way into the tracks and then got carried around beneath the track guards. Often bits of wood or lumps of rock would get caught in it, and there are instances on record of the mud scraper, just ahead of the drive sprocket, being torn off. These foreign objects, lodged in the mud, could literally rip segments of the tracks guards off or at least bulge them to the point where they jammed up against the turret ring. Attempts were made to improve matters by altering the arrangement of the skid rails beneath the top run of the tracks but the preferred option was to take off part of the track guard, especially the bit either side of the turret, and operate the tank without it.

The kind of interesting detail which can be gleaned by careful study of the regimental war diaries is typified by entries in the document compiled by 145RAC at this time. In July they announced that their headquarters squadron would comprise one Mark IV Churchill fitted with two Number 19 (tank to tank) and one Number 18 radio sets; a Mark V Churchill with one Number 19, one Number 18 and a Number 38 (Infantry) set along with two Mark I close support tanks each with one Number 19 set. During a long move by tank transporters in August one rig slipped off the road on a corner and a Churchill rolled off, crushing two members of its crew. But it is a comment entered on 10 August that probably sums up the reality of good tank warfare. The regiment is to be ready for battle by the 12th and while making his notes the adjutant remarks; "it's the little things that count ... all hatches and pistol ports working efficiently, guns and W/T (radio) in perfect working order, all spares present and stowed by the book, all ammunition checked and rechecked for loose rounds". On 11 August six of the regiment's Churchills were fitted with frames to carry fascines.

The first action involving 12th RTR was in support of the Canadians on the Argilla River. Things did not go according to plan and a composite squadron, which had made the crossing, had to be supported as it withdrew by three troops of

A Mark IV Churchill of C Squadron, North Irish Horse on the approaches to the Hitler Line. The instruction *not to be stowed on deck* suggests that it has only recently arrived but the damaged remains of the track guards suggests some hard usage.

tanks, presumably Churchills and Shermans, along with a troop of M10 tank destroyers. Although basically Royal Artillery weapons, and therefore more properly described as batteries rather than troops, there is a lot of evidence of M10s operating closely with tank regiments on certain occasions when their greater anti-tank firepower would have been welcome. A curious entry in the 48th RTR War Diary for about this time refers to tanks by code names; Shermans are Spiders, M10s Mosquitos and Churchills are Beetles.

Early September 1944 found the Allies up against another formidable German position known as the Gothic Line. A subsequent action, involving A Squadron 142 RAC is well documented. The squadron comprised three troops of 6-pounder Churchills, two troops of Shermans and a Headquarters Troop of four NA75 Churchills with 75mm Sherman guns. The objective was a German position on Monte Colombo and the tanks were to support the 2/5th Battalion, the Leicestershire Regiment of 46th Infantry Division. As usual, in the Italian campaign, it was the infantry that led the way, moving forward at 0700 hours on 14 September, followed by two troops of Churchills with the third troop and the two Sherman troops in support.

The leading troops lost three Churchills almost at once; one bellied in soft ground, another rolled over in a ditch and a third lost a track. The three remaining tanks now combined and, with the Shermans firing in support, soon had their infantry onto a ridge about half way to the objective. Here they were seen by enemy observation posts on Monte Colombo and soon came under heavy fire. Further progress was made, but only with the greatest difficulty. Only three tanks could manage to occupy firing positions at any one time and soon they were virtually trapped. One, indeed, had its track broken by enemy fire and it took the crew two hours, working under intense shellfire, to replace ten track plates. The objective was taken after some fifteen hours difficult fighting but it is interesting, in the summary, to learn that the squadron officers believed "that the chances of success would have been considerably lessened if they had had Shermans only". Which at least is a modest compliment to the Churchill.

The same regiment reported that in the October fighting they encountered many more Panther turrets, which they left it to the artillery to deal with. They even reported an emplaced Tiger turret but were unable to confirm it. Meanwhile 145 RAC report, in the wake of their Gothic Line actions, that from 17 September to 9 October their tanks averaged 60 miles on their tracks and only reported one engine failure, but whether this was a Churchill or Sherman is not recorded. This regiment also operated a Panther tank, named *Deserter*, which was captured by the Seaforth Highlanders of Canada who presented it to the regiment. It served with C Squadron but they soon discovered that the tank was too wide to use British Bailey Bridges, which rather hampered its usefulness.

A trial involving Churchill tanks that was carried out at Riccione on 4/5 October 1944 revealed the range of ammunition now available to British tank gunners. Two targets were provided. One was a 12 foot square canvas screen, the other a rather battered old Tiger tank that had been dragged down to the beach which became the trial range. This Tiger was without its gun or mantlet and, having been burned out there was some doubt as to the hardness of its armour. At least three tanks were used. A Mark IV with a long (Mark V) 6-pounder fitted with free elevation; another Mark IV with the

Churchills of 51st RTR forming up in line abreast during fighting on the Gothic Line. Notice how, as in north west Europe, camouflage is particularly applied to the turret.

Another view of 51st RTR on the Gothic Line with a Daimler Dingo scout car in the immediate foreground. The tanks are well dispersed, suggesting a threat of air attack, but it all looks peaceful enough.

Tanks as artillery. Five Churchill NA75s and one with a British 75mm firing a concentrated barrage of high explosive.

shorter Mark III gun with geared elevation and a Mark IV NA75 with the American 75mm weapon. Ranges, on average, were about 650 yards, which would have been very dangerous if the Tiger still had its gun and an aggressive crew.

The only round that actually penetrated the 102mm front plate was from the long 6-pounder firing Sabot. The others did little more than scratch the armour. The 75mm high explosive round chipped away chunks of armour where it hit, but the charge within the shell did no damage at all. A Sabot round fired from the short 6-pounder also went through the frontal armour. Firing at the canvas screen revealed that Sabot ammunition was not as accurate as APCBC and tended to strike about two or three feet higher up the target.

Among the rounds tried was 6-pounder APCBC, APHV (Armour Piercing High Velocity), AP Discarding Sabot and for the 75mm just M61 High Explosive shells.

An interesting action carried out over the period 24 October to 9 November 1944 involved 51st RTR supporting British 10th Infantry Brigade. The object was to cross the river Ronco, establish a bridgehead on the other side and then capture the enemy airfield at Forli. Despite the fact that the infantry and tank men knew one another so well that they were all said to be on first name terms, the initial stage seems to have been a bit of a mess. The infantry waded the river and set up a defended perimeter but, when the tanks tried to find a crossing, the bank proved so high they reckoned a Sherman dozer tank would be needed to prepare a route. Why this had not been done before the infantry moved off is not explained. However another young officer believed he had found a suitable crossing place about 1,000 yards up river. It would not be ideal, it meant that, having crossed the river the tanks would still have to fight their way down the far bank to meet up with the infantry.

Even so it was decided to use this route and a troop from C Squadron moved off at first light. One tank was disabled on a mine before it reached the river and, when Lieutenant Neill's tank took the plunge it was immediately hit in the turret by an enemy anti-tank round. All of the turret crew died at once and the co-driver was wounded but, worse still, the crossing was now blocked. Then the river rose, no more tanks could get across and the infantry, under heavy counter attack, had to be pulled back. By watching the abandoned Churchill in the river the troops were able to monitor the water level. When it started to go down a new assault was planned. In an attempt to gauge the level of opposition it is said that the infantry brigade commander went down to the river bank and waved a white handkerchief. Since the enemy took no notice they came to the conclusion that the opposite bank was not now well defended.

This naïve piece of reconnaissance seemed to work since the tanks and infantry crossed the Ronco without opposition, although one tank got stuck. Meanwhile the Germans shelled another potential crossing point further north. The next day, however, the infantry ran into opposition which the tanks sorted out and then plans were laid to hit the airfield. This was known to be well defended and each tanks was given an extra load

Locals smile and applaud as an unidentified Churchill Mark IV leads a convoy of American trucks through a liberated Italian town.

of stores, including ammunition, grenades and spare radio batteries for the infantry. There was a lot of artillery fire about and the crews had to live inside their tanks. B Squadron was earmarked for the direct attack on the airfield and their crews were brewing breakfast inside their tanks at 0315 in the morning. Churchills from two troops moved forward while the third remained in reserve. Two of the tanks were towing 2-pounder anti-tank guns for the infantry. They sat on the start line until signals from a flare pistol announced that the infantry were in possession of some buildings on the far side. Now the tanks rolled across the airfield, met up with the infantry and dropped off the anti-tank guns. The tanks now began systematically shelling the buildings and they had soon collected a good crowd of prisoners. The little anti-tank guns succumbed to artillery fire very early on, and one is left to wonder what anyone hoped to achieve with this pre-war weapon on a 1944 battlefield with Tigers and Panthers roaming about. By the end of the action nine enemy troops had been killed and 27 captured. The attacking force suffered no casualties at all.

Meanwhile A Squadron advanced on the left of the airfield. They record one unusual incident when a troop sergeant's Churchill was in action. The breech block being opened to accept a new round a piece of shrapnel came down the barrel, into the turret and into the sergeant's arm. By 6 November A Squadron had been living in their tanks continuously for thirteen days. The men were exhausted and it was essential to relieve them. However any attempt to move the tanks out was likely to attract enemy attention so it was decided that crews from C Squadron should go up on foot and change over with the A Squadron men. This caused a great deal of anguish. Despite their weariness the crews were very reluctant indeed to hand over their tanks to strangers, even though they were men from the same regiment.

The summary of this action reveals that over the fifteen day period five Churchills suffered ignition problems from the rain, three were mined, seven suffered minor mechanical

Some Mark VII Churchills arrived in Italy towards the end of the war. This one demonstrates its ability to climb over a Sherman Octopus ARK.

faults and three of the 75mm tanks were rendered unserviceable due to splash. We know that when Morrell modified the Churchill IV turret he pointed out the advantage of fitting an external mantlet. One is therefore surprised to learn that the turret could suffer from bullet splash to the extent that it was impossible to operate. Since no further explanation is given, and since the same complaint is not reported on conventional Mark IVs, one assumes that splash affected the gun mounting (but if so why not on the Sherman?) or possibly the sights.

As 1944 came to an end there were changes in the air for the two tank brigades. The 25th was to be disbanded and there would be changes to the regiments. Both 142 and 145 RAC were due to be disbanded but, before they went, they left a few useful details in their war diaries. Staring with 142 RAC they note that early in November 1944 three of their close support tanks (presumably Mark Is) were sent to base workshops for conversion to ARKs. These were the turretless ramp tanks used for filling ditches and fording rivers. The close support tanks were replaced by 95mm Mark Vs and they also added four Churchill ARKs to their establishment. It is interesting to note that the regiment did not believe in the idea of Royal Engineers operating these specialised tanks. In their view the Sappers could never be trusted to turn up where they were needed, when they were needed and preferred the idea of managing their own specialised armour. By December, in addition to M3A3 and M5A1 Stuart tanks, M4A1 Shermans and one remaining Churchill Mark I, the regiment had nine Churchill fascine carriers, three Mark I and six Mark IV ARKs and thirteen NA75 Churchills. It had also been playing with Churchill Crocodile flamethrowers but, all this useful experience notwithstanding it was disbanded in January 1945.

Much the same can be said for 145 RAC. Early in November they carried out tests using Churchills in the indirect fire role. On the 11th their captured Panther went into action for the first time, shelling an enemy observation post in an occupied village. Unfortunately the Panther's gunner dismounted at one point and trod on a Schu-mine, as a result of which he died. During a shoot on 14 November two 95mm Churchills, acting in support of infantry, managed to hit selected houses at ranges up to 5,000 yards and they were firing an average 200 rounds per day. During one patrol at this time an enemy telephone was found still intact, but nobody answered when they tried to make contact. Towards the end of November there was a change to the squadron organisation. It was now based upon two 95mm Churchill V, two Sherman, six Churchill III/IV and two troops (six tanks) which were either Shermans, Churchill NA75s or Mark VIIs. Although disbandment was announced on 1 December the regiment continued to receive Mark VII Churchills to replace the Shermans and on the 22nd they carried out a trial to compare the performance of a Churchill, Panther, and a Sherman with end connectors fitted to its tracks, over soft ground. Unfortunately the results of this trial were not recorded. Disbandment was complete by 17 January 1945.

While this was going on 51st RTR was withdrawn to emerge in an entirely different form. Shortly after the action at Forli it starts testing Crocodiles and from December it was in the process of converting to an armoured engineer regiment with two squadrons of Crocodiles and one of Sherman Crab flail tanks. It remained active on the Italian front until the end of the war but now as part of a combined Royal Armoured Corps/Royal Engineers armoured engineer brigade.

Meanwhile 21st Army Tank Brigade was reformed with 12th RTR as the senior regiment, followed by North Irish Horse and 48th RTR. We shall concentrate on the senior regiment for this last phase, since it has no other published history. Going back to September 1944 it is interesting to note that the war diary lists six Mark I Churchills, 23 Marks III/IV, six Mark V and six NA75. These were supported by twelve Shermans, eleven Stuarts and ten scout cars. In December there are references to a 75mm troop of A Squadron enjoying a shoot against a church tower west of the River Lamone, which was thought to be an enemy observation post. The tower was duly destroyed. On another occasion a troop of 95mm equipped tanks carries out harassing fire against enemy positions with the aid of an airborne observation post, which is an unusual practice for tanks.

In January 1945 there are many more references to indirect shoots as, for instance, during the attack out of Granarolo against enemy position on the east bank of the river Senio. Two 95mm Churchills from Headquarters Troop, supplemented by two troops of 95mm Churchills and one troop of 75mm Shermans act as artillery, in support of an infantry attack. In February B Squadron try something even more exotic, indirect fire with BESA machine-guns firing tracerless ammunition. In this case no comment can be found to explain whether it worked or not.

At the end of March three Shermans of B Squadron are replaced by one Churchill NA75 and two Mark VIIs, the British 75mm version of A22 now starting to arrive in the Italian theatre. In April, as part of Operation Buckland C Squadron, now apparently equipped with Crocodile flamethrowers, along with the Crocodiles of 51st RTR and some Wasp flamethrowers based on Universal Carriers, set up a mass flame attack against the German bank of the river. The attack his highly successful and the infantry roll into action in Kangaroo personnel carriers of the 4th Hussars.

On 20 April the regiment takes part in a rapid advance that cuts through a mass of German armour. The war diary describes how Panzer IV tanks crossing a field 'ceased to exist as Panzer IVs after concentrated fire from three troops of Churchills'. In all 20 enemy tanks were destroyed by the combined effects of tank gunfire, artillery and cabrank fighter bombers. On the following day one Churchill is hit three times by *Faustpatrone* rounds without recording any damage or casualties. Right up to the end the Churchill was maintaining its reputation as a tough fighter.

12

The Last Battles

Meanwhile in northern Europe great things were happening. Even before the attack on Le Havre began British troops had liberated Brussels (3rd September) and Antwerp on the following day. Not that the capture of Antwerp in itself was much help. The port lay a long way from the sea, and the mouth of the Scheldt river, along with its northern shore, was still firmly in enemy hands. Until the waterway was free the port facilities would remain useless.

The Germans understood this every bit as well as the Allies, and they also knew that while supplies from Britain had to come into Europe via western France it would inevitably slow General Eisenhower's advance. It followed that any other Channel port between Le Havre and Antwerp would be an asset to the Allies and therefore something the Germans should hang on to for as long as possible. There were, in fact, four important sites along the Channel coast; the ports of Bolougne, Calais and Dunkirk along with the massive gun batteries on Cap Gris Nez. General Montgomery passed responsibility for clearing these to General Crerar's 2nd Canadian Division which set about its duty in great style. Even while the attack against Le Havre was getting under way Canadian forces were making their presence felt all along the coast. On 6 September they appeared before Bolougne, by the 9th they had arrived in front of Dunkirk. Then, having gathered their strength the attacks went in. The battle for Bolougne was fought between 17th and 22nd September and that for Calais from the 25th to the 30th. The only Churchill tanks to take part were specialised armour types of 79th Armoured Division which fall outside the scope of this study. None of the ports so far attacked had given up easily, and in every case there was a considerable risk to the civilian population. Dunkirk promised to be another tough nut but for some reason Montgomery ordered only that it be masked and not attacked. Where masking ends and a siege begins is not entirely clear except that the former sounds like a more passive activity which, if Dunkirk is any indication, appears to generate more aggression on the part of the defenders.

After the Le Havre operation the four surviving Churchill regiments not in Guards Armoured Brigade pulled out of the line for a period of rest and maintenance. The first to move was 7th Royal Tank Regiment which, towards the end of September, advanced eastwards in a series of 30 mile bounds until it reached a location due south of Dunkirk where it came under the command of the Czech Independent Armoured Brigade Group. This British administered force had just taken over responsibility for masking Dunkirk from the Canadians and already had the city well covered. On the night of 7/8 October 7th RTR moved into a location at Loon Plage, to the west of the port. Here they joined up with the Czech Reconnaissance Squadron.

It was not going to be a peaceful stay. The very next night the Germans subjected the area to heavy shelling and launched an infantry attack which drove some of the Allied infantry from their posts. There was little that the tanks could do so the crews fought dismounted. On this occasion it was mainly a matter of protecting the tank park from infiltrators but on the following night they went onto the offensive. On the 10th tank crews from A Squadron, fighting on foot, joined dismounted personnel from the Czech recce squadron and Free French Infantry in a counterattack which, as the war diary remarks, was the first attack of its kind ever undertaken by the regiment. It was successful. There follows a long period of investment in which such attacks become commonplace. Sometimes the tanks serve as artillery to hammer identified German positions, at other times they move out to patrol roads around the perimeter. On one night, early in November, German aircraft were heard overhead, dropping supplies to their beleaguered troops. One or two packages fell into Allied hands and were found to contain newspapers, mail and what the report describes as Bazookas, meaning the equivalent German hand held anti-tank weapons. On 5 November a Churchill tank, out on patrol, came under fire on a stretch of road and counted seven Bazooka type rounds fired at it. Only one hit and that did no damage beyond dislodging a spare track plate welded to the turret. Since a number of self-styled experts in the British Army decried this habit of adding bits of iron to tanks, arguing that in practice they did not do any good, it is interesting to note this one unequivocal example of its value. The regiment assumed that the weapons were part of the air-dropped consignment and put their poor effort down to lack of training.

The fighting on both sides was remarkably aggressive. During an action in November 7th Royal Tanks claim that they captured a German gun and mortar. This latter may well be one of the multi-barrelled *Nebelwerfers* because the regiment certainly had one by December and was having fun using it against its former owners. Even so no effort was made to attack Dunkirk and in fact the port remained invested virtually until the end of the war. It is interesting to note that the Czech Brigade was still surrounding the port in May 1945 when General Patton was moving into Czechoslovakia itself. Long before then 7th RTR had left. At the end of January 1945, having withdrawn from Dunkirk, it was in the process of converting from regular Churchills to Crocodile flamethrowers.

With the departure of 7th RTR from the scene we are left with just six regiments of Churchill gun tanks in Europe. These are; 9th RTR, 107RAC and 147 RAC in 34th Tank Brigade and, of course, those battalions of the Grenadier, Coldstream and Scots Guards that comprised 6th Guards Tank Brigade.

Following their actions in Normandy the Guards spent more than a month resting and doing maintenance work on their tanks. The rapid advance to Brussels and Antwerp was perceived as work for armoured divisions which could cover the ground faster, always assuming that they did not run into any stiff opposition. It is quite clear, from reading the war diaries of other Churchill Regiments, that they believed the Guards Brigade received preferential treatment. The implication being that they lived in the lap of luxury and were only let out to fight carefully orchestrated battles. In practice, of course, they fought as hard as anyone else with no advantages in terms of equipment or weapons; their Churchills were no better than those in 34th Tank Brigade. On the other hand

Caution exemplified. A Churchill noses cautiously round a bend in the road, covered by an infantryman from his foxhole who is guarding the corner with a PIAT (Projector Infantry Anti-Tank).

there were some very wealthy officers in the Guards, and if they located a few dozen bottle of champagne they tended to buy them. And it probably does not help to dispel their reputation for exclusiveness to read their war history which, when it is not telling of their own deeds, somehow seems to suggest that the only other tank force doing anything in Europe was the Guards Armoured Division.

The Allied advance across France could not be sustained. Tanks can only run so far without maintenance and men can only stand so much; but in any case there comes a point when supply lines are stretched so thin that fuel and ammunition is only trickling through. Indeed while Eisenhower and Montgomery continued to disagree over future strategy quite a serious situation was developing where supplies were concerned. The next major obstacle was clearly the river Rhine but until supply lines could be secured and shortened there

Sporting an Allied star on the turret the Coldstream Guards' captured Panther *Cuckoo* joins one of the regiment's Churchills in action in the Maastricht area.

The Churchills in this case are Crocodile Flamethrowers but this one of very few photographs showing Churchill tanks operating with Ram Kangaroo armoured personnel carriers.

A Squadron, 9th Royal Tank Regiment near Overloon, Holland in February 1945. A Universal Carrier is parked outside the building.

was a delay which enabled the Germans to reorganise and stiffen their defences. The attempt to bounce the Rhine by an airborne coup at Arnhem failed in September and the only option now, at least in the north, was a hard slogging match through Belgium and Holland. It was time, once again, to send for the Churchills.

Sixth Guards Tank Brigade arrived in Holland at the end of September. They were soon in action, with the Grenadiers and Coldstreams attacking south of Nijmegen while the Scots Guards acted as a back stop. They had gained a healthy respect for the German 88mm in Normandy but now its seems they were coming up against Panthers more frequently, and they soon discovered that its 75mm gun was just as lethal. In a firefight near Overloon a Churchill of No 2 Squadron, Coldstream Guards encountered a Panther just 500 yards away, but was obliged to move out from cover in order to get a shot at it. Four rounds of 6-pounder APDS followed in rapid succession, each striking the German tank on its front plate and each bouncing off without doing any damage at all. The Panther responded to this provocation with four rounds of 75mm AP, all of which penetrated the British tank but, miraculously, did not harm the crew. It is also worth noting that when the survivors if this squadron pulled back from what had been a very uncomfortable position after dark they employed smoke to disguise the exhaust fumes that might tell the enemy that they were on the move.

It was while they were in Overloon that the Coldstreams acquired their Panther tank. It was found, abandoned in a barn, nicknamed *Cuckoo* and taken into service. The crew appointed to the German tank clearly liked it but if one feature appealed to them more than any other it was the superb quality of the sights. Germany had always been noted as a source for high quality optics such as binocular and camera lens but it is clear that high standards were maintained right through to the end of the war. British manufacturers, by comparison, never seem to have achieved these standards at any time during the conflict. It was during the attack on Tilburg, later in the month, that the Guards worked with infantry in Kangaroo personnel carriers for the first time. The extra mobility conferred on the infantry seemed to promise greater opportunities for exploitation and when it was decided to try and capture a bridge over the river Aa, Left Flank Squadron of the Scots Guards with 7th Battalion the Seaforth Highlanders made the attempt. Most of them got snarled up in a traffic jam from which just three Churchills emerged. These made a dash for the bridge, or at least as close to a dash as Churchills could make, only to discover that it had been blown.

The three regiments of 34th Tank Brigade arrived in Holland early in October. While 147 RAC supported the RAF Regiment protecting an airfield near Westerloo the other regiments were in the Eindhoven area, mainly mounting raids, on what appears to have been a daily basis. On 20 October the brigade was involved in a new offensive, the object of which was to break into an occupied area from which the east side of the Scheldt could be cleared. Under overall control of 49th Infantry Division the spearhead of the attack was to be Clarkeforce which comprised 107 RAC, 49th Reconnaissance Regiment, infantry from the Leicestershire Regiment and a troop of self-propelled M10 17- pounder anti-tank guns. The whole under the command of Brigadier W. S. Clarke, the commander of 34th Tank Brigade.

Using Churchill tanks in this way was unusual. Normally they would operate, with infantry, over relatively short distances against defined objectives. Now they were to set out

A Dingo scout car burns on the road so this Churchill of Guards Tank Brigade is forced to make a detour through the trees.

on an expedition into enemy held territory with the general objective of going as far as they could, roughly in a north easterly direction. It was the sort of work one would normally hand to a Sherman regiment but they were fully committed elsewhere. The main body of 49th Division, along with the other two Churchill regiments of 34th Tank Brigade, would hold open the corridor that Clarkeforce created. The action began on the afternoon of the 20th and lasted through ten action packed days. Although the total distance achieved was no more than 25 miles it must be borne in mind that the countryside was flat as a billiard table, with a succession of waterways and thick woods which the enemy defended tenaciously. The infantry normally travelled on the tanks of the reserve squadron while the anti-tank guns kept up with the leading Churchills in order to deal with enemy armour. Most of the opposition seems to have come from enemy anti-tank weapons of the self-propelled variety that took every advantage of the flat landscape. While Clarkeforce forged ahead the other two regiments were just as heavily engaged, fighting off German attempts to break the corridor they had created.

A typical action is recounted in 147 RAC War Diary. On 21 October 11 troop was advancing down the Wuestwezel - Breda road when the leading Churchill was hit by a 75mm round. It immediately pulled back and the troop took up a fire position astride the road. Soon they were joined by the rest of C Squadron and a reconnaissance revealed many Clarkeforce vehicles knocked out further ahead. It was also possible to see enemy artillery and self-propelled guns moving around a nearby wood. The main attack was expected from the north and half the squadron tried to get into fire positions around some nearby buildings to cover this. The other half moved into a farm on the south side. Unfortunately the ground was very wet and most of the tanks on the northern front got bogged down. As the first German SP approached there were only two tanks in position to deal with it. The rest were either stuck, or helping to pull the others out of the mud. As the Churchills struggled into position four more German guns rolled up. The leader, described as a long 75mm on a Panzer III chassis was knocked out by 11 Troop at 150 yards, the second succumbed to the combined efforts of 11 Troop, 14 Troop and the Forward Observation Officer's tank.

Some impression of Churchill casualties at this time may be judged from the following report, compiled by Brigadier Clarke and issued on 7 November 1944. The report lists 31 Churchill and four Stuart tanks knocked out between 20th October and 30th October 1944 and it provides a fascinating picture of the variety of Churchills in service with the brigade along with a wealth of incidental information:

In winter conditions rapidly got worse but the crew of this Churchill Mark VI of the Grenadier Guards seem to be highly amused as their tank ploughs through a small stream near Overloon.

DATE	UNIT	TYPE	WD No.	EXTENT OF DAMAGE
21 Oct.	9 RTR	Mk VI 75mm	T251680	Shell damage.
		Stuart	T156915	Brew-up.
		Stuart	T156821	Ammunition fire - turret brew up.
22 Oct	147 RAC	Mk III 6-Pr	T32422	75mm shell in turret.
		Mk III 6-Pr	T68581	Turret badly gouged - no penetration.
		Mk III 6-Pr	T31723	Brew-up.
		Mk III 75mm	T67998	Gun split by shell & bad hull damage.
		Mk III 75mm	T31627	High Explosive - superficial damage.
		Mk IV 75mm	T68773	Turret roof penetrated - brew-up.
		Mk IV 75mm	T31109	Shell damage to horns & idler.
		Mk VI 75mm	T172541	75mm through ammunition bin - brew-up.
22 Oct	107 RAC	Mk III 6-Pr	T68578	75mm through front of hull.
		Mk VII 75mm	T251513	Engine and gearbox only brewed-up.
		Armoured Recovery Vehicle		Severe mine damage - complete write-off.
		Stuart	T156753	Brew-up.
		Stuart	T156857	Brew-up.
		Stuart	T156718	Frontal penetration.
26 Oct	107 RAC	Mk VII 75mm	T252024	Turret penetration at junction of roof and side by 75mm - repaired in unit lines.
27 Oct	9 RTR	MkIV 75mm	T31009	Shell damage to bogies. Hull distorted
27 Oct	107 RAC	Mk IV 75mm	T172714	75mm frontal penetration.
		Mk IV 75mm	T251691	Shell damage to turret ring.
		Mk IV 75mm	T251699	Bogie hit and hull floor damage.
28 Oct	9 RTR	Mk III 75mm	T31023	Cupola & turret roof smashed by mortar.
		Mk VII 75mm	T251520	75mm - hits in nine places - no penetration but considerable damage.
		Mk IV 6-Pr	T31777	Hull penetrated - brew-up.
		Mk IV 75 mm	T68618	Gun mantlet damaged and idler smashed.
		Mk IV 75mm	T172704	Hull penetrated - brew-up.
		Mk IV 75mm	T251711	Hull penetrated.
		Mk VI 75mm	T251676	Three turret and one hull penetrations, partial brew-up.
		Mk VI 75mm	T251958	Turret holed three times, partial brew-up.
		Mk VI 75mm	T251678	Hull penetrated, turret and idler damaged, brew-up.
28 Oct	107 RAC	Mk III 6-Pr	T31968	Shell penetrations but only partial brew-up.
		Mk VI 75mm	T251688	Shell damage to gun and mounting.
		Mk VI 75mm	T251920	Gun mantlet penetrated.
		Mk VII 75mm	T173186	Hull penetrated and idler smashed.
29 Oct	107 RAC	Mk III 75mm	T31634	Turret penetration and hull damage.

1. Of these 35 tanks (all types) knocked out in action, only the following are believed to be completely irrecoverable:
   - 7 Churchills.
   - 1 ARV
   - 3 Stuarts

2. Other tanks incidentally damaged by falling into pits, by their guns striking obstructions etc., were:
   7 Churchills
   1 Stuart
These have been recovered.

3. Personnel casualties were as follows:

Killed;	6 officers	25 other ranks.
Wounded;	8 officers	55 other ranks.
Missing;		1 other rank.

STATEMENT OF SP GUN CASUALTIES INFLICTED ON THE ENEMY BY 34 TANK BRIGADE DURING OPERATIONS 20TH OCTOBER TO 30TH OCTOBER 1944.

The following equipments were viewable at the locations given on 2 November 1944 and are verified by Commander, 34 Tank Brigade, personally.

88mm Jagdpanther	Brew-up, all crew buried.
105mm	Rounded mantlet.
105mm	Rounded mantlet.
105mm	Square mantlet.
75mm	Nine listed, three with rounded mantlet.

1. All the above 75mm and 105mm equipments on the PzKpfw III chassis, some have the new rounded type and some the old square type gun mantlet. All of them have full head cover and the majority the new all-round vision cupola; many have concrete reinforcement to the front slopes of the superstructure.

2. Most of them appear to be brand new, with low mileage readings on the speedometers, several recording under 150 kilos (one as low as 87 kilo). When seen some of them appeared to be only slightly damaged and could have been made fit to fight again after minor repairs.

3. Undoubtedly equipments other than those listed above were knocked out and claimed by regiments but were subsequently recovered by the Bosche; marks of recovery work and towing activity are visible at the locations given by the regiments when stating their claims. Every SP in the above list can be seen at the locations given, unless since moved by other units.

4. CONCLUSIONS. In these operations, which resulted in an advance of 16 miles, the relation of total losses suffered by us, vis-a-vis the enemy, will be noted as relatively creditable to the gunnery of this formation, in view of the fact that the enemy was operating over known ground and we were compelled to expose ourselves constantly in attacking.

Signed W.S. Clarke
BRIGADIER
Commanding 34 Tank Brigade

• • •

The winter of 1944/45 was tough on both Churchill brigades, as it was on everyone engaged in the struggle. A period of heavy rain with interminable mud was followed by a deep freeze and heavy falls of snow which turned the ground to iron. Writing to his regiments in January Brigadier Clarke offered congratulations and ended up by saying ' the Great Toboggan rides into Namur and Liege will not soon be forgotten!' He referred to the Churchill's unfortunate habit of sliding about all over the place in snow or icy conditions. The Coldstream Guards noted it too. Their Churchills would slither off the road at every bend but what made them really mad was the way in which their captured Panther *Cuckoo* just rolled along as normal, showing no regard for the conditions at all.

In the middle of all this came the German irruption in the Ardennes. Although directed against US 1st Army it involved everyone to some extent and had the distinct disadvantage that it came so close to Christmas, a holiday everyone had been looking forward to. Most of 34th Tank Brigade was deployed in blocking positions, covering the liberated part of Holland, but in fact the attack ran out of steam just before

Part of the great toboggan ride. Guards' Churchills, with limited snow camouflage, tackle an icy road between Birgden and Waldenrath.

Christmas and things settled down for a while. Guards Tank Brigade did very well for themselves. For some time they had been working closely with the United States Army which, they found, was well supplied with turkeys but rather short on festive spirit. With two Scottish battalions in the brigade there was more than enough Scotch to go round and a brisk trade ensued. Although we are not committed to examine specialised armour in this study a brief exception might be made for an experiment carried out by 9th RTR at about this time. They helped to develop an attachment for Churchill tanks that would detonate Schu mines, which were a serious problem for the infantry. Lengths of Churchill track, hanging guide horn down, were suspended ahead of each track, with links of Universal Carrier track attached to the ends. This swept the ground in front of the tank, set off the Schu mines - which could do no harm to the tank - and left a safe path for the infantry to follow.

Early in February 1945 34th Tank Brigade was instructed to prepare for Operation Veritable, the clearing of the Reichswald Forest. This had been planned, as Operation Gatwick, back in October but cancelled while other matters took priority. The Reichswald covers an area of about 45 square miles, being about nine miles long west to east and five miles wide at its widest end. It is also higher to the west than the east. Despite the impression given elsewhere it is no jungle, but a well managed forest, mainly of pine but with deciduous trees at its western end. Within the entire forest is divided into square stands of timber, like city blocks, measuring 800 yards by 400 yards and separated by a criss-cross pattern of straight tracks, on average about eight feet wide. These tracks are used to haul timber out, but they were never designed to take heavy traffic, and certainly not tanks. The fight to clear the Reichswald is so intimately bound up with the combat career of the Churchill tank that it justifies a more substantial study.

In 1945 the Reichswald more or less marked the northern limit of the German Siegfried Line, although this extended a bit further north to cover a neck of land down which ran the main road from Nijmegen to Cleve. Beyond that the land was flooded, almost as far as the Rhine itself. However before venturing into the forest with 34th Tank Brigade let us examine the activities of 6th Guards Tank Brigade which was instructed, in conjunction with 15th Scottish Division, to break through the Siegfried defences between Nijmegen and Cleve and then to drive the enemy out of Cleve. The plan was to have the Scots and Coldstream Guards work down both sides of the road with their infantry brigades while the Grenadiers followed up and then led the way through the Siegfried Line. Known minefields in the area were to be cleared by flail tanks but before they had gone very far most had succumbed to the mud. The Germans had already flooded the area and now, as the water rose, the mud rapidly got worse. Soon it was so bad that the Coldstreams and Grenadiers were trapped. It was no place for Churchill tanks and to make matters worse a massive amount of wheeled traffic, moving with them, also got stuck. However the Scots Guards, still staying on the road, had more luck. True the road surface was deteriorating rapidly but all the while their Churchills could keep rolling the Jocks kept going and headed for the German defence line. They even had the foresight to bring a bridgelaying tank with them to tackle the anti-tank ditch but found, to their surprise, that the road passed over the ditch, and so did they.

Churchills of the Scots Guards with infantry of 2nd Argyll and Sutherland Highlanders forming up for an assault on the Siegfried Line, 8 February 1945. The tanks have been supplied with fluorescent material to drape over their turrets so that they can be recognised by friendly aircraft.

Churchills passing a Sherman Crab flail tank in the advance towards the Reichswald Forest.

Churchill Mark VI *Lyon* of Left Flank Squadron, 3rd Battalion Scots Guards during the advance to the Siegfried Line north of the Reichswald. Behind the tank is a Universal Carrier belonging to the infantry, then a Stuart, or Honey light tank of the recce troop and more Churchills.

The Guards Brigade historian believed that the German defenders along the road had been stunned by the ferocious preliminary bombardment by heavy bombers and artillery, but it seems incredible that they should leave such a gap in a vital part in their main defence system open at such a crucial time.

Although the Grenadiers and Coldstreams later managed to break through the Siegfried Line the latter had so many tanks bogged in the mud that they could make no further progress. Meanwhile the Grenadier Guards, along with the Scots, joined in the attack on Cleve which was finally occupied on the 12th. Now, however, the water had risen to a point where no traffic could get through and the force of British and Canadian troops in Cleve looked as if it was stuck there without supplies. After attempts to outflank the flooded zone failed it was decided to bring in a fleet of DUKWs and M29c Weasel amphibians which managed to maintain supplies to a sufficient level.

The forming up point for 34th Tank Brigade ahead of Operation Veritable was a wood west of the town of Groesbeek but the start line for the attack was within the Reichswald itself. Most of 8 February was taken up with the advance to the start line through pouring rain over atrocious ground that was a worse hindrance to the tanks than a minefield. The order of march for 147 RAC, which was to skirt the northern edge of the forest, was as follows; Reconnaissance troop led the way in their Honeys, followed by A Squadron's Churchills and a Sherman containing the Royal Artillery Forward Observation Officer or FOO. Next came C Squadron, Westminster Dragoons with their Sherman flail tanks, then Regimental HQ of 147 RAC followed by C Squadron and another FOO, B Squadron Westminster Dragoons, some Churchill AVREs and finally B Squadron of 147 RAC.

A Squadron, supporting 4th Battalion Royal Welsh Fusiliers found the going very bad indeed. Most of the flail tanks got bogged but for the first 1,000 yards of their advance there was no opposition from the enemy at all. Some enemy activity was noted during the next stage but it did not slow down the advance and the squadron gained its objectives and stopped for a meal. B Squadron, advancing two troops up with the third in support was working with the 1st Battalion, the Highland Light Infantry. They made quick progress to the anti-tank ditch which was quickly bridged. However as more tanks rolled across it the bridge started to break up and the squadron commander decided to try his tank at the ditch without the aid of a bridge. Having managed to struggle across the remaining tanks tried and found that, as long as they each went over at a different place the ditch proved to be no great problem. C Squadron, operating outside the forest along the line of the Nijmegen to Cleve rail-

In the thick of the Reichswald, a Churchill of 107 RAC smashes its way through some tall but slim timber. However the mess it leaves in its wake could prove impassable to infantry.

way, seems to have struck some of the worst ground. Soon all the flails, AVREs and the FOO's Sherman had bogged and then the Churchills began to drop out. In the end their infantry, 1st Battalion the Oxfordshire and Buckinghamshire Light Infantry, went on and took the objectives by themselves while the tanks rallied. Any attempt to continue the attack next day was defeated by traffic congestion as the mud and floods took hold. C Squadron finally managed to catch up with its infantry and support them in an attack from which they emerged with no casualties, while A Squadron spent the day helping their infantry to suppress enemy machine-gun positions firing from the fringes of the forest.

Yet it was in the heart of the forest that the real struggle would take place, and it seems an unlikely place to commit tanks. It would be gloomy enough anyway, beneath the trees, but we must remember this was February when the days are short and it was raining most of the time. It would have been overcast, to add to the gloom, but even when the rain stopped water would continue to drip down on everyone. Within no time at all the forest rides, churned up by tank tracks, would be impassable to wheeled vehicles so there would be problems just getting supplies through. As if all this were not enough elements of the Siegfried Line actually ran through the forest near its widest point and we should also remember that most German infantry now had access to hand held anti-tank weapons such as the *Panzerfaust* which could, with luck, destroy virtually any tank. And luck was on the side of such weapons with deep undergrowth for the men to hide in, until their victims came very close indeed.

Ninth Royal Tank Regiment estimated that they would need two weeks in which to train with their infantry on the difficulties of forest warfare. They got one week. Having broken into the forest, leaving most of their Stuart reconnaissance tanks bogged down outside, the Churchill tanks moved up to the start line in pitch darkness. It is worth noting that they did not follow the forest rides but moved through the trees. No explanation is given but, reading between the lines of surviving documents, it seems that the infantry commander had banned tanks from the rides in case they messed them up. That way, presumably, he hoped to keep his supply vehicles running. But driving through the trees was no joke. Troop leaders led the way on foot, guiding their tanks around those trees that were too big to push over.

The 9th RTR had already done some trials on forest fighting and learned a good deal more in the Reichswald. They found that in general conifers were easier to push over than deciduous trees and since the forest was mainly conifers this helped to some extent. A Churchill, they discovered, could push over a pine tree with a trunk two feet in diameter, but the best a Stuart could manage was a one foot diameter tree. Also they found that when a Churchill pushed at a tree it simply snapped off near the base while a Stuart tended to uproot it. This, they reckoned, was due to the shape of the tanks. The flat front of the Churchill gave the tree no option but to break while the sharp nose of the Stuart caused the tank to ride up the tree until it was pushed over by sheer weight. Now it had also been agreed that with the risk of ambush being what it was, the drill in the forest should be that infantry preceded their tanks, but the regiment now laid down a rule that the infantry should always be at least thirty yards in front of the tanks to avoid the risk of being hit by falling trees.

Even so it was necessary to maintain close liaison between the tanks and their infantry and to this end the Ninth recommended that the men wear distinguishing marks on their backs. The tanks, they said, were almost blind in the forest either by day or night and without the infantry they would be in serious difficulties. On one occasion, during the Reichswald fighting, one group of infantry identified themselves by attaching white, enamel drinking mugs to their backs and then arranging to advance in bounds of just eighty yards at a time. Having made the move they would then turn and flash red torch lights at the tanks which came steadily forward and prepared for the next move.

It was a slow process and it took four days to pass right through the forest. Opposition on the first day within the forest, 9 February, was minimal and all three squadrons had reached their objectives by the end of the day. Even so the war diary emphasises how gloomy the forest was in the rain and how eerie it must have been to hear bursts of machine-gun fire echoing all around. Most of the following day was taken up in waiting for supplies of petrol, oil and water to keep the tanks going. This problem had been anticipated and some of the tanks had been provided with special supply sledges which they towed behind them. In practice these proved too difficult to haul through very bad going and in amongst the felled trees they must have been impossible. Sledges fitted with rollers proved somewhat better. At one point someone had the idea of loading supplies onto the recce troop Stuarts and this seemed to be a promising idea until someone else, further up the chain of command, put a stop to it. In practice the most successful method was to have two Churchills from regimental headquarters, each towing a pair of three ton trucks and this provided enough fuel until a Royal Army Service Corps column got through to the tanks in the evening. Even then the only way to distribute the stores was to load up four tanks and set off around the squadrons like a milkman doing his rounds.

When the advance resumed on the morning of 11 February the regiment was down to just 29 tanks, and eleven of them were with B Squadron, so B Squadron led the way. At one point a tank in 10 troop was attacked by a *Panzerfaust* team that got so close to the tank that it could not depress its gun sufficiently to deal with them. The ground was so soft that the tank could not turn either so another tank fired a 95mm high explosive round near it and solved the problem that way. The ensuing blast prompted the immediate surrender of most of the nearby infantry. By 10am B Squadron was deployed in an advanced position when a Jagdpanther appeared, heading for their location. Without a turret the big SP must have been at even greater disadvantage in the forest than the tanks, but it could easily outgun the Churchills and was only driven off by some M10 tank destroyers which were operating in support.

Late in the day there were indications of an enemy counterattack but when it came in, with artillery support, the tanks held it off. On 12 February, as they neared the eastern end of the wood the regiment noted a tremendous increase in the use of *Panzerfausts* but by the end of that day it was clear that enemy morale was dropping fast. They emerged into the daylight on the 13th and next day, their part in the battle over, the survivors of 9th RTR went back to Groesbeek for a rest. Looking back on the battle the regiment made a series of observations, many of which have been noted above. They reckoned that the presence of tanks in a forest was probably of more moral than practical value. During night time advances, for instance, they made it a practice to remove tracer rounds from machine-gun ammunition belts and then spray the tree-tops with fire, above the heads of their own infantry. This, they said, had a certain horrific value, presumably on the enemy. They had also discovered that bits of wood got wedged under the track guards, forcing up the central section and jamming the turret. Just like the regiments in Italy, 9th RTR responded by removing the centre section on each side. They explained that an adequate number of bulldozers, capable of clearing up the deadwood, would be a better way of dealing with the problem.

Indeed they regarded machine-guns as of far more use than main armament in wooded country. Firing high explosive rounds, except in clearings, tended to be as much of a risk to ones own infantry as the enemy and they reported that consumption of BESA ammunition far exceeded all other types. They also advocated firing machine-guns into the surrounding shrubbery whether enemy troops could be seen or not since, if there were any around they invariably got hurt as they tried to retreat. Night time was clearly the most trying. Practice in 9th RTR was to pull back about 20 yards behind the infantry but not to rely on the infantry to protect them. The tanks should be guarded by alert crew members equipped with Sten guns and all movement within the leaguer area should be forbidden. Anyone heard moving about should be shot without waiting for identification. They recorded at least two instances of enemy troops penetrating a leaguer and doing serious damage.

The third element of the brigade, 107 RAC, had far more diverse objectives and remained in action far longer than the other two regiments. To begin with it had to fight its way down the western edge of the Reichswald and then take two routes. It was to support 51st Highland Division and it is interesting to note that in this action each squadron of the regiment was supporting an entire infantry brigade whereas 147 RAC, on the opposite side of the forest, only had a single regiment allotted to each squadron.

***Dundee*, a Churchill of the Scots Guards, eases its way across a Churchill tank bridge, laid over a crater. There is precious little room to spare and the infantry passengers do not appear to be entirely sure of the driver's skill.**

Even in the first day's fighting, working down the end of the forest, 107 RAC had problems. The ground conditions were so bad that all tracked vehicles had to remain on the road and this was not just the regiment's tanks, it included flail tanks of the Lothian and Border Horse along with AVREs of 79th Armoured Division and a self-propelled anti-tank battery of the Royal Artillery. The AVREs helped them through a network of defences south of Breedeweg which included objectives that some poetically inclined soul had named Shelley, Tennyson and Wordsworth. Upon turning the south west corner of the forest 107 RAC now found itself technically in Holland once again, without any frontier formalities, since the border between Holland and Germany ran along the edge of the wood at this point.

Now, however, the regiment split up. A and B Squadrons had been leading the way until now and the former had lost a lot of tanks on mines and in the mud. Thus B Squadron, less one troop, headed south west along a causewayed road leading to Gennep on the river Maas while C Squadron came up and headed east along the southern side of the forest with the remnants of A Squadron and the detached troop from B Squadron. The fighting in this area soon became as confused as the squadron organisation. What had begun as B Squadron pushed south through Gennep, crossed the river Niers and then swung east towards the Rhine. The main thrust, based on C Squadron, continued eastwards to join the assault on Goch but both arms soon began to feel the pinch for want of ammunition since supplies were not getting through, due to the condition of roads within the Reichswald.

The fall of Goch, on 20 February, marked the end of the last major assault before the Rhine crossing a month later. But it did not mean the end of fighting. Some of the pillboxes that formed the outer ring of Goch's defences were still holding out for some days afterwards. Working in conjunction with 79th Armoured Division 107 RAC worked out an effective method of dealing with them. Stage one was a bombardment by 75mm and 95mm gun Churchills which, with luck, brought forth the white flag. If not stage two involved bring up a couple of AVREs to hit the thing with Petard rounds, but it was necessary to keep Churchill gun tanks nearby because the AVREs had to advance close up to the pillboxes to score a hit. If this also failed to induce surrender it was time to send for the Crocodiles and if the occupiers did not give up after the first burst of flame they never got another chance. Watching, in a sense, from the sidelines when the specialised armour went in, 107 RAC noted that applying flame without first involving an AVRE did not always work. The occupants of the pillbox only had to retire behind a door into the inner chamber and wait until the flaming stopped. But the detonation of a Petard or two invariably knocked this door off its hinges and guaranteed success.

The advance to the Rhine marks the last combat actions of 34th Tank Brigade. The three regiments remained in the area and retained their tanks for quite a while but they now had a new role. They became, in effect, occupation force policemen, moving into zones in occupied Germany which they would systematically clear of all weapons and munitions that they duly stockpiled. In addition they were ordered to

Emerging from the confines of the Reichswald a well-protected Churchill Mark V passes a column of Loyd Carriers. Notice how a track link has been hinged to drop over the driver's visor when it is shut.

round up displaced persons and send them to holding areas, to take care of released Allied prisoners of war and deal with any Germans who still showed signs of giving trouble. It was uncongenial work for men who had fought so well and they found it rather boring but the war in Europe was coming to an end and at least it was relatively safe.

The war was also changing again and for the last time, and this became obvious to the last Churchill crews when they were notified, at the time of the Rhine crossing, that they should now style their formation the 6th Guards Armoured Brigade. While no particular reason was given for this at the time it is possible, with hindsight, to detect at least one line of thinking which led to it.

Until Field Marshal Sir Bernard Montgomery took over command of the Eighth Army in the summer of 1942 he had not had very much to do with tanks. As an infantry officer in the First World War he was probably only just aware of them, and even in the formative years between the wars it was the exception, rather than the rule, for an infantry officer to get involved seriously with tanks. Even as a divisional commander in France in 1940 he would have had little direct experience of them and the period he spent as a corps commander in Britain would not have improved the situation a great deal. It is only after he is elevated to General Officer Commanding South East Command in Britain that we find him writing, in his normal evangelising style, on the proper use of tanks. At such an exalted level tanks would now come more directly under Montgomery's view but there is a world of difference between a command where the only objective is training, and one that involves combat.

Like all army commanders Montgomery would have had a Royal Armoured Corps advisor on his staff who, in theory at least, kept his superior up to date on tank matters and endeavoured to prevent them from being misused through inexperience, but there was little risk of that with an officer as cautious as Montgomery. Whether he had realised it before or not, the desert situation confirmed to Montgomery that tanks were a vital part of his army. He would learn fast and soon came to realise that they could break through enemy lines, that they were more effective if employed in large numbers and that, generally speaking, the best way to deal with enemy tanks was to use your own against them. Yet they also bewildered him. He was unable to make any sense at all out of the various types and in particular he came to distrust the accepted wisdom of having different tanks to fill different roles. In Montgomery's view the idea of having special tanks to work

Sixth Guards Tank Brigade yet again, entering the ruined town of Cleve. All three tanks appear to be conventional Mark IVs mounting 6-pounder guns.

with the infantry, while others operated as cruisers like the cavalry was a nonsense. His own logic told him that it must be possible to combine the best features of both into one design that could do most things; what he described later as the universal tank.

Events in Italy, for the short time he was there, can only have served to confirm him in this view because, as we have seen, the Allies managed well enough with only the Sherman for a good deal of the Italian campaign. Yet in Normandy the dichotomy was as strong as ever and it must have irritated him. By careful planning it was possible, in an area like Normandy, to find work that suited both types of tank but again, as we have noted, when the advance through France began the Churchills were almost redundant. The thought that six or more good tank regiments were simply resting while a crucial attack was mounted was galling to everyone and clearly, by the time they reached the Rhine, the entire scheme had become pointless.

Thus we see that by March 1945 at the latest all of the surviving Churchill regiments except three had been stripped of their tanks and put to other work. The fact that the three survivors were from Guards regiments, considered as elite troops and with more influence in their officers' messes than the rest of the Army put together, may well have a bearing on this. But in order to participate in this last act of the war in Europe they had to adapt. By that time they had been out of the line for three weeks and would probably have settled for anything; as it was they got an offer that was impossible to refuse.

Despite the unfortunate result of the Arnhem operation the Allies were anxious to use airborne troops again during the Rhine crossing. The plan was to land glider borne troops and drop parachutists on the eastern side of the river with a view to trapping those forces who were expected to retreat in the face of the direct Allied assault across the river. It was a lot to ask of airborne troops. For one thing they would lack heavy support equipment while the enemy was between them and the river and for another they would be fighting some of the best troops in the German army, on their home soil and anxious to break clear of any trap. On the positive side the parachute troops would come from British 6th and American 17th Airborne Divisions and as soon as it proved possible they would be joined by Churchill tanks from Guards Armoured Brigade.

The Rhine is not a river, as that word is understood in Britain; just a narrower version of the English Channel. So the first phase of the attack was carried out as a full scale amphibious operation with landing craft, tracked amphibians and DD tanks. Once a bridgehead had been secured on the enemy's side work began on building bridges; but in the meantime a ferry service was established. This involved shifting powerful winches across the river and setting them up on the far bank while similar winches were emplaced on the near side. With cables stretched the full width of the river, attached to rafts made from sections of Bailey bridge resting on pontoons, a regular service began. The Scots Guards crossed on the night 24/25 March. The Jocks, in their Churchill tanks, had been listening in to radio messages from 6th Airborne Division since the paratroops landed and now they were on their way to join them. As soon as they got ashore Left Flank Squadron led the way on a four mile drive to link up and, as the brigade history says, the Churchills arrived within twelve hours of the airborne troops landing.

However plans change in battle almost as soon as they are made and the Rhine Crossing was no exception. The collapse of German resistance on the east bank, largely as a result of having tough airborne troops in their rear, rendered the original plan pointless, now there was nothing for it but pursuit and for two battalions of the Guards this was the most welcome opportunity they could imagine. The new plan was for the Churchill tanks of the Scots Guards, joined by those of the Coldstream Guards which had come over on one of the first bridges to be completed, to join up with 513 Regiment in US 17th Airborne Division and race for the German city of Munster, 50 miles to the north east. They would be accompanied by 3rd Reconnaissance Regiment and Royal Artillery batteries of field, medium and anti-tank guns. In attempting this the Churchills would be working to the same timetable as the armoured divisions with their Cromwells and Shermans. The heavy tanks would be operating out of their league; it would be interesting to see how they stood up to it.

The tanks met up with the Americans just outside Wesel. The Scots Guards were matched with 2nd Battalion of the 513th while the 3rd Battalion joined the Coldstreams. The plan was to have the paratroops ride on the Churchills until action was imminent and then dismount. If the tanks were ambushed by enemy infantry with *Panzerfaust* type weapons then the soldiers would jump down and deal with them. If it was heavier stuff, such as eighty-eights, then the troops took cover while the Churchills sorted it out.

At 3 o'clock on the afternoon of 27 March the Churchills, with their passengers aboard, were lined up on the Munster road and ready to start. The full account of that amazing drive can be found in the Brigade history and there is only space here to record the odd incident. Arriving in the town of Haltern on the second night, trying to find a bridge over the river Lippe which the Germans had not blown up yet, the Coldstream Guards were so far ahead of their Brigade HQ that they had lost radio contact. Yet anxious to push on they despatched a squadron of Churchills to the next town. Hammering down the road in the dark they found that they were being overtaken by another column, led by a tank which turned out to be a Panther; a type with which they were already familiar although their own pet *Cuckoo* had expired from a broken fuel pump some weeks earlier. This Panther, however, was still in service with the German Army although it now found itself leading a British tank column without realising it.

If the Panther has a vulnerable spot it is the back, and as it drew ahead of the leading Churchill it received two rounds in this sensitive area and brewed up. Other elements of the

The Yanks are coming. Men of 17th Airborne Division riding Churchills of 6th Guards Armoured Brigade on the advance through Germany.

Another view of the Guards' Churchills with their American passengers. The man closest to the camera is not wearing a special camouflage suit, its is a mark on the photograph.

An M10 (17-pounder) SP anti-tank gun of the Royal Artillery guards a cross roads in Germany as Churchill tanks roll by carrying men of British 6th Airborne Division.

German column were dealt with in the same way until the Churchills were ordered to draw off the road and wait for the rest of their squadron to catch up.

By the late afternoon of 2 April 1945, which was just after Easter that year, the Churchills with their paratroopers, had fought their way into the city centre of Munster. It had taken seven days of hard driving and a lot of equally hard fighting to get there and the Guards were highly impressed with the fighting qualities of the 513th. If an average advance of seven miles each day does not sound impressive, when compared with the famous run across France, one must first acknowledge the scale of the resistance, the state of some of the ruined towns through which they had to navigate and the endless diversions caused by the need to find alternative routes around wrecked bridges. It is unlikely, under these circumstances, that a Sherman regiment could have done much better.

While the Coldstream and Scots Guards had been charging across Germany with the Americans the Grenadier Guards had not been idle. Having been held outside Wesel for a day they joined up with 6th Airborne Division for an attack on the town on Koesfeld, 28 miles away. Although each squadron of the armoured regiment was allocated to a brigade in the airborne division they each carried one battalion in this advance. It was achieved against virtually no opposition within the day. Next morning two squadrons were selected for what was described as a lightning dash to capture a bridge over the river Ems at Graven. The idea of anyone suggesting that Churchill tanks could make a lightning dash anywhere seems somewhat ridiculous but in the event they reached Graven just as it was getting dark, only to capture the wrong bridge and have the real objective blown up as they watched.

To chronicle the last few weeks of the war in any detail is merely to repeat what has already been written and create an unnecessary triumphal paean. It would tell us nothing new about the Churchill tank. Suffice it to say that when VE Day came the brigade was already established on the Baltic coast and the Coldstream Guards was given the special task of making a formal entry into the city and naval base of Kiel, in an attempt to overawe the German Navy which, at this point, had not actual made a formal declaration of surrender. The tanks had all been stripped of the extra track links welded to their turrets, repainted and cleaned for the parade, largely by captured German soldiers. The long line of tanks wound its way through the city and their history remarks that the route was so narrow, through the rubble of bombed buildings, that had one tank broken down it would have brought the entire column to a halt. The fact that none of them did is really a tribute to the high standard of maintenance practised by the crews. If any of those men who had nursed the early Churchills through their long period of teething troubles could have seen them now, running smoothly in a formal parade after weeks of continuous high speed driving - at least high speed Churchill style - they would have been more than just surprised.

After a few more weeks touring around the countryside so that the local population could be reminded just what war looks like from close to, the Brigade returned to the Cologne area in June and held its final parade. It had already handed over its tanks at Hamburg and was now destined to revert to its traditional infantry role. There had been rumours of a final campaign in Norway where the Churchills would operate against German die-hards believed to be planning a last ditch campaign in the mountains. Many in Germany scoffed at the idea of these ponderous tanks operating in mountainous territory but had they been able to talk to Churchill crews from Italy they would have learned the truth. As it was this expedition proved unnecessary, if it had ever been planned, and by now all the Churchills had finished with engines.

On VE Day all tanks in 107 RAC, except those of B squadron which was away on detached duty, fired a twenty one gun salute.

At the end of May 1945 9th RTR was down to six Churchills and six Honeys but they were still engaged upon security duties. They were informed that in June they would be joined by 4th Royal Tank Regiment which had been selected to convert to Churchills and prepare to go out to the Far East. Their

Above: A white flag signifies surrender as Guards' Churchills with British paratroops on board enter the German town of Billerbecke at the end of the war.

Below: *Iron Duke* the Churchill Mark VII of Lieutenant Colonel P N Veale, commanding 9th Royal Tank Regiment, is prepared to take part in the Paris Victory Parade.

comrades in the 9th were to assist with their conversion and training in the infantry tanks. In the event it came to nothing and 9th RTR was disbanded in January 1946.

The 147th RAC had already gone by then, having been disbanded in the previous October and of the three regiments in 34th Army Tank Brigade it was 107 RAC that lasted longest. In June 1945 they were told that they would go to the Far East as a Light Tank Battalion, which meant a form of organisation, not a regiment equipped with light tanks. Changes in the army, resulting from the German surrender, saw many men leaving the regiment and others being switched around so that, for instance, at one time there were more ex-147 RAC men than original 107 personnel in the regiment and the dilution continued as men were released from the colours or transferred. The need for tanks in the Far East rapidly diminished and 107 RAC seems to have become something of a forgotten regiment. When its history was written, probably in January 1946, the commanding officer, Lieutenant Colonel R H Taite, could not say whether they were headed for disbandment or suspended animation. But his last entry in the war diary, dated 2 March 1946, deserves to be quoted in full;

> '*We have today received official written information that the 107 Regiment RAC (King's Own) ceased to exist on 4 February 1946. I give up! I could say and write much, but on second thoughts I feel that my - our - sentiments had better remain unuttered and, like the regiment itself, pass quietly into suspended animation.*'

Meanwhile there were interesting changes in Britain, one of which concerns Churchills. Among the regiments that never went overseas one was 1st Battalion The Royal Gloucestershire Hussars. Throughout the war they acted as a training regiment, equipped with a variety of tanks, and in February 1945 they were to be found at Crowborough, in Sussex. Here they were charged with retraining men who had been wounded in action and hardening them up for a return to front line service. In June they learned that they would be reorganised as an Armoured Regiment (Light) to be equipped with Churchill tanks.

Whether this presaged a move to the Far East is not certain, but they came under command of 35th Tank Brigade and soon received 21 Churchill tanks from the Guards Armoured Training Wing which also supplied instructors to teach them the rudiments of the new tank. By July they had moved to Fornham Camp East, at Bury St. Edmunds in Suffolk where, according to their war diary, they had the following peculiar collection of Churchill tanks; thirteen Mark IV, one Mark V, thirteen Mark VI and one Mark VII along with three Churchill 3 inch Gun Carriers which should have vanished long ago and two more described as Churchill Mark IIIB (headless). The suffix B might indicate one of the mechanical alterations but the term headless is impossible to comprehend, unless it means turretless. In August, still at Bury St. Edmunds, they acquired ten more Mark VI, nine Mark VII and seven Stuart V light tanks along with a motley collection of Jeeps, trucks, motorcycles and five Daimler Scout Cars.

In the event nothing came of this. Within months all of these vehicles had been taken away and the regiment had sailed for Austria, via Italy, where it operated armoured cars in an internal security role.

13

A Hero Retires

With the war in Europe finally over, serious consideration was given to the defeat of Japan. But at the same time there was a great urge to take stock and absorb the lessons of the recent campaign before it was too late. One manifestation of this was the publication, in January 1946, of a document entitled *A Survey of Casualties Amongst Armoured Units in North West Europe.* It was compiled by two Royal Army Medical Corps Captains named Harkness and Wright and perhaps one should explain at once that the armoured units involved were exclusively British and they did not include any Churchill battalions, at least not in any detail.

It was a prodigious work of research which attempted to analyse such things as where a tank was hit and by what sort of projectile, what effect the hit had on the crew and how their various injuries compared. Indeed the amount of detail available is staggering but not if one is studying Churchills. The reason for this, Harkness and Wright explained, was due entirely to the difference in organisation and operation between armoured brigades in armoured divisions and tank brigades attached to infantry divisions. In simple terms it came down to the fact that an armoured brigade generally functioned as an entity. Its own personnel recovered damaged tanks and reported on them, it processed its own killed and wounded and maintained comprehensive records to a high standard. Such was not the case with a tank brigade. It was rare for regiments to work together and quite often one found independent squadrons out on their own. Damaged tanks were left scattered across the landscape to be recovered by whoever could reach them and wounded would go back to the dressing station of the infantry division they were working with; so it soon became impossible to keep track of them.

Nevertheless they did have some luck. Enquiries revealed that 4th Grenadier Guards and 107 RAC had maintained fairly comprehensive records, but for the others they had to rely on hearsay. It makes a poor sample when compared with the hard evidence available from the armoured brigades but at least it gives some small insight into the situation as far as Churchill crews are concerned. Generally the scale of casualties, calculated as the ratio between those killed and the total of killed and wounded taken together, was more or less the same for Churchills, as it was for Cromwells and Shermans; that is something around .48 to 1 inside tanks and .19 to 1 outside. Churchills appeared to have less incidence of fire than the other tanks, at least to judge from the casualties, but this may have been due to the side doors of the infantry tank which increased the chances of escape. There were also fewer reported cases of men being killed or wounded when partly exposed, meaning with head and shoulders protruding from the turret hatches. This was put down partly to the added protection provided by some regiments, in the form of an armoured plate behind the commander's hatch, but mainly to the fact that the All Round Vision Cupola was more common on Churchills than on other tanks.

One of the most interesting factors, and again one that was common to all types of tank, was the low percentage of kills due to the use of hand held, hollow charge weapons such as the *Panzerfaust.* Far more crewmen were killed when a tank was hit by a conventional armour piercing round then ever occurred from strikes by hollow charge projectiles, yet this weapon seemed to have the psychological edge. Nobody likes being in the lead tank which comes around a corner only to be stopped in its tracks by a strike from an 88. The results were almost always fatal for some of the crew at least, yet this seems to have been regarded as a normal hazard of tank life. It was not dreaded in the same way that men felt about the sneak attack delivered by a man, hiding behind a wall, with a glorified drainpipe under his arm. True the effect of a strike by a hollow charge round could look terrible when the tank was examined afterwards but the horror of it seems to have been the very idea that one human being, carrying a frail weapon in his arms, could creep up to anything as solid as a tank and disable it with one shot.

As a result experiments were carried out to find a means of defeating it, although the principle was already well understood. The trick was to cause the hollow charge projectile to explode before it touched the armour of the tank and the best way to do this was to add an extra screen of armour, spaced away from the hull, to set the thing off. The result, in the case of the Churchill, was the production of a kit which attached to sections of the hull and turret. It was demonstrated on a Mark VII with the improbable nickname of *Gnat* and consisted of a set of expanded metal screens in steel frames, fixed to the tank by brackets. The photograph shows the main protected areas. The use of expanded metal, rather than armour, was clearly done in order to keep weight down but there is no reason why it should not have been just as effective. In the event it appears to have come too late to fit onto tanks serving in Europe and was never required for the Far East.

Another reason for making the screens from expanded metal was that it was less attractive to magnetic grenades. These weapons had been exercising the minds of British experts since the latter half of 1944 when it was know that both Japan and Germany had developed them. Of course the Germans, having designed the magnetic grenade, also produced the antidote, a putty like compound known as *Zimmeritt* which was coated over the armour of a tank and allowed to harden. The Shell Mex Petroleum Company in England already produced a similar substance but the authorities were unwilling to adopt it without testing everything else first and this included wire netting, plywood and hessian sheets impregnated with paint among others. These experiments were not limited to the Churchill by any means but it is worth recording that the Mark VII gun tank now on display in the Tank Museum, and reputedly the last one to be built, has wire mesh panels over the engine decks which are never seen on service tanks and this may well be a feature adopted for those tanks destined for the Far East.

The glacis plate of a Mark VII Churchill of 141 RAC penetrated by a 75mm German projectile, fired from an unknown range but which struck at an angle of 80 degrees.

An unfortunate choice of name for this 9th RTR Churchill which has taken an incoming round through the appliqué armour just below the turret on the right side.

This uparmoured and upgunned Churchill Mark III was penetrated by a round entering the right side of the hull with such force that it has bulged a panel of appliqué armour on the opposite side and forced it away from the hull.

In due course the research team came to the conclusion that an anti-magnetic paste of some sort was the best solution but they pointed out some drawbacks. In order to make a thick enough covering it had to be applied in layers over a period of three days. It could be vulnerable to rubbing by shrubbery, chipping from soldiers' boots or even coming off in great chunks when the tank fired. On the other hand it took paint well and did reduce the metallic gleam given off by a

A Churchill Mark VII named *Gnat* displays spaced panels of steel mesh as protection against magnetic grenades and hollow charge rounds.

tank, even one that had been sprayed with matt finish paint. In June 1945 21st Army Group in Europe conducted trials on a coated Churchill Mark VI. They found that even on a bright day the coated tank did not stand out against the background. They even tried mixing straw with the compound to give it texture and then tried paint. The tank was first coated with a German paint known as *Spezial Grun* which is described as a bright yellow/green camouflage shade. This was over painted with patches of a black emulsion and the result was excellent. None of these experiments resulted in any of these materials being adopted but by then the war was virtually over so it did not matter.

An unusual test of this paste is mentioned in an undated report by the Department of Tank Design. It concerns two Churchill tanks, one coated with anti-magnetic paste and the other without, which were subjected to attack from a Wasp flamethrower. Most of the flame shots were aimed at the front of the tank and measurements were taken electrically so that they could be read immediately outside the tank. The results were presented in tabulated form and no attempt was made to evaluate them but it does seem clear that in general the unprotected tank got a lot hotter than its coated counterpart. In most cases where the coated tank was hit the temperature did not reach an unbearable level but on the unprepared tank, at one point, they obtained an internal reading of 600 degrees centigrade which would probably have been fatal to the crew and set off the ammunition. Whether there was much risk of this occurring in action is uncertain but one imagines that it would have been quite unlikely.

On 2 May 1945, with the war against Germany in its last few days, Claude Gibb, Director General of the Fighting Vehicle Division at the Ministry of Supply visited the Newton Chambers factory in the Chapeltown district of Sheffield. He was there on behalf of Field Marshal Sir Alan Brooke to congratulate the company on its efforts to build Churchill tanks during the war. At some point a new Mark VIII was rolled out with the legend CHURCHILL 1,000 painted on the sides and, when a large Union flag fell away from the turret side it revealed the United Kingdom and United States flags painted on the turret. The number 1,000 was inevitably a fiction and its significance was not even made clear. It could hardly be the 1,000th Churchill to be built, but rather the 1,000th to be completed by Newton Chambers.

The workers and management of Newton Chambers cheered Gibb and congratulated themselves on what had been achieved, but now that the war was ending demand for tanks would fall off dramatically. Even so there are no indications of large scale cancellations from the surviving contract cards and there is evidence to show that the rework programme was continuing. As we have already seen the main production run of the Churchill ended with the 95mm gun Mark VIII. Up to that point, except for the Mark VI which is little more than an upgunned Mark IV, each type was a distinct combination of gun and turret. Yet various publications appearing at the end of the war carry the range of Churchills through to the Mark XI.

According to a supplementary handbook first published in September 1945 Marks IX, X and XI were mechanically identical to Marks IV, V and VI but they carried the heavy, composite turret. A more detailed descriptions further on in the book proves that this means the cast and welded A22F turret. The book goes on to say that these tanks also have extra applique armour on the sides, thicker front plates, reinforced roof (presumably to the hull) and skirting plates to the hull. The difference was that the Mark IX still carried the 6-pounder, making it an improved Mark IV, the Mark X a 75mm being an improved Mark VI and the Mark XI the 95mm naturally an improved Mark V. Unfortunately the book fails to give any weights, but one imagines they must have been a considerable increase on the original and a good deal more than the suspension was designed to bear. The final six-monthly report on progress in the Royal Armoured Corps, covering the six months up to 30 June 1945 gives a few more details. The rework programme involved adding applique armour 20mm thick to the sides, replacing the existing visor plate with a Mark VII pattern, complete with ball machine-gun mounting and round driver's visor; fitting heavy suspension units and the H41 gearbox.

What we should be looking for is the combination of the original type hull, with its square side doors but new driver's visor and machine-gun mounting, carrying the later style cast turret with the heavy lip around its base. At the time of writing

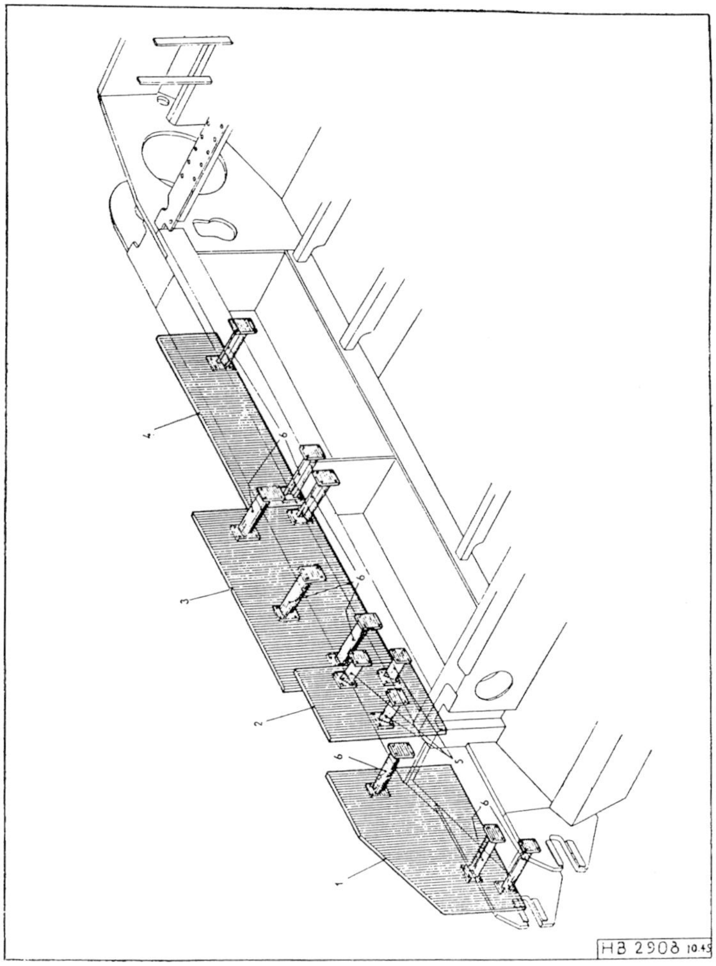

A drawing of the skirting plates designed for use on Churchill tanks.

Below: A Churchill Mark VI experimentally coated with the British version of *zimmeritt* and then painted sand and black. Even in this unlikely setting the camouflage seems remarkably effective.

this combination has never been identified in a single photograph or document. It is not even clear what is meant by skirting plates unless this refers to the spaced panels mentioned earlier. Applique armour on the sides is clear enough and it should not be too difficult to thicken up the roof armour, at least over the engine if not the front The six-monthly report does not show how many reworked tanks had been ordered but it is clear that none were completed by the end of June and the chances are that the programme was cancelled in its early stages. The only evidence to suggest otherwise is a note in a DRAC Liaison Letter of 1945 which claims that some Marks X and XLT had been delivered, but that the rework programme was now being phased out and he anticipated that no Marks IX or XI would appear. The reference to XLT introduces three other marks, the IXLT, XLT and XILT where LT stands for Light Turret. Of these the final RAC report states they 'are reworked Churchills brought up to the heavy Churchill standard of full immunity but retaining their current Light Turrets as there are not sufficient heavy turrets' however a further note explains that none of these were produced because production of heavy turrets exceeded the number of tanks being reworked.

In 1948 the War Office introduced a new registration system for military vehicles. This meant replacing the wartime WD number with a new style of registration which also served as a road license plate. In the case of tanks it involved painting out the old T series number and replacing it with a new serial painted front and back. The new system took the form of two digits, two numbers and two digits; 23ZR64 being a typical example. The original cards which recorded these changes list just 32 Mark X Churchills which were probably all that were ever converted.

As soon as the War Office started to concentrate its collective mind on the problems of fighting in South East Asia it initiated the usual bureaucratic nightmare, issuing papers on a variety of subjects in vast quantities. From the RAC point of view the main effort appears to have been reinforcement, with the emphasis on specialised armour. The new forces would include two armoured brigades, qualified with the word Light in parenthesis which is never fully explained. an Indian armoured division, presumably extracted from Italy and special regiments dedicated to CDL, personnel carriers, amphibians, airborne tanks and general specialised tanks. Churchills would also be used but the idea seems to have been to convert an armoured brigade already in Burma, and the formation chosen was 254 Indian Armoured Brigade which comprised 3rd Dragoon Guards (The Carabiniers), 149 RAC (7th Battalion the King's Own Yorkshire Light Infantry) and 150 RAC (10th Battalion the York and Lancaster Regiment). Up to this time these regiments had been fighting mainly in US M3 Lee and Grants, supplemented by Shermans, but now they would change.

According to RAC records in Britain 85 Churchill tanks and three Churchill bridgelayers had been shipped to India by June 1945 and these were to equip 149 RAC. The regiment is shown as having left the brigade in Burma by May but it would not have needed all 85 tanks at once. Assuming the normal establishment, about 50 tanks would be needed at once to complete the regiment so the remainder would presumably be held as a reserve. The RAC explained that the 254th would become a Churchill brigade as soon as sufficient tanks were available and records show that the brigade had moved back to India, from Burma by the end of June 1945. The Carabiniers, in their history, say that as they prepared to leave Burma for India after the fall of Rangoon they were told that they would be re-equipped with Churchills and transferred to the newly created XIIth Army for the final battles. Events elsewhere ensured that no Churchills would fight in the Far East during the Second World War.

The precise number of Churchills shipped out to India is difficult to judge. Another RAC report, listing tanks by type and quantity world wide, claims that there were 308, including Crocodile flamethrowers, in South East Asia Command (SEAC) by the end of June 1945 while an Australian report dated one month later, says that the British have shipped 'large numbers' of Churchills to SEAC. This would provide a Churchill brigade with near enough one hundred per cent reserve which, considering the distances involved, would seem sensible. The Australians themselves, the first country since Canada to show an interest in the tank, ordered 310 and the similarity between the two figures is suggestive. In the event a report from Major A R Nairn of the Royal Australian Engineers, then stationed in London, reported that ten Mark VI, 30 Mark VII and 15 Mark VIII were ready for shipment by July. Fifty one of these tanks finally arrived in Australia.

Major Nairn's main concern while he was in London was to make absolutely sure that the British made a good job of preparing the tanks for shipment. From the early days of the war the problem of shipping tanks long distances overseas had been largely ignored, with the result already recorded to the first two Churchills shipped out to North Africa. However when discussing the Far East and Australasia it was not just a question of distance, great as that was, but climatic conditions. The tropics could do amazing things to machines, even something as solid as a Churchill tank. Condensation would build up in enclosed spaces, turning everything to rust, while instruments would steam up and become breeding grounds for insects and electrical components would corrode. Fabrics just rotted away. As it was Major Nairn need not have worried. British manufacturers had already learned their lesson and experts at REME had carried out extensive investigations into all problems associated with delivering, storing and using vehicles in the tropics.

Even so the list of tasks to be undertaken when tanks were shipped so far was extensive. First the turret had to be traversed to six o'clock and the gun supported in a wooden cradle. Then very orifice had to be packed with grease or covered in anti-corrosive paint while a wooden bung went into the end of the gun barrel and special fabrics were used to seal other areas. It is interesting to read, in conjunction with Major Nairn's report, further comments on the anti-magnetic compound mentioned earlier. It seems that on-going trials had revealed that if it was applied with a ripple finish, as the Germans did with their *Zimmeritt*, it was difficult to fit waterproofing gear. Since this was considered important Britain had now decided to adopt a smooth finish for the paste but after further thought the Australians elected to have their tanks delivered without the covering and apply the paste later if they wished.

Just to complete the story of the Australian Churchills the first ten, all Mark VII, sailed from Liverpool aboard the SS *Empire Abercorn* on 10 July1945. Other sources suggest that the initial order for 310 was later increased by 150 and there are further references to an order for 100 which may be the same thing. However we are talking about 460 tanks or 560 at most. In practice, with the delivery limited to 51, it would just about suffice for one regiment without any reserves.

According to Major-General Ronald Hopkins in *Australian Armour* (Australian Government Publishing Service 1978) the Australian General Staff had formed an unaccountable preference for the Churchill over the Sherman in 1943. Whether this had anything to do with the success they had enjoyed with Matildas in New Guinea, compared with American M3 Stuarts on Buna, is not clear, but it is difficult to imagine how a tank as heavy as the Churchill could be used effectively in the Pacific, or indeed on the Australian mainland itself. The Churchills were subsequently issued to the 1st Armoured Regiment, Royal Australian Armoured Corps at Puckapunyal, Victoria in 1949. Some remained in service until 1956 when they were finally replaced by Centurions.

Australian troops watch with their usual mixture of cynicism and disbelief as a Churchill Mark VIII negotiates a muddy jungle track. Australia received 15 Mark VIII out of a total of 51 Churchills.

A pair of immaculate Churchills pass the saluting base in London during the Victory Parade. His Majesty King George VI takes the salute, Queen Elizabeth, the present Queen Mother stands beside him. To the left of the dais one can distinguish Prime Minister Clement Attlee, bare headed, then Winston Churchill and, on the end of the line in uniform the South African General Jan Christian Smuts who, like Churchill, was a veteran of the Boer War, albeit on the other side.

The end of World War II, with the signing of surrender documents by a Japanese delegation aboard USS *Missouri* in Tokyo Bay on 2 September 1945, posed numerous problems for the Royal Armoured Corps. A large number of the tanks then in service came from the United States under the Lease Lend agreement and there was every reason to believe that most of them would be reclaimed. The RAC was most anxious to retain as many light tanks, Stuarts and Chaffees, as possible since there was no British equivalent. If the Shermans had to go then it was calculated that there were enough Cromwells, Comets and Churchills to fill the gap until Centurion came on stream in sufficient quantities to replace them.

Those police duties that fall to most victorious armies after a major war were seen, by the British, as a modest threat most easily contained by wheeled armoured vehicles while the tank regiments stationed abroad would be furnished with Comets and Cromwells, the latter mainly because some of them carried the 95mm close support weapon. There did not seem to be a place in all this for the Churchills.

In fact the Royal Armoured Corps was in the process of reorganising itself. Most of the special wartime regiments had already been disbanded, while others would revert to the Territorial Army leaving a core of regular regiments serving in Britain or abroad with tanks or armoured cars. None of the surviving documents appear to specify exactly which regiments would have particular types of tanks but the general rule seems to have been that those regiments serving abroad ran Cromwells and Comets while Churchills were only allocated to some regiments in Britain. Meanwhile those tanks not required for active duty would have to be stored and the Director, Royal Armoured Corps, Major General Raymond Briggs, was at pains to explain that those tanks in store must be maintained and preserved in the best possible condition.

Another innovation at this time was the multi-barrelled smoke discharger which was, essentially, an improved version of a device that had been carried on German tanks since about 1942. The British model consisted of six barrels in a cluster on a common mounting, each pair of barrels set at a slightly different angle to the adjacent pair. A mirror image set was mounted on each side of the turret. The barrels were set at a fixed elevation to give a range of about 50 yards and they were splayed out in such a way that the projectiles spread out in an arc with about 20 degrees between each pair.

The projectile was the No. 80 smoke grenade, which was fired electrically by a push button system inside the turret. Each button controlled a vertical bank of three grenades. A scheme was organised to fit them to most British AFVs, including the Churchills, the only exception being the Shermans whose future was still uncertain. Subsequent trials indicated that the device was very vulnerable to small arms fire and until this had been investigated no more were issued. This meant that a number of Churchill tanks that should have had them fitted probably never did. In the event it proved impossible to overcome this unreliability and they were reinstated.

A rather battered looking Churchill Mark V apparently serving with 4th Royal Tank Regiment in Libya after the war.

Generally speaking it is difficult to find the Churchill mentioned in official documents for the immediate post-war period. The implication, clearly, is that it was the least considered of all wartime types. It was still the mainstay of specialised armour units, and of course the Crocodile flamethrowers, but as a gun tank it was disappearing rapidly from the scene. Even so, when the five surviving RAC Training Regiments were reviewed in 1946 the 57th was shown as the training regiment still devoted to instructing on Churchill. But a later entry explains that when the 57th changed to training on cruiser tanks no Churchill trained personnel would be available after 29 July 1946. The writing was now clearly on the wall.

A review of tank holdings, published in the first quarter of 1946, listed the following holdings of Churchill gun tanks; Mark V - 197, Mark VI - 594, Mark VII - 896, Mark VIII - 173, Mark X - 33. They were described as being maintained against the requirements of the post-war Army, both active and auxiliary, but a footnote explained that production of Centurion would gradually eliminate the older types. However when the Victory Parade was held in London on 6 June 1946 two Mark VII Churchills took part. Cleaned, painted and polished to the highest standard they rolled past the elaborate dais on which

Sacrifice; the battered carcass of a Churchill Mark IV on a target range in Britain.

His Majesty, King George VI, attended by other members of the Royal Family and various notables, took the salute. Among those seated alongside the saluting platform one can identify the distinctive form of Winston Churchill, now leader of His Majesty's opposition, who leans forward and, no doubt, permits himself a wry smile as these tanks trundle by. He, who had used all the power at his command to get the tank built, and supported it against a lot of professional opposition, had some cause for self-congratulation in the end.

The Territorial Army was officially reinstated on 1 January 1947. It was to consist of two armoured divisions, three independent armoured brigades, three armoured car regiments, three divisional regiments and one armoured delivery regiments. The armoured divisions and divisional regiments would be equipped with Comets, Cromwells and Stuarts while the independent brigades would have Churchills as gun and close support tanks and Stuarts for reconnaissance. Each brigade would have four armoured regiments but they would only have a handful of tanks on their regular establishment. Others would be drawn, as necessary, for annual camps and exercises. However it seems that many of the regiments listed in the original table soon reverted to other roles so it is difficult to say how many actually acquired Churchill tanks in the end.

The diminishing requirement for Churchills in the British Army meant that many could go for scrap, unless anyone else wanted them. However there was a lot of competition. The United States had thousands of spare Shermans, which were far more suitable for most resurgent armies than the temperamental old Churchill, while the Soviet Government was busy distributing redundant T-34 tanks to those country already under its sway, or those that it hoped to influence. The only country which may have employed Churchills in any quantity appears to have been India, although the evidence

A fate worse than death. A Churchill Mark IV in Ward's scrap yard amidst the tower blocks of Sheffield. This tank sports the post-war style of numbering 12ZY48,

Churchill tanks at the Royal Armoured Corps Training Centre in Germany in October 1947. The tall officer is Lieutenant General Sir Richard McCreery, General Officer Commanding the British Army of the Rhine. He is greeting Lieutenant Colonel H H K Rowe RTR.

A Churchill Mark VII of the Royal Jordanian Army. How many they received and how long they remained in service is not clear.

is slim. In fact it takes the form of a handbook, issued by the Indian Armoured Corps in December 1946. It was printed in Roman Urdu, the western alphabet version of Hindustani, and covered Marks IV to VIII. It seems reasonable to assume that the tanks in question were those shipped out to India in the last months of the war for use in Burma, but how many served with the Indian Army, and for how long they were retained has proved impossible to confirm.

One country that did acquire Churchills was Ireland, or Eire as it was known at that time. The Irish Free State had operated armoured cars as part of its armed forces since its formation but it had formed a Tank Cadre between the wars, to provide a tank trained nucleus of personnel and this was revived in 1949. Four Irish Army officers had been trained on Churchills in Britain and in due course four tanks arrived in Dublin. The first, which was delivered on Christmas Eve 1948, was recovered from the hold of a ship which had subsequently been loaded with agricultural fertiliser but it was duly rescued and cleaned out. The four tanks have been identified as type XLT, but a frontal view of two of them shows the original square driver's visor and machine-gun mount which, for all practical purposes, should make them the Mark VI type with appliqué armour and all-round vision cupolas. The tanks remained in service for many years although it proved increasingly difficult to keep them mobile. At one point a Meteor engine is said to have been fitted in one of them, which resulted in an amazing improvement in its performance, but it must have been an incredibly tight fit and probably meant a considerable reduction in fuel capacity. The last survivor was kept, in running order, on a firing range although its gun was removed after each firing demonstration and stored at the Curragh. This practice ended when the supply of ammunition ran out in about 1970. At the time of writing one Irish Army Churchill remains in the Museum collection at the Curragh.

In 1949 one Churchill Mark VII, described as unserviceable, was subjected to a battering such as no other tank of its class had ever received. In fact the trial was not specifically related to the Churchill, which was already considered obsolete. But since a Churchill was involved it justifies inclusion

This Irish Army Churchill has been identified as a Mark XLT yet, as we have seen, the existence of this type is denied by official records. To comply with specifications a Mark X should have the round driver's visor and BESA mounting of the A22F which this tank clearly does not have.

here. The tank was uparmoured to a maximum 235mm with patches of appliqué armour welded to the sides of both turret and hull. and then fired at with armour piercing capped rounds of 128mm from the ridiculously short range of 100 yards. The gun in question was either the PaK44 l/55 as fitted to the huge German Jagdtiger (SdKfz 186) self propelled gun or the KwK44 L/55 of the superheavy tank Porsche Type 205 *Maus.* The actual gun is not specified but it is interesting to speculate upon whether, if it was the PaK 44, an actual Jagdtiger was used or if the gun was mounted on a range mounting. If it was the KwK44 then it would have to be on a range mount.

Three rounds were fired; two at the hull sides, one at the turret. Those which struck the hull gouged great chunks out of the appliqué armour and cracked many welds but did not penetrate. The one that struck the turret virtually tore it off. The appliqué plate was smashed to pieces and the turret side bulged inwards but again there was no actual penetration. Even so many internal fittings were torn loose and all the hatches were ripped off the turret, including the entire commanders cupola. There seems little doubt that any tank, hit this hard, would be totally disabled by the blows. To discover how the crew might have fared two rabbits were placed inside the tank, each in its own little wire cage, but they survived all three rounds and, according to the report, appeared to be unaffected.

What happens when a German 128mm armour piercing round, fired at a mere 100 yards range, strikes the turret of a Churchills Mark VII. Witness the extra panels of armour on the hull and turret sides which gave a maximum thickness of 235mm.

By 1949 it must have been a reasonable bet that the Churchill had had its day. As a concept the infantry tank was now discredited, and Vauxhall's 1940 design, which had hardly been in the forefront of technology then, was now looking decidedly long in the tooth. Added to which the peacetime establishment of the Royal Armoured Corps did not call for so many tanks, and the numbers required could be filled by surviving Cromwells and Comets until the new Centurion replaced them. The only place you would find Churchills now was with specialised armour units and there were not many of them. But among them was 7th Royal Tank Regiment.

We left the 7th outside Dunkirk at the time when it learned that it was going to convert to Churchill Crocodiles. It had done very well in its new role but among its less pleasant tasks was the destruction of Belsen Concentration Camp. From January 1946 until September 1947 it had been stationed in India, operating light tanks and armoured cars on internal security duties and when it arrived home, in October, it was located at Wrottesley Park near Wolverhampton and it was here that it learned of its next role. It would move down to Tidworth, in Hampshire at the end of the year and join the Specialised Armour Establishment, operating a wide range of vehicles, many based on Churchill chassis, for trial purposes.

By 1950 the regiment was split up; Regimental Headquarters, along with A Squadron was at Bovington Camp. C Squadron was in Norfolk and D Squadron at Gosport in Hampshire. The big news item that summer was of unrest in a far off country that none of them had probably ever heard of, called Korea. One month later they were told to raise a squadron to go to this place and operate Churchill Crocodile tanks. C Squadron was selected, but like most elements of the post war British Army it was desperately short of men. It was made up to strength by calling up reservists - men who had already seen service during World War II - along with National Servicemen, a few volunteers and a core of experienced regulars. Many of the tanks were stored in Germany.

The squadron sailed from Southampton on the transport *Empire Fowey* on 12 October, after two hectic months working on the tanks, many of which had been standing idle for five years with very little care and maintenance. At the same time new tank crews were being put through a rapid training programme. The *Empire Fowey* arrived at Pusan on 15 November 1950 and the 7th then sat down to wait for their tanks. These arrived in installments, usually in poor condition, and as each troop was completed it prepared to move north. Their

A Churchill Mark VII, with Crocodile flame gun in place, is unloaded onto the dockside in Korea.

destination was Pyongyang but the first troop to arrive discovered a chaotic situation as a massive invasion by the Chinese Army threatened United Nations positions and prompted a rapid evacuation. No. 6 Troop thereupon turned south and over the next few days covered 200 miles on their tracks over indifferent roads, an achievement for the Churchill which the regiment regarded as without parallel.

By December the squadron was located eight miles north of Seoul, acting as cover for retreating UN forces and their main task was to guard crossing points on the Han River. To this end they were shorn of their flame trailers and employed as gun tanks, which is why they qualify for inclusion here, and they never did operate as flamethrowers. For some time one troop retained its flamethrower trailers, just in case, but they were more of a nuisance than they were worth. C Squadron had been supplied with AEC Matador artillery tractors to tow the trailers on route marches, but since the trailers had no brakes, and were exceedingly heavy, accidents were quite common. In fact the squadron had been sent out in a support capacity. The main British armoured force was headed by the VIIIth King's Royal Irish Hussars, a full regiment of Centurion tanks, but 7th RTR now learned that the Centurions would operate further south. The reason given was that Centurions were too heavy to operate north of Seoul because the bridges would not take them. But Major A J D "Jack" Pettingell, commanding C Squadron 7th RTR believed, rightly as it turned out, that the real reason was security. One Centurion had already been lost at Pyongyang and the authorities did not wish to lose any more of these brand new machines to the Communists.

Their first major action took place on 2 January 1951 near a village known as Chunghung Dong. With the departure of the Centurions, British armour in Korea was represented by C Squadron, 7th RTR and the Cromwells of the reconnaissance troop of 8th Hussars, along with similar Royal Artillery OP tanks under the command of Captain Astley Cooper. Early on the morning of 2 January the Cromwells, along with 1st Battalion the Royal Ulster Rifles were over run by Chinese troops but the tanks managed to restore the situation. Next 1st Battalion the Royal Northumberland Fusiliers found themselves under attack and Pettingell was asked to provide help. One troop of Churchills, along with half of squadron headquarters, made their way in single file along a gully towards the threatened position. Progress was slow, the ground frozen and the tanks slipping as they went so it took something like eight hours to work their way up to the scene of action.

Firing high explosive rounds the tanks managed to help the infantry retake one village and then on to the second. It was all achieved with virtually no casualties, although the enemy, who persisted in using mass infantry techniques, suffered heavily. Pettingell was worried that his tanks might be asked to stay forward after dark and he had great difficulty in persuading the Brigadier in charge to release them. Just before the troop pulled out it received a request for help from the infantry and agreed to stay until the light faded. Meanwhile the general situation was becoming desperate all along

All that remains of a Churchill of C Squadron, 7th Royal Tank Regiment, blown apart during the premature evacuation of Seoul.

the line and orders were received to pull back south of the Han River. Pettingell responded by sending back as many tanks as he could spare to a leaguer position but all the while he had tanks committed he had to maintain a supply of fuel and ammunition, plus maintenance support from REME.

Finally, as darkness fell, the tanks and infantry returned with great care. It took six hours, with tank commanders guiding their driver's on foot down the narrow track. Pettingell had done an excellent job, getting all his tanks back with no casualties and yet covering his infantry up to the last moment. But this was done with a liberal and intelligent interpretation of orders in a very difficult situation. He describes the final wait with his half-track and a troop of Churchills, until the last company of Northumberland Fusiliers came out, as very trying. The Chinese used what he described as 'toy bugles' to communicate and the sound of these, drawing ever closer, caused him and his tank crews to 'have kittens' while they waited. All was not so well with the Cromwells. Cooperforce, as it was called, was ambushed and lost all of its tanks in the night due, according to Pettingell, to the fact that they were not allowed to withdraw once it got dark.

Despite a statement that Seoul would be held, it was evacuated shortly afterwards. Pettingell could not understand what all the haste was about and tried to move out in a dignified fashion through the burning city, but orders to hurry resulted in the destruction of one tank and an armoured recovery vehicle which were not ready to run. He later learned that two days had been allowed for the withdrawal which meant that he probably could have saved the tanks. He gives some graphic descriptions of the conditions. The smell caused by primitive methods of fertilisation made everyone ill and it became so cold that Pettingell had to warm up the ink in order to write his reports. He could detect no enthusiasm on the part of the high command to stand and fight and he regarded the lack of accurate information as chronic.

In a later report Major Pettingell describes some of the problems encountered when operating Churchills in Korea. Much of the trouble was due to bitter winter weather conditions which caused moisture in the air servo systems to turn to ice and the grease in turret rings to freeze, jamming the turrets. The solution to this problem was to lubricate the ring with gearbox oil and pack straw around the outside. Most of the tanks had done between 350 and 500 miles at the time of this report and forded many rivers. This caused another problem. Water would get into the gearbox and freeze under the oil. The tank had to sit with its engine running for thirty minutes before the gearbox became warm enough to thaw out and if the tank had been parked in gear it meant sitting there with the clutch pedal pressed hard down for the same period of time. Standard practice, if there was time, was to chock up the suspension of each tank and park it in neutral overnight.

The Churchill's behaviour on ice, according to Pettingell, terrified everyone. One tank, he explains, slid sideways for 45 feet during one journey. It is something one recalls the Guards complaining about in Holland. The cold also affected track pins which broke continually and crews had to work hard in dreadful conditions to replace pins or even refit thrown tracks if the break was not detected in time. One thing Pettingell does not explain is whether the Churchills, having been restored to the gun tank role, had their flame guns replaced by BESA machine-guns. It would be an obvious thing to do, and not particularly difficult since the hull mounting was adaptable to either weapon. Tank commanders also used a Bren gun from the turret top although under mass attack by enemy infantry most found it safer to retire inside the tank, close all the hatches and invite a neighbouring tank to spray them with machine-gun fire.

In February the United Nations troop started to fight their way northwards again, and 7th RTR went with them. Opposition was minimal and Jack Pettingell attributed this to the massive firepower employed against the enemy. Arriving once again on the Han River nobody could find a valid use for the Churchills so they were tasked with mounting patrols along with infantry platoons travelling in Oxford Carriers. Their commanding officer regarded this as misuse of his slow and mechanically vulnerable tanks but on one occasion they were employed to fire against enemy positions north of the Han and during this period he notes with some pride that they carried out shoots of ranges up to 5,500 yards, which was extremely good for the 75mm gun firing high explosive. Obviously the tanks' own sights were not much use in these conditions but the squadron employed an air observer. This was achieved by negotiating the unofficial loan of a Stinson L5 aircraft and pilot from the US Army, fitting it out with a British observer and radio set, and using it to spot the fall of shot. The same practice, it will be remembered, had been tried in the last months of the war in Europe.

Following these operations the squadron was withdrawn to a reserve area, based on Suwon. Many of the tanks were in dire need of maintenance. Most engines were beyond repair and had to be replaced while sprockets and tracks were starting to break up at an alarming rate. The Chinese had started using crude but effective anti-tank mines and there were suggestions that the squadron should be re-equipped

The almost unpronounceable *Gynaecolater II* caught in a rare moment of leisure on a warm day in Korea.

with flail tanks. Indeed Pettingell believed that they should have been sent out as a composite squadron, with troops of flame tanks, flails and bridgelayers. After all they had been trained in specialised armour, but it was too late to worry about that now. By June they were enjoying the Korean summer in the Inchon area, aware that they would not be called upon to go into action unless an emergency developed. Pettingell noted that they would not be getting any new tanks, by which he presumably meant more Churchills, but he managed to borrow a Centurion from the Hussars for training.

By September they knew they were going home. In the meantime they had received a generous compliment upon their good work from Major-General John W O'Daniel commanding I Corps. They sailed from Pusan on 8 October 1951 aboard the transport *Empire Orwell*, leaving their tanks with an Ordnance Depot for later shipment to Britain. So ended the active service life of one of Britain's most famous tanks and the reason, apart from age, is summed up in a document issued by the Fighting Vehicle Design Establishment in 1951.

This document reviewed the immunity of five British tanks to a variety of Russian weapons. Two of them; Cromwell and Comet, could be written off as vulnerable in all respects while Centurion and the new FV214 Conqueror had high standards of immunity. Churchill came somewhere in the middle. In respect of the Soviet 76mm gun Churchill was immune at all ranges up to 2,000 yards. Against the 85mm gun it was critical up to 500 yards but immune at 1,000 yards and above but when it came to considering the new 100mm and 122mm weapons the poor old Churchill was considered vulnerable at all reasonable fighting ranges and only immune above 3,000 yards. This was the future.

A final statement said that from the date of this document the A22 was excluded from all future consideration, except as an armoured recovery vehicle.

14

Black Prince: the Big Gun Churchill

Although it is perfectly legitimate to employ the adjective lamentable when discussing Britain's efforts to produce a reasonable tank during the Second World War, it need not be entirely unqualified. There were undoubtedly those, and some in very high places, who made sensible proposals with the very best of intentions. But we all know where good intentions can lead. The difficulty seems to have been in the translation. The ideas were sound enough, the requirements fully understood and most of the difficulties appreciated, but it was somewhere after that, when these concepts were translated into reality that things started to go wrong and this seems to have happened time and time again. This was compounded by another trait which meant that, even if a tank appeared to be unsuitable at an early stage of its development, production continued until the contract was honoured.

Sometime in the latter half of 1941 the British General Staff came to the conclusion that they needed a new range of tanks mounting bigger guns. The weapon they had in mind was the new 76.2mm anti-tank gun, known as the 17-pounder from the weight of its projectile. Perhaps suggesting a range of tanks is to over state the case but the General Staff certainly wanted one in the cruiser category and one in the infantry class which covers the two main British types. The cruiser emerged, ultimately, as the hybrid A30 Challenger about which little need be said here but the infantry tank seems to have been given a lower priority. When it was discussed it appears that there was some doubt as to whether it needed a turret or not, and it was partly as a result of these deliberations that the unlovely Churchill three inch gun carrier was born.

The idea of a turreted infantry tank with a 17-pounder gun appears to have been revived in 1943 when preliminary approaches were made to Vauxhall Motors. Why Vauxhall is not clear. Granted their work on the Churchill was almost done and production of the improved Mark VII was under way but there seems to be little doubt, to judge from previous experience, that if you asked Vauxhall Motors for a tank you would inevitably get something that looked like a Churchill, and there were those who believed that the basic design of that tank, never mind most of its details, were well and truly obsolete. Thus you might be able to produce a new tank, with a bigger gun and thicker armour, but if it still looked more or less like a Churchill then you had not achieved very much. However in 1943 this was not a foregone conclusion.

In December of that year the Director of Tank Design gave the matter some thought and wondered, on paper, whether one could risk losing the BESA machine-gun in the front of the tank in order to improve protection. What he had in mind was a tank with a sloped front, somewhere between 45 and 55 degrees, which would provide sufficient protection from the strike of a German 88mm. Why he believed that you could not have sloped armour and a hull machine-gun is not clear. The German Panther had both of these features and DTD actually cited the Panther as his example. So did the Sherman. Possibly it was simply the BESA that was the problem. Perhaps its bulk meant that it would not fit comfortably into a sloped plate. Yet the Sherman was proof that the Browning could be used in this way, so if the BESA would not do, why not use the Browning?

Of course the key factor was the turret. It had to be large enough to accommodate the 17-Pounder, and that meant a larger turret ring, a wider tank and so on. Having more or less settled on this the usual mischief maker arrived and asked if, perhaps, an even bigger gun might not be better. If one works on the basis that the 17-pounder was first considered in 1941 then there is nothing wrong with suggesting something larger in 1943. At least there would not be anything wrong if we were discussing anywhere but Britain. Having taken two years to generate enough enthusiasm for an infantry tank with a 17-pounder gun, was it wise to jam a spanner in the works by suggesting something else? Was one not automatically inserting another two years into the gestation period? The gun up for consideration was a modified anti-tank variant of the classic British 3.7 inch anti-aircraft gun; a 94mm weapon

Black Prince prototype Number 3 photographed at Lulworth Camp. The wider turret with its new style mantlet, shrouded in canvas, is the best way to distinguish it from a Churchill.

The same tank viewed from the side. Now one can appreciate the great size of the 17-pounder gun when compared with the Churchill's 75mm.

known as the 32-pounder. It was said to be capable of a penetration performance 25 per cent better than that of the 17-pounder. But the 32-pounder was a lot bigger than the 17-pounder and it would require a larger turret. That would mean a larger turret ring, probably a wider hull with stronger suspension, more powerful engine and so on. In other words, with a simple question about a gun our mischief maker was really postulating an entirely new tank.

It is worth remarking at this point that there was a third 17-pounder gun tank in the design stage at this time, the A41 Centurion. So we had the A30 Challenger now entering production, the Centurion in a fairly advanced state of design and now Vauxhall Motors was tinkering about with an enlarged version of their Churchill. The Centurion had first been considered in September 1943, and it had also been subjected to indecision over the gun. Yet design work on the new infantry tank must have gone ahead at a rare pace because Vauxhall Motors had a set of turret drawings in December 1943 for what would be known as A43, the General Staff designation for their heavy infantry tank. The A41 designers actually had two sets of turret drawings, both of which were seen by the Vauxhall design team. One was relatively conventional in that the main gun and machine-gun were on a stabilised co-axial mounting; the other had the BESA machine-gun mounted independently of the main gun. DTD thought that until the matter had been decided for A41 he would not go any further on A43 and it was probably left up to Vauxhall in the end.

We come now to what has to be the most astounding decision made with regard to A43; the plan to use the original Bedford engine fitted to A22. This, we recall, was a twelve-cylinder, horizontally opposed petrol engine rated at 350bhp. It had been designed at the outset for a tank that was planned to weigh just 32 tons and was later adopted for the early Churchills which weighed 38 tons and, of course, was used in the final model that weighed 40 tons. Now, it seems, the manufacturers were proposing to fit it into a tank which was expected to weigh something not far short of 50 tons. There were plenty of people, from Churchill drivers to military engineers, who were quite prepared to admit that the tank was already underpowered but everyone knew that the engine bay in the Churchill was simply too constricted to take anything else; yet that could hardly be true of the new tank which it was now decided should be called Black Prince.

In fact, inch for inch, there is not a tremendous difference between Churchill and Black Prince but it was sufficient to create enough space for a bigger engine and the obvious choice was the 600bhp Rolls-Royce Meteor which would easily fit. The Meteor was in production, it was a well tried and much liked power unit with a high standard of reliability, but if it was considered officially for the Black Prince there is precious little evidence. Certainly in May 1944 a Mr A R Code, from the automotive division of the Ministry of Supply, did a comparison on paper between Black Prince and the German tanks Tiger and Panther. He justified the comparison on the grounds of firepower and protection but was ready to admit that the German designs were a good deal older. In the light of that, he asked, why was the more modern Black Prince so weak in terms of mobility? Working from his own figures he calculated that the respective power to weight ratios of the

Viewed from above, notice that the commander is now on the right of the turret, with the gunner, while the loader serves the gun from the left. Also visible are the air intakes, relocated on the engine deck.

The true bulk of Black Prince's turret is best appreciated from the rear. How much better it might have looked with a Centurion turret instead.

three tanks were; 15.3 to 1 (Panther), 11.5 to 1 (Tiger) but just 7 to 1 in the case of Black Prince. Code also advocated the Meteor. But was it ever done?

Surviving records on Black Prince are few but with one exception they state unequivocally that the Bedford engine was used. Furthermore the one surviving tank has a Bedford engine in it. The exception is a typescript summary of the tank's development history, issued under the auspices of the War Office in August 1947. The data sheet that rounds off this document claims that the engine was a Rolls-Royce Meteor but this is not fully confirmed by the text. Having recorded Code's comments the paper goes on to state that his department was made responsible for investigating the possibility of fitting the Meteor engine. Indeed it was hoped to install the entire engine/transmission assembly from the Cromwell but there were problems with the final drive, which was probably too narrow. There was also a size problem with the engine, in terms of headroom, and Code reckoned this could be overcome by sloping the engine by just four degrees. One other drawback was the fact that the starter motor would protrude into the fighting compartment, but all the other factors were positive. There would be an increase in fuel capacity and improved power to weight ratio of 12.5 to 1 while the top speed was estimated at 22.2mph. All the evidence suggests that this work was never done, but that does not help to explain why it was not decided to design the tank around the Meteor engine in the first place.

There is, theoretically, another objection to the Meteor, one that was made against a whole range of power units which, like the Meteor, had derived from aircraft engines. There were those who argued that such engines, designed to run at high speed in the air, did not deliver sufficient torque at low revolutions to power a tank, and there was some truth in this. However the same complaint could easily be levelled against many tank engines which had not spent their lives in the air, and many of them had lower horsepower ratings to boot; the Bedford flat twelve for one. Another problem that worried the designers was the clutch. The plan was to use the same eighteen inch, single plate type already provided for the Churchill. But claims were made that there was insufficient allowance for stress, caused by rough driving conditions, in the clutch as fitted to the Churchill and that fitting it into a 49 ton tank would be courting disaster. However this problem soon got swept up in other matters.

It had already been agreed to abandon the Lockheed air assisted steering system employed on Churchill, in favour of a manually operated system and Vauxhalls were in the process of designing a five speed gearbox which they hoped would overcome the clutch worry. We should remember at this point that some of the first Churchills built had a five-

This rear three-quarter shot of another Black Prince prototype reveals the two rear smoke emitters and infantry telephone box. It also provides a very good view of the heavy duty suspension.

speed box. Anyway at just this time the General Staff announced that they wanted two reverse speeds, the normal ratio and another which would enable the tank to move backwards more quickly.

In order to speed up the trial process Vauxhall Motors installed the new gearbox into a standard Churchill tank, with the 350bhp engine and eighteen inch clutch, but with extra weights fitted to bring it up to 49 tons 17 cwt. Trials were carried out over a sea of thick mud on the Vauxhall testing ground with a variety of drivers and all declared themselves quite happy with it. The ratios in the five-speed box had been thought by some to be too close but in practice they made driving a lot easier, despite the fact that it was now an entirely manual operation. A single shift from one gear to the next could be achieved in .75 of a second. Mind you that did not leave much room for error. If the tank was travelling at 3.5 mph and one wished to change gear it did not do to bungle it. The tank was so sluggish that in 1.5 seconds the speed would drop from 3.5mph to zero. Vauxhall also noted, to their delight, that in 60 miles running the clutch on the over weight tank only had to be adjusted three times, and that other Churchill tanks being run for comparison were just as bad.

Before launching into a general description of the new tank there is one other feature that warrants attention. Again it displays a curious pattern of indecisiveness in the way it is handled. The Department of Tank Design, which was still saying that the turret should be able to accept a bigger gun, then swung around the other way and asked why it was that A41 (Centurion), which had the same turret ring diameter and armament as A43, had a different style of turret. In the view of the Director of Tank Design there was obviously something wrong with one or other of the turrets. It was then explained that the original design for A41's turret was square, with slab sides, like Black Prince, but it had since been altered. The reason given was that, after giving the matter considerable thought, and even taking account of the psychological implications, a slope sided turret was chosen because it gave 'an exaggerated impression of immunity'. It was an explanation of sorts, at least for A41, but the documents do not go on to say why it was not adopted for Black Prince. Was it, perhaps, that infantry tank crews felt secure enough behind their thick armour not to need the added psychological fillip of sloping sides? One doubts it.

As the illustrations show Black Prince did indeed look just like a big Churchill. There are differences of course. The track was wider and of a new pattern. The suspension included one more wheel station on each side and the ten central sets of load carrying rollers were paired with heavy duty brackets. The rollers themselves were still steel rimmed and the springs made a good deal heavier but the basic principle

(text continues on page 208)

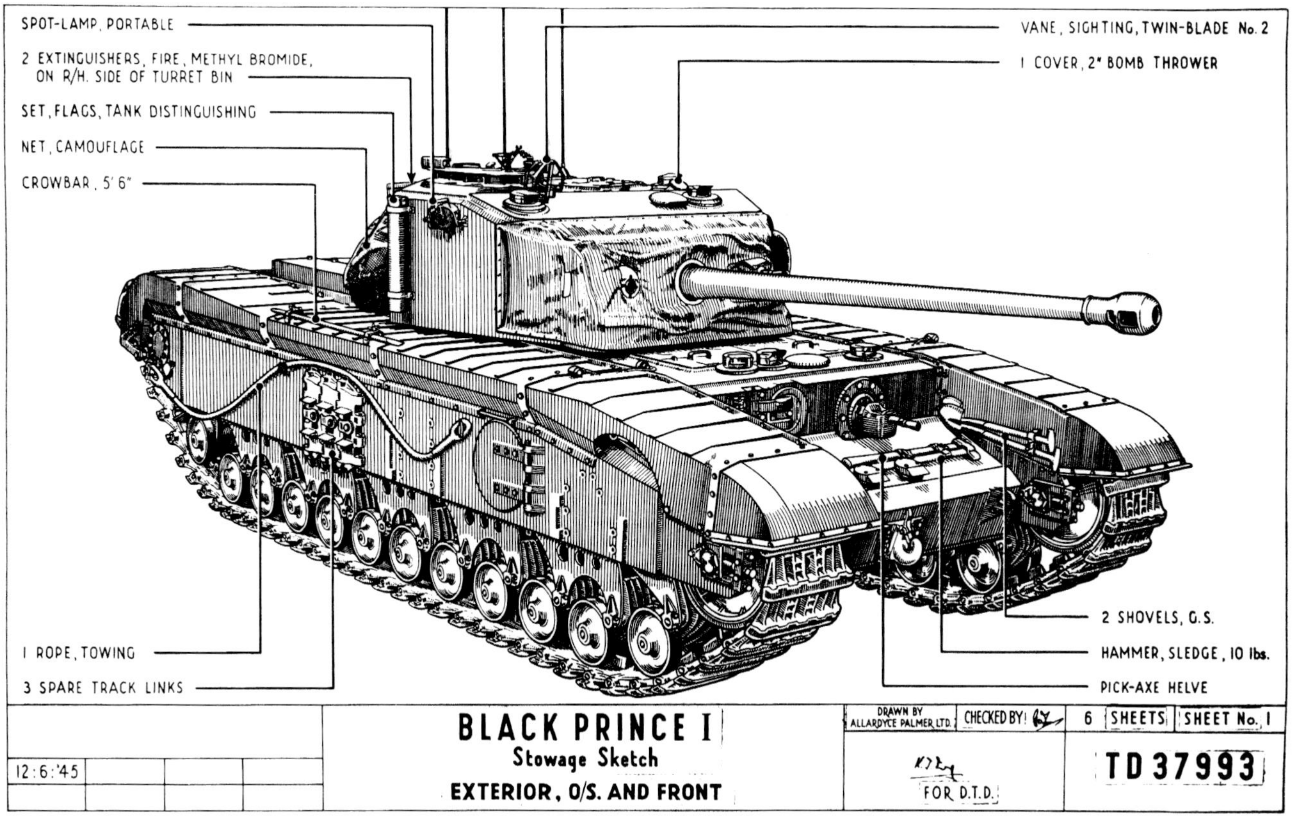

Pages 205-207: Stowage diagrams for Black Prince

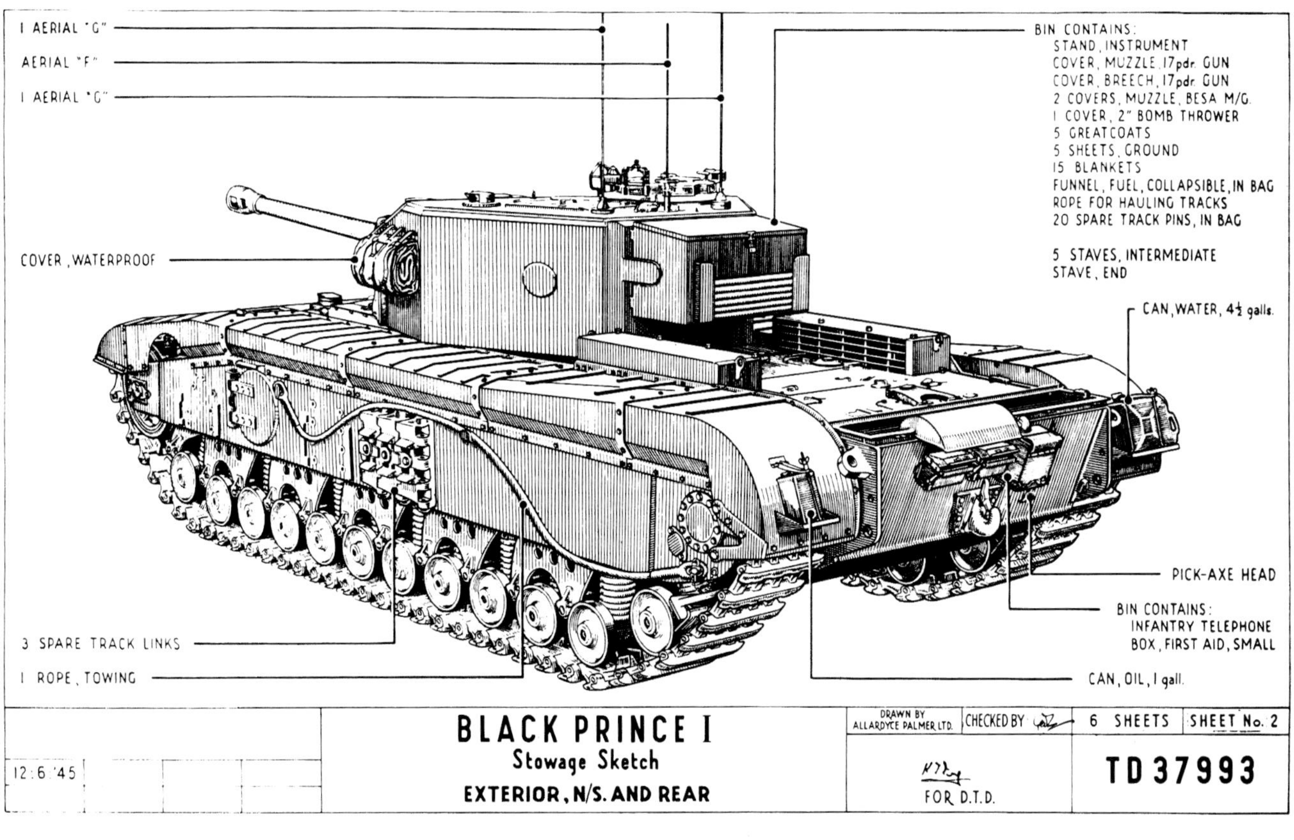

BOMB THROWER, 2"
LANYARD, COCKING
1 PERISCOPE, No. 6, Mk. I
1 BOX, BESA AMMUNITION IN FEED TRAY
BIN CONTAINS: PLUG, CLEARING, BESA M/G. CAN, OIL, M/G., Mk. II CAN, GREASE, M/G., Mk. I
11 BOMBS, SMOKE, 2"
BRUSH, CLEANING BORE, 2" BOMB THROWER
9 BOMBS, SMOKE, 2"
CAN, OIL, "WESCO", 1pt., FOR BUFFER OIL
L/H. PANNIER BIN CONTAINS: 28 CARTRIDGES, A.P.C.B.C., A.P.C., H.E. OR A.P.D.S.
READY BIN CONTAINS: 15 CARTRIDGES, A.P.C.B.C., A.P.C., H.E. OR A.P.D.S.
KEY, REMOVING JAMMED CARTRIDGE
BAG, SPENT BESA M/G. CARTRIDGE CASES AND BELTS
14 BOXES, BESA AMMUNITION
1 BESA M/G., WITH DEFLECTOR CHUTE
GUN, Q.F., 3" 17 pdr., Mk. VI
1 PERISCOPE, No. 6, Mk. I
CASE, RANGEFINDER
2 PERISCOPE PRISMS AND 1 BRUSH
1 WATER BOTTLE
1 TELESCOPE, SIGHTING
BOARD, MAPS
ALTERNATIVE STOWAGE POSITION FOR BREN M/G.
R/H. PANNIER BIN CONTAINS: 28 CARTRIDGES, A.P.C.B.C., A.P.C., H.E. OR A.P.D.S.
CARRIED IN UNIT TRANSPORT: SUIT, ANTI-GAS, IN VALISE CASE, SPARE W/T. VALVES
SPENT CARTRIDGE CONTAINER
12:6:'45
BLACK PRINCE I
Stowage Sketch
TURRET, FRONT
DRAWN BY ALLARDYCE PALMER LTD
CHECKED BY
FOR D.T.D.
6 SHEETS
SHEET No. 3
TD 37993

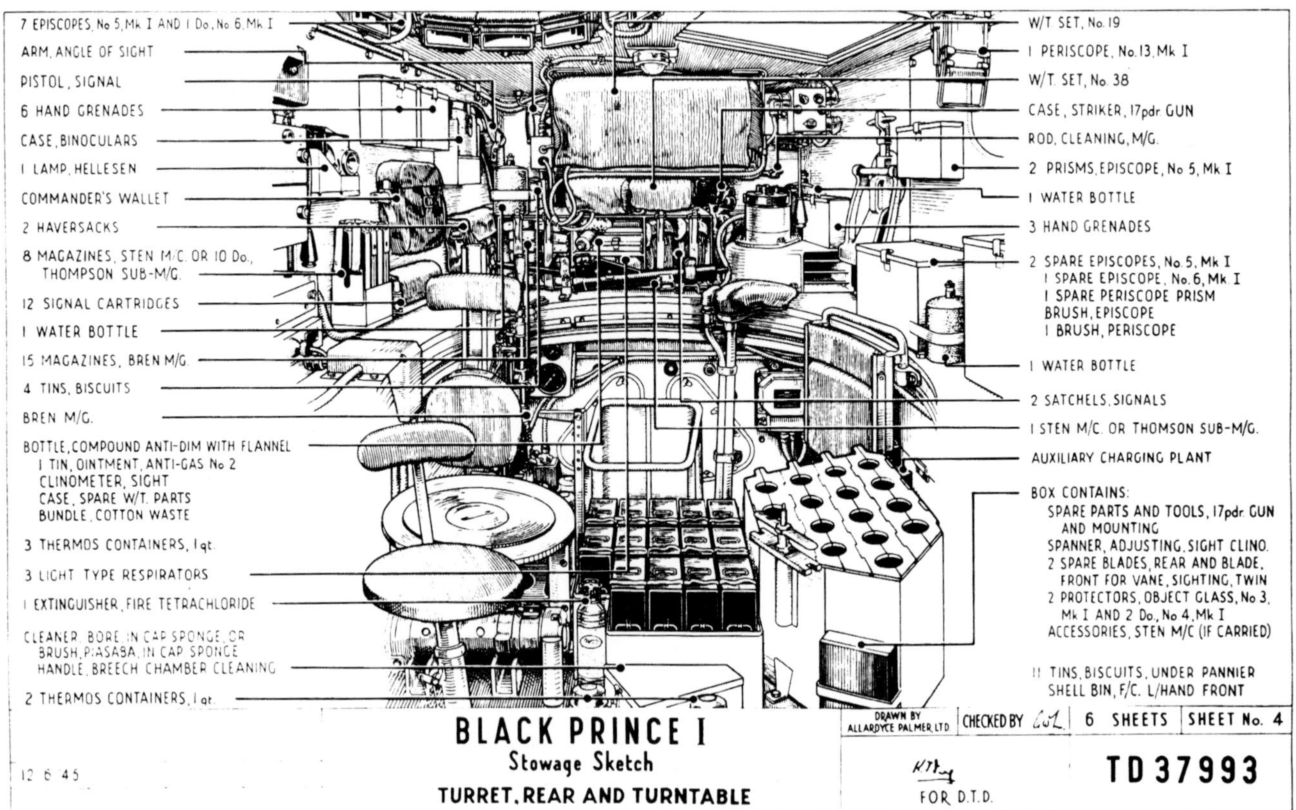
7 EPISCOPES, No. 5, Mk. I AND 1 Do., No. 6, Mk. I
ARM, ANGLE OF SIGHT
PISTOL, SIGNAL
6 HAND GRENADES
CASE, BINOCULARS
1 LAMP, HELLESEN
COMMANDER'S WALLET
2 HAVERSACKS
8 MAGAZINES, STEN M/C. OR 10 Do., THOMPSON SUB-M/G.
12 SIGNAL CARTRIDGES
1 WATER BOTTLE
15 MAGAZINES, BREN M/G.
4 TINS, BISCUITS
BREN M/G.
BOTTLE, COMPOUND ANTI-DIM WITH FLANNEL 1 TIN, OINTMENT, ANTI-GAS No. 2 CLINOMETER, SIGHT CASE, SPARE W/T. PARTS BUNDLE, COTTON WASTE
3 THERMOS CONTAINERS, 1 qt.
3 LIGHT TYPE RESPIRATORS
1 EXTINGUISHER, FIRE TETRACHLORIDE
CLEANER, BORE, IN CAP SPONGE, OR BRUSH, PIASABA, IN CAP SPONGE HANDLE, BREECH CHAMBER CLEANING
2 THERMOS CONTAINERS, 1 qt.
W/T. SET, No. 19
1 PERISCOPE, No. 13, Mk. I
W/T. SET, No. 38
CASE, STRIKER, 17 pdr. GUN
ROD, CLEANING, M/G.
2 PRISMS, EPISCOPE, No. 5, Mk. I
1 WATER BOTTLE
3 HAND GRENADES
2 SPARE EPISCOPES, No. 5, Mk. I 1 SPARE EPISCOPE, No. 6, Mk. I 1 SPARE PERISCOPE PRISM BRUSH, EPISCOPE 1 BRUSH, PERISCOPE
1 WATER BOTTLE
2 SATCHELS, SIGNALS
1 STEN M/C. OR THOMSON SUB-M/G.
AUXILIARY CHARGING PLANT
BOX CONTAINS: SPARE PARTS AND TOOLS, 17 pdr. GUN AND MOUNTING SPANNER, ADJUSTING, SIGHT CLINO. 2 SPARE BLADES, REAR AND BLADE, FRONT FOR VANE, SIGHTING, TWIN 2 PROTECTORS, OBJECT GLASS, No. 3, Mk. I AND 2 Do., No. 4, Mk. I ACCESSORIES, STEN M/C (IF CARRIED)
11 TINS, BISCUITS, UNDER PANNIER SHELL BIN, F/C. L/HAND FRONT
12 6 45
BLACK PRINCE I
Stowage Sketch
TURRET, REAR AND TURNTABLE
DRAWN BY ALLARDYCE PALMER LTD
CHECKED BY
FOR D.T.D.
6 SHEETS
SHEET No. 4
TD 37993

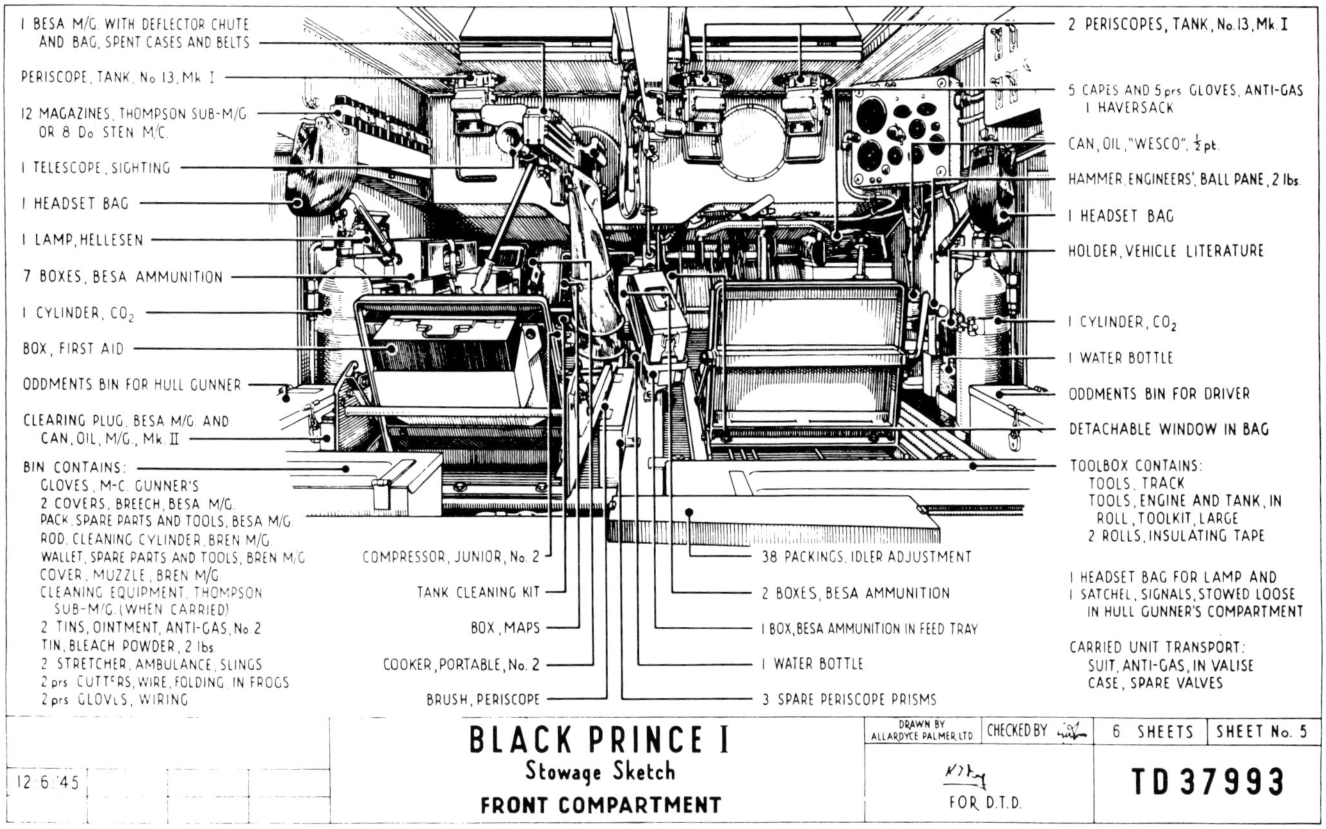

1 BESA M/G. WITH DEFLECTOR CHUTE AND BAG, SPENT CASES AND BELTS
PERISCOPE, TANK, No 13, Mk I
12 MAGAZINES, THOMPSON SUB-M/G. OR 8 Do STEN M/C.
1 TELESCOPE, SIGHTING
1 HEADSET BAG
1 LAMP, HELLESEN
7 BOXES, BESA AMMUNITION
1 CYLINDER, CO_2
BOX, FIRST AID
ODDMENTS BIN FOR HULL GUNNER
CLEARING PLUG, BESA M/G. AND CAN, OIL, M/G., Mk. II
BIN CONTAINS:
GLOVES, M-C. GUNNER'S
2 COVERS, BREECH, BESA M/G.
PACK, SPARE PARTS AND TOOLS, BESA M/G.
ROD, CLEANING CYLINDER, BREN M/G.
WALLET, SPARE PARTS AND TOOLS, BREN M/G.
COVER, MUZZLE, BREN M/G.
CLEANING EQUIPMENT, THOMPSON SUB-M/G. (WHEN CARRIED)
2 TINS, OINTMENT, ANTI-GAS, No 2
TIN, BLEACH POWDER, 2 lbs
2 STRETCHER, AMBULANCE, SLINGS
2 prs CUTTERS, WIRE, FOLDING, IN FROGS
2 prs GLOVES, WIRING
COMPRESSOR, JUNIOR, No 2
TANK CLEANING KIT
BOX, MAPS
COOKER, PORTABLE, No. 2
BRUSH, PERISCOPE
38 PACKINGS, IDLER ADJUSTMENT
2 BOXES, BESA AMMUNITION
1 BOX, BESA AMMUNITION IN FEED TRAY
1 WATER BOTTLE
3 SPARE PERISCOPE PRISMS
2 PERISCOPES, TANK, No.13, Mk I
5 CAPES AND 5 prs GLOVES, ANTI-GAS
1 HAVERSACK
CAN, OIL, "WESCO", ½ pt.
HAMMER, ENGINEERS', BALL PANE, 2 lbs.
1 HEADSET BAG
HOLDER, VEHICLE LITERATURE
1 CYLINDER, CO_2
1 WATER BOTTLE
ODDMENTS BIN FOR DRIVER
DETACHABLE WINDOW IN BAG
TOOLBOX CONTAINS:
TOOLS, TRACK
TOOLS, ENGINE AND TANK, IN ROLL, TOOLKIT, LARGE
2 ROLLS, INSULATING TAPE
1 HEADSET BAG FOR LAMP AND
1 SATCHEL, SIGNALS, STOWED LOOSE IN HULL GUNNER'S COMPARTMENT
CARRIED UNIT TRANSPORT:
SUIT, ANTI-GAS, IN VALISE
CASE, SPARE VALVES
12.6.45
BLACK PRINCE I
Stowage Sketch
FRONT COMPARTMENT
DRAWN BY ALLARDYCE PALMER LTD
CHECKED BY
6 SHEETS
SHEET No. 5
FOR D.T.D.
TD 37993

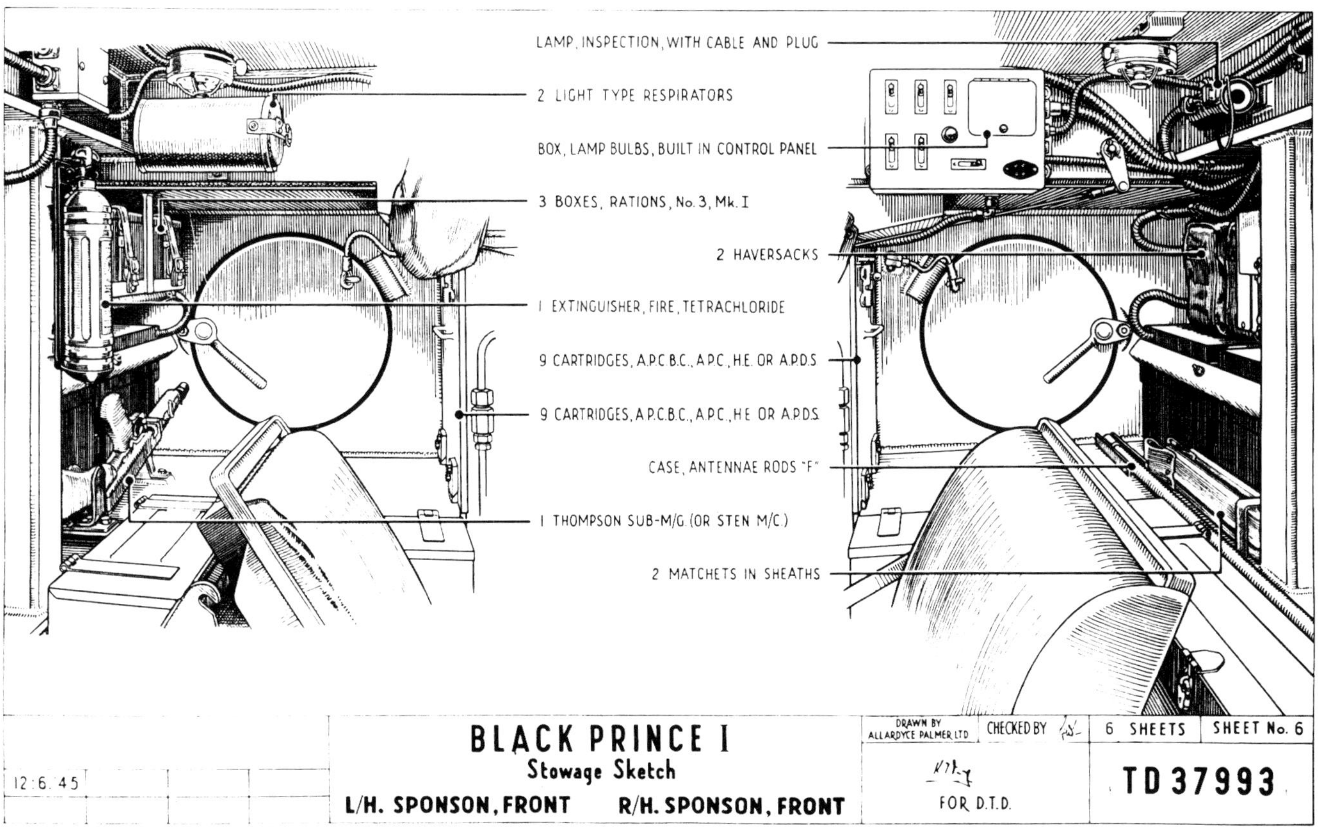

LAMP, INSPECTION, WITH CABLE AND PLUG
2 LIGHT TYPE RESPIRATORS
BOX, LAMP BULBS, BUILT IN CONTROL PANEL
3 BOXES, RATIONS, No. 3, Mk. I
2 HAVERSACKS
1 EXTINGUISHER, FIRE, TETRACHLORIDE
9 CARTRIDGES, A.P.C.B.C., A.P.C., H.E. OR A.P.D.S.
9 CARTRIDGES, A.P.C.B.C., A.P.C., H.E. OR A.P.D.S.
CASE, ANTENNAE RODS "F"
1 THOMPSON SUB-M/G. (OR STEN M/C.)
2 MATCHETS IN SHEATHS
12.6.45
BLACK PRINCE I
Stowage Sketch
L/H. SPONSON, FRONT
R/H. SPONSON, FRONT
DRAWN BY ALLARDYCE PALMER LTD
CHECKED BY
6 SHEETS
SHEET No. 6
FOR D.T.D.
TD 37993

of the suspension remained the same. The hull was shaped like the Churchill with the stepped front but there had been a conscious effort to improve the driver's view by lowering the height of the front idlers. The hull was of welded construction and generally well sealed for wading. The air intakes had been moved from the sides to the top of the engine deck and the outlet was at the rear. One source explained that all upper surface hatches and hinges had been strengthened when compared with Churchill, to give greater resistance to blast, suggesting an otherwise unrecorded fault with the latter.

All experienced tank crew who had anything to do with Black Prince remarked on the roominess and comfort of the crew stations, but these were otherwise laid out like the Churchill. The turret by comparison was an odd affair. With the exception of the mantlet it was little more than a large, crude, welded box like a scaled up versions of the old Churchill III turret. Apart from its thick armour it had no ballistic properties at all and looked like an afterthought. The mantlet, on the other had, was quite sophisticated. Created by the old engineering firm Stothert and Pitt of Bath it consisted of two huge castings with semi-exposed trunnions. The same design had already been adopted for the Comet cruiser tank and the A41 would be similar. The top of the turret was crowned by an All Round Vision cupola for the commander.

In May 1944 it was decided to build six prototypes of Black Prince. This contrasts sharply with Centurion for which 20 prototypes had been ordered in February. Granted Centurion prototypes would be required to test different weapon mountings and gearboxes but even so six was regarded as insufficient and somebody must have said so, because the War Office summary echoes the complaint. One can only guess at the reason but it does not seem to be too far fetched to suggest that even now, at the design stage, those in authority realised that the Centurion was a far more significant design than Black Prince ever could be. The surprise is not that they only built six of them, but that the bothered to build any at all.

Whatever the reason, six prototypes were built to be distributed as follows:

> Two would remain with Vauxhalls for development work, one would suffer firing at trials another would go to Lulworth for firing trials, yet another to the Fighting Vehicle Proving Establishment for acceptance field trials and the last would be used for wading trials.

Another curiosity surrounding the British method of producing tanks in World War II comes out of the War Office report on the development of Black Prince. This is the strange practice of allowing supposed experts with particular obsessions to comment on the design even after one would imagine that these matters had all been settled. As an example in April 1944, shortly before the order for prototypes was placed, and at a time when one might suppose that most of the design work was done, the General Staff announced requirements for a fully rotating turret smoke mortar, pressure ventilation, rear mounted smoke dischargers and provision for conversion to a command tank. They also stipulated wading

Prototype Number 4, the surviving Black Prince, seen here at full speed during trials. Now one can see the new mantlet without its canvas cover.

to 7 feet and stowage for 90 rounds of ammunition. Others chipped in with demands for greater depression of the main gun and a lower overall height. These last requirements, it was claimed, conflicted directly with one another while most of the others suggested a lack of co-ordination which simply had to stop. Yet a month later we find Mr Code once again demanding that the tanks should all be fitted with Meteor engines. When the subject of a command tank version was raised again in June Vauxhalls were told to go ahead with their programme and ignore it. The contract card, which is undated, refers to six Churchill "A43" Black Prince prototypes which would carry the WD numbers T353280 to T353285. A second contract card, dated 30 June 1944, refers to 300 Churchill "A43" (to which someone has added "Black Prince" in pencil) but gives no WD numbers and, of course, was never fulfilled.

Apparently the original six tanks were due to be built in armour plate, which is unusual for prototypes. At least this may be deduced from a comment dated 28 July 1944 to the effect that the War Office was under the impression that these tanks, now being referred to as pilot models, would be made in mild steel. Normally, when this was done, the tank in question was fitted with small, red, cast metal triangles bearing the words CAUTION UNARMOURED. This, presumably, was to prevent them from being sent into action by mistake, although that would appear to be a highly unlikely contingency. Certainly the surviving example of Black Prince, Pilot Model No. 4, is not marked in this way.

A mock-up of the tank was shown to various interested parties in September 1944, including representatives from 21st Army Group fresh from combat experience in Europe. Various matters were raised and certain features criticised but nothing, apparently, of a serious nature and it was all soon put right. Vauxhalls had hoped to have the first pilot model ready by October 1944 but this proved impossible. For one thing Stothert & Pitt had been sent the wrong kind of 17-pounder barrel (Mark II instead of Mark VI) so they could not complete the design of the mantlet until that was put right. Thus it was early January 1945 before the first tank ran out of the Vauxhall plant to undergo trials, which included a road run to FVPE in Surrey.

Shortly before the first tank was completed someone pointed out that it was too heavy for any of the existing road transporters. The standard rig for heavy tanks was the American Diamond T tractor hauling a three-axle drawbar trailer with a carrying capacity of 40 tons. Other, heavier, trailers were in the design stage so for the present it was regarded as acceptable to let the Diamond T take the load, allowing for the fact that Black Prince was also a good deal wider, and they seem to have managed well enough.

The trial road run showed that the tank could manage a steady 11 mph average speed and during cross-country runs at Chertsey it often managed to remain at 7.5 mph for a reasonable distance. This was put down to the fact that much of the ground was frozen at this time so the tank ran a lot in 4th and 5th gears but it was a good omen. The tracks stood up well and the tank was considered remarkably free from pitching which indicated that it would be a good gun platform. The suspension also worked well, even on hard ground the tank showed no tendency to bottom on its springs. The driver reported that the controls were light and positive while the only complaint, that the engine tended to over run when going down hill, was blamed on poor governors. Over running was indicated by flashes and loud bangs from the exhaust system and it was a fault which Churchill crews were reporting from Europe.

In this view the last example of the Churchill family strikes a fitting pose alongside the shot up remains of an old Mark I, first of the line.

The most ominous problem, towards the end of this trial, was a gearbox failure. This immediately revived all the old nightmares that beset the Churchill in its early days but Vauxhall announced that they had already taken steps to strengthen it.

In February 1945 Pilot No 3 was sent down to Lulworth for firing trials while the next tank was prepared for shipment to Shoeburyness in Essex for shooting at trials. This was followed by the wading trials pilot which went to Westward Ho! in North Devon. No reports appear to have survived of the Shoeburyness trial but there is some commentary on the wading tests. The welded hull was regarded as virtually watertight but it is interesting to note that the tank was fitted with an inflatable rubber ring which sealed the joint between turret and hull during wading. This had been a feature of the early Tiger tanks that had been designed to wade, and may well have derived from that. Black Prince was too wide to pass through the bow of a Landing Craft Tank (LCT) Mark 3 so it was tested with an LCT 4 fitted with extensions to the ramp. Conditions must have been rough because the report claims that water sometimes splashed right over the turret and it was later found that three gallons of water was sloshing about inside. However since this had not come in through the hull the Combined Operations Experimental Wing declared the trial satisfactory.

Full circle. One of the last production Churchill Mark VIIs is posed alongside a newly built Vauxhall saloon. In this publicity shot the company hopes to establish its wartime credentials with the post-war motoring public.

A preliminary report, issued by the commanding officer at Lulworth Camp in March 1945 on Pilot No. 3 is remarkable detailed, considering that it was compiled in haste. The opening remarks are worth quoting; "it is disappointing to see the standard Churchill vertical front plate perpetuated. It is hoped that vertical front plates will not appear in future AFV designs". And it is worth emphasising that this vertical plate was 152mm thick which was no different to that on a Mark VII Churchill. In general, however, the report is quite positive although toxicity levels inside the turret were higher than they should be and rapid escape, through the commander's hatch with its All Round Vision cupola, was impossible for all but the skinniest men. It also seems that the rubber air seal around the turret ring proved to be a bit of a nuisance, especially when the tank was traversing its turret on the move.

A report, issued by the Army Operational Research Group (AORG) in June 1945, carried details of gun loading trials. Black Prince was designed to carry 89 rounds of 17-pounder ammunition and as far back as April 1944 it had been agreed that it would only carry three natures of ammunition; Armour Piercing Capped, Ballistic Capped (APCBC), Armour Piercing Discarding Sabot (APDS) and High Explosive (HE). Fifteen rounds would be stowed in a ready-use bin on the turret floor while the rest was loaded in three racks on each side within the hull panniers.

The AORG trial involved three men. An experienced gunner from the Royal Artillery, who presumably had not had much to do with tanks before. A trooper from the Royal Armoured Corps who had experience as a tank gunner in North Africa and Italy and a Corporal who normally drove the tank but was roped in to the loading trials for the very reason, or so it seems, that he knew nothing whatever about handling ammunition or loading guns. Even the height of each man was recorded; they ranged from 5ft 8ins to 5ft 11ins in their army boots. In order to make the trial as difficult as possible each man had to load all fifteen rounds from the ready-use bin and he was only provided with APCBC rounds because they were the longest and heaviest; each round weighed 37 lbs. The report is generally complimentary, claiming that an average loader could lift a round from the bin and slam it into the breech in 5 seconds with ease. It is worth reporting that when these trials were over, and the tank was stripped down for inspection, it was revealed that the final drive was on the point of collapse. The unfortunate tank probably had no more than a few miles left in it.

The fact that these trials were carried on at all is a reflection of relaxing attitudes as the war wound down to its end, since on 30 May 1945 it was stated officially that the tank would not go into production. Today just one example survives, Pilot No. 4 was moved to the Tank Museum in 1947 and remains today a rather impressive but forlorn exhibit. In its final summing up the War Office sounded a weary note. It pointed out that by the time the project was abandoned two years had been spent in evolving a heavily armoured carrier for the 17-pounder gun which was already two and a half years overdue (presumably a comparison with Tiger) when it was first built. And all this for a tank which was incapable of expansion in armour, weapons or motive power. All it really proved was that it was possible to develop the standard Churchill, which was already on the point of being declared obsolete, but that parallel developments (undoubtedly a reference to Centurion) would finally render two classes of tank unnecessary.

Abbreviations

ACIGS	Assistant Chief of the Imperial General Staff
AFV	Armoured Fighting Vehicle
AORG	Army Operational Research Group
APCBC	Armour Piercing Capped Ballistic Capped (ammunition)
APDS	Armour Piercing Discarding Sabot (ammunition)
APHV	Armour Piercing High Velocity (ammunition)
ARV (i)	All Round Vision (cupola)
ARV (ii)	Armoured Recovery Vehicle
AVRE	Armoured Vehicle Royal Engineers
bhp	brake horsepower
Bn	Battalion
BSA	Birmingham Small Arms
cc	cubic centimetres
CDL	Canal Defence Light
CGRD	Controller General of Research & Development
CIGS	Chief of the Imperial General Staff
CO	Commanding Officer
CTA	Cemented Tank Armour
DAFV	Director of Armoured Fighting Vehicles
db	decibels
DGTT	Director General of Tanks & Transport
DRAC	Director Royal Armoured Corps
DTD(i)	Department of Tank Design
DTD(ii)	Director of Tank Design
FDS	Field Delivery Squadron
FOO	Forward Observation Officer
ft/lbs	foot/pounds
FVPE	Fighting Vehicle Proving Establishment
GHQ	General Headquarters
GM	General Motors
GRC&W	Gloucester Railway Carriage & Wagon
GS	General Staff
H&W	Harland & Wolff
HE	High Explosive (ammunition)
HEAT	High Explosive Anti Tank (ammunition)
HMS	His Majesty's Ship
hp	horsepower
HV	High Velocity
IT	Izod Test
LAD	Light Aid Detachment
LCT	Landing Craft Tank
LNER	London & North Eastern Railway
MEE	Mechanisation Experimental Establishment
MGO	Master General of Ordnance
MoS	Ministry of Supply
mph	miles per hour
MQ	Machineable Quality (armour)
MV	Medium Velocity
NCO	Non Commissioned Officer
OP	Observation Post
pdr	pounder
QF	Quick Firing
RAC	Royal Armoured Corps
REME	Royal Electrical & Mechanical Engineers
RTR	Royal Tank Regiment
SEAC	South East Asia Command
STT	School of Tank Technology
TOG	The Old Gang
WD	War Department
WVEE	Wheeled Vehicle Experimental Establishment

NOTES

NOTES

NOTES

NOTES

NOTES